Lecture Notes in Physics

Springer-Verlag Berlin Heidelberg GmbH

The Editorial Policy for Proceedings

The series Lecture Notes in Physics reports new developments in physical research and teaching – quickly, informally, and at a high level. The proceedings to be considered for publication in this series should be limited to only a few areas of research, and these should be closely related to each other. The contributions should be of a high standard and should avoid lengthy redraftings of papers already published or about to be published elsewhere. As a whole, the proceedings should aim for a balanced presentation of the theme of the conference including a description of the techniques used and enough motivation for a broad readership. It should not be assumed that the published proceedings must reflect the conference in its entirety. (A listing or abstracts of papers presented at the meeting but not included in the proceedings could be added as an appendix.)
When applying for publication in the series Lecture Notes in Physics the volume's editor(s) should submit sufficient material to enable the series editors and their referees to make a fairly accurate evaluation (e.g. a complete list of speakers and titles of papers to be presented and abstracts). If, based on this information, the proceedings are (tentatively) accepted, the volume's editor(s), whose name(s) will appear on the title pages, should select the papers suitable for publication and have them refereed (as for a journal) when appropriate. As a rule discussions will not be accepted. The series editors and Springer-Verlag will normally not interfere with the detailed editing except in fairly obvious cases or on technical matters.
Final acceptance is expressed by the series editor in charge, in consultation with Springer-Verlag only after receiving the complete manuscript. It might help to send a copy of the authors' manuscripts in advance to the editor in charge to discuss possible revisions with him. As a general rule, the series editor will confirm his tentative acceptance if the final manuscript corresponds to the original concept discussed, if the quality of the contribution meets the requirements of the series, and if the final size of the manuscript does not greatly exceed the number of pages originally agreed upon. The manuscript should be forwarded to Springer-Verlag shortly after the meeting. In cases of extreme delay (more than six months after the conference) the series editors will check once more the timeliness of the papers. Therefore, the volume's editor(s) should establish strict deadlines, or collect the articles during the conference and have them revised on the spot. If a delay is unavoidable, one should encourage the authors to update their contributions if appropriate. The editors of proceedings are strongly advised to inform contributors about these points at an early stage.
The final manuscript should contain a table of contents and an informative introduction accessible also to readers not particularly familiar with the topic of the conference. The contributions should be in English. The volume's editor(s) should check the contributions for the correct use of language. At Springer-Verlag only the prefaces will be checked by a copy-editor for language and style. Grave linguistic or technical shortcomings may lead to the rejection of contributions by the series editors. A conference report should not exceed a total of 500 pages. Keeping the size within this bound should be achieved by a stricter selection of articles and not by imposing an upper limit to the length of the individual papers. Editors receive jointly 30 complimentary copies of their book. They are entitled to purchase further copies of their book at a reduced rate. As a rule no reprints of individual contributions can be supplied. No royalty is paid on Lecture Notes in Physics volumes. Commitment to publish is made by letter of interest rather than by signing a formal contract. Springer-Verlag secures the copyright for each volume.

The Production Process

The books are hardbound, and the publisher will select quality paper appropriate to the needs of the author(s). Publication time is about ten weeks. More than twenty years of experience guarantee authors the best possible service. To reach the goal of rapid publication at a low price the technique of photographic reproduction from a camera-ready manuscript was chosen. This process shifts the main responsibility for the technical quality considerably from the publisher to the authors. We therefore urge all authors and editors of proceedings to observe very carefully the essentials for the preparation of camera-ready manuscripts, which we will supply on request. This applies especially to the quality of figures and halftones submitted for publication. In addition, it might be useful to look at some of the volumes already published. As a special service, we offer free of charge LATEX and TEX macro packages to format the text according to Springer-Verlag's quality requirements. We strongly recommend that you make use of this offer, since the result will be a book of considerably improved technical quality. To avoid mistakes and time-consuming correspondence during the production period the conference editors should request special instructions from the publisher well before the beginning of the conference. Manuscripts not meeting the technical standard of the series will have to be returned for improvement.

For further information please contact Springer-Verlag, Physics Editorial Department II, Tiergartenstrasse 17, D-69121 Heidelberg, Germany

Harald Grosse   Ludwig Pittner (Eds.)

Low-Dimensional Models in Statistical Physics and Quantum Field Theory

Proceedings of the
34. Internationale Universitätswochen
für Kern- und Teilchenphysik
Schladming, Austria, March 4–11, 1995

Springer

Editors

Harald Grosse
Institut für Theoretische Physik, Universität Wien
Boltzmanngasse 5, A-1090 Wien, Austria

Ludwig Pittner
Institut für Theoretische Physik, Universität Graz
Universitätsplatz 5, A-8010 Graz, Austria

Supported by the Bundesministerium für Wissenschaft und Forschung, Vienna, Austria

Cataloging-in-Publication Data applied for.

Die Deutsche Bibliothek - CIP-Einheitsaufnahme

Low dimensional models in statistical physics and quantum field theory : proceedings of the 34. Internationale Universitätswochen für Kern- und Teilchenphysik, Schladming, Austria, March 4 - 11, 1995 / Harald Grosse ; Ludwig Pittner (ed.).

(Lecture notes in physics ; Vol. 469)
ISBN 978-3-662-14068-0 ISBN 978-3-540-49778-3 (eBook)
DOI 10.1007/978-3-540-49778-3
NE: Grosse, Harald [Hrsg.]; Internationale Universitätswochen für Kern- und Teilchenphysik <34, 1995, Schladming>; GT

ISBN 978-3-662-14068-0

Originally published by Springer-Verlag Berlin Heidelberg New York in 1996
Softcover reprint of the hardcover 1st edition 1996

Typesetting: Camera-ready by the authors
SPIN: 10520052 55/3142-543210 - Printed on acid-free paper

These proceedings are dedicated to the memory of Professor Paul URBAN, founder and long-time director of the Schladming Winter School, who died at the age of ninety a few days before the start of this year's School.

Preface

This volume contains the written versions of the invited lectures and some of the seminars presented at the "34. Internationale Universitätswochen für Kern- und Teilchenphysik" in Schladming, Austria, which took place from March 4th to March 11th, 1995. The main contributions are centered around "Two-Dimensional Models of Quantum Field Theory and Statistical Physics". The lively developments within these subjects were also reflected by the many interrelated contributions.

The main items can be grouped into integrable (quantum) spin systems, which lead in the continuum limit to (conformal invariant) quantum field theory models and their algebraic structures, ranging from the Yang - Baxter equation and quantum groups to noncommutative geometry.

A review of integrable quantum spin models based on the reflection equation is given by P. Kulish from the St. Petersburg group. L. Faddeev uses this technology to quantize the Liouville model. V. Pasquier solves Calogero - Moser - type models of various forms. H. Grosse gives an introduction to this rapidly developing field. The Luttinger model as well as spin models such as the XX, XY, XXX and XXZ models are treated and their solution and interrelations sketched.

Nonrelativistic problems, from stability of matter to magnetic field problems and the Quantum Hall Effect (QHE), are treated by J. Fröhlich. In particular the fractional QHE is of great insterest. Theoretical ideas behind it are connected to the Chern - Simons model, and edge currents are connected to conformal invariant models.

A. Alekseev tells us of a method to quantize the Chern - Simons model. Within the Hamiltonian approach, a two-dimensional system is obtained. The Poission - Lie algebra of monodromies leads after deformation to a quantum group-invariant lattice model, whose representations are studied. Fusion rules lead to Verlinde's algebra.

The conformal symmetric two-dimensional models are reviewed by I. Todorov. Virasoro algebra representations were classified long ago, but the richness of these models is still not fully uncovered. Recently even number theoretic aspects have been used to study solutions of the Knizhnik - Zamolodchikov equation, describing the monodromies for n-point functions.

Since our low-dimensional models sometimes allow us to connect to physically realizable situations, we also invited E. Gornik to report on fascinating mesoscopic experiments. Modern technology and experiments with quantum wires and quantum dots are reported. Through tunneling, individual energy levels can be measured. The Luttinger model allows us to describe transport properties of quantum wires.

The algebraic aspects behind integrability led to the introduction of Lie - Poisson algebras, quadratic algebras, the classical Yang - Baxter relations, and deformations of these structures. N. Reshetikhin went through all these developments.

On any associative algebra, there exists (at least one) differential calculus. This interesting observation leads to many recent developments. It applies to quantum groups, but also to matrix algebras, as they occur within the Standard Model of elementary particles. The Higgs - effect is incorporated in a natural way. These developments are explained in detail by R. Coquereaux. The different approaches are compared.

Not only integrable models but also others are treated. K. Gawedzki tells us how to use the functional integration method to study problems of turbulence. A hierarchy of scales shows up. W. Thirring explains how the notion of Lyapunov exponents can be used for quantum systems too.

This year's Schladming School indicated lively development. A large number of physicists made this an impressive event.

In addition, excellent seminars were presented at the school, but due to space limitations not all of them could be included in these proceedings. However, a list of seminar contributions is given at the end of this volume, so that interested readers may request information or pertinent material directly from the authors.

Finally, we would like to express our thanks to the lecturers for all their efforts, to the sponsors of the school, above all the Austrian Ministry of Science and Research and the Government of Styria, for providing generous support, to our colleagues in the organizing committee for their assistance, and to Mrs. E. Neuhold for her help in bringing the files prepared by the authors in TEX or LATEX into their final form.

Graz, January 1996

H. Grosse and L. Pittner (Directors of the School)
V. Winkler (Scientific Secretary)

Contents

Universality in Turbulence: An Exactly Solvable Model

By *K. Gawedzki* and *A. Kupiainen*

Introduction to Integrable Models of Statistical Physics

By *H. Grosse*

Yang - Baxter Equation and Reflection Equations in Integrable Models

By *P.P. Kulish*

Scattering Matrices and Affine Hecke Algebras

By *V. Pasquier*

Some Algebraic and Analytic Structures in Integrable Systems

By *N. Reshetikhin*

Monodromy Representation of the Mapping Class Group $\mathcal{B}_n$ for the su_2 Knizhnik - Zamolodchikov Equation

By *Y.S. Stanev* and *I.T. Todorov*

What are the Quantum Mechanical Lyapunov Exponents?

By *W. Thirring*

Area-Preserving Structure of Massless Matter-Gravity Fields in 1+1 Dimensions

By *G. Amelino - Camelia, D. Bak,* and *D. Seminara*

Solitons in the Calogero - Sutherland Collective-Field Model

$W_{1+\infty}$ Minimal Models and the Hierarchy of the Quantum Hall Effect

Creation and Annihilation Operators for $b - c$ Systems on General Algebraic Curves

Interacting Electrons and Localized Spins: Exact Results from Conformal Field Theory

On the U(1) Problem of QED_2

The Finite Gauge Model on the Truncated Sphere

By *H. Grosse, C. Klimčík,* and *P. Prešnajder*

Local Operators in Massive Quantum Field Theories

By *A. Koubek*

On Anomalies and Noncommutative Geometry

By *E. Langmann*

Quantum Deformations of Space-Time Symmetries with Fundamental Mass Parameter

By *J. Lukierski*

Magnetic Field-Induced 2*D* Wigner Solid versus Incompressible Quantum Liquid

By *G. Meissner*

List of Contributors

A.Yu. Alekseev
Institut für Theoretische Physik
ETH Zürich
Hönggerberg
CH-8093 Zürich
Switzerland
alekseev@itp.phys.ethz.ch

R. Coquereaux
Centre de Physique Theorique
CNRS - Luminy, Case 907
F-13288 Marseille, Cedex 9
France
coque@cpt.univ-mrs.fr

L.D. Faddeev
St.Petersburg Branch of
Steklov Mathematical Institute
Fontanka 27
St.Petersburg 191011
Russia
Research Institute for
Theoretical Physics
University of Helsinki
Siltavuorenpenger 20 C
SF-00014 Helsinki
Finland
faddeev@lomi.sph.su

K. Gawedzki
I.H.E.S., C.N.R.S.
F-91440 Bures-sur-Yvette
France
bongo@ihes.fr

H. Grosse
Institut für Theoretische Physik
Universität Wien
Boltzmanngasse 5
A-1090 Wien
Austria
grosse@pap.univie.ac.at

P.P. Kulish
St.Peterburgs Branch of
Steklov Mathematical Institute
Fontanka 27
St.Petersburg 191011
Russia
Depart. de Física Teór., IFIC
Centro Mixto Universidad
de Valencia - CSIC,
46100 Burjassot (Valencia)
Spain
kulish@evalvx.ific.uv.es

V. Pasquier
Service Physique Theorique
CEA-CEN Saclay
F-91191 Gif-sur-Yvette
France
pasquier@amoco.saclay.cea.fr

N. Reshetikhin
Department of Mathematics
University of California
Berkeley, CA 94720
USA
reshetik@math.berkeley.edu

W. Thirring
Institut für Theoretische Physik
Universität Wien
Boltzmanngasse 5
A-1090 Wien
Austria
fwagner@pap.univie.ac.at

G. Amelino-Camelia
Center for Theoretical Physics
MIT
Cambridge, MA 02139
USA
amelino@mitlns.mit.edu

I. Andric
Ruder Boskovic Institute
POB 1016
41001 Zagreb
Croatia
andric@thphys.irb.hr

A. Cappelli
INFN
Largo E. Fermi 2
I-50125 Firenze
Italy
cappelli@andrea.fi.infn.it

C. Gattringer
Max Planck-Inst. f. Physik
Föhringer Ring 6
D-80805 München
Germany
chg@dmumpiwh.bitnet

H. Johannesson
Inst. of Theoretical Physics
Chalmers Univ. of Technology
S-41296 Göteborg
Sweden
tfkhj@fy.chalmers.se

I.T. Todorov
Institute for Nuclear Research
Bulgarian Academy of Sciences
Tsarigradsko Chaussee 72
BG-1784 Sofia
Bulgaria
todorov@bgearn.bitnet

A. Koubek
Dept. of Appl. Math.
and Theor.Physics
University of Cambridge
Silver Street
CB3 9EW Cambridge
United Kingdom
a.koubek@damtp.cambridge.ac.uk

E. Langmann
Department of Theor. Physics
Royal Institute of Technology
S-10044 Stockholm
Sweden
langmann@theophys.kth.se

J. Lukierski
Institute of Theor. Physics
PI. Maksa Borna 9
University of Wroclaw
50204 Wroclaw
Poland
lukier@proton.ift.uni.wroc.pl

G. Meissner
Inst. f. Theor. Physik
Postfach 15 11 50
Universität des Saarlandes
D-66041 Saarbrücken
Germany
meissner@usb612.phys-th.uni-sb.de

P. Presnajder
Dept. of Theoretical Physics
Comenius University
Mlynská dolina
SK-84215 Bratislava
Slovakia
presnajder@fmph.uniba.sk

A. Protogenov
Inst. of Appl. Physics
Russian Acad. of Sciences
46 Ulyanov Street
603600 Nizhny Novgorod
Russia
alprot@appl.nnov.su

P. Pyatov
Bogoliubov Lab. of
Theoretical Physics
JINR
141980 Dubna
Russia
pyatov@thsun1.jinr.dubna.su

J. Sobczyk
Institute of Theor. Physics
PI. Maksa Borna 9
University of Wroclaw
50204 Wroclaw
Poland

Th. Strobl
Inst. f. Theoret. Physik
RWTH-Aachen
Sommerfeldstr. 26-28
D-52056 Aachen
Germany
tstrobl@pluto.physik.rwth-aachen.de

B.M. Zupnik
Bogoliubov Laboratory
of Theoretical Physics
JINR
141980 Dubna
Russia
zupnik@thsun1.jinr.dubna.su

The Chern-Simons Theory and Quantized Moduli Spaces of Flat Connections

Anton Yu. Alekseev[1] and Volker Schomerus

[1] Institut für Theoretische Physik, ETH-Hönggerberg, CH-8093 Zürich, Switzerland

[2] II. Institut für Theoretische Physik, Universität Hamburg, Luruper Chaussee 149, 22761 Hamburg, Germany.

1. Introduction

The main topic of this lecture series is the program of Combinatorial Quantization of the moduli space of flat connections. In the language of Physics we call the same subject operator quantization of the Chern-Simons theory. This most naive and old-fashioned approach to quantization was not applicable to the moduli space until recently, when the r-matrix description of the corresponding Poisson structure has been suggested [1]. The quantization program has been carried out in [2], [3]. One of the main results presented in these lectures is the classification of unitary representations of the moduli algebra [4]. Their representation spaces are natural candidates for the role of the Hilbert space of the Chern-Simons theory. And indeed, one can check that the dimensions of these spaces are given by Verlinde numbers.

Another point of view on the same algebra comes from the dimensional reduction. An algebra of observables of the 3-dimensional theory (Chern-Simons) may be viewed as a noncommutative algebra of dynamical variables of a certain 2-dimensional statistical model. This model proves to be a lattice gauge theory with a quantum gauge group. This idea has been suggested in [6]. The simplest algebra which possesses such a symmetry and corresponds to a 1-dimensional lattice was suggested in [7]. The moduli algebra has been treated from this point of view in [8]. There one finds the first comparison of the results with the Verlinde formula.

Let us also mention the perturbation theory approach to quantization of the moduli space [9], which relates this problem to the theory of Vassiliev invariants.

On the other pole of the Mathematical Physics globe are attempts to apply the lattice gauge models with quantum gauge group to regularization of gauge theories in 4 dimensions [10]. We are not able to comment on this issue.

In these lectures we extensively use the results and ideas of [1], [2], [3], [4], [8], [3], [4], [5].

2. Chern-Simons Theory and the Moduli Space of Flat Connections

The Chern-Simons theory is constructed by the following data. We pick up some semi-simple Lie algebra $\mathcal{G}$, a coupling constant k, and a 3-manifold M. The Chern-Simons action is a functional of a $\mathcal{G}$-valued gauge field A on M and has the form

$$CS(A) = \frac{k}{4\pi} Tr \int_M (A \wedge dA + \frac{2}{3} A^3). \tag{2.1}$$

We do not describe truly exceptional properties of this action and of the corresponding theory, and refer the reader to the original papers and numerous reviews. Let us only mention that, for $\mathcal{G}$ being a compact Lie algebra, k is required to be integer in order to ensure the global gauge invariance.

When the manifold M has a structure of a direct product of a 2-dimensional surface and a segment of the real line, the Chern-Simons theory admits a Hamiltonian interpretation. As in any topological theory, the Hamiltonian is equal to zero. The action (2.1) induces a symplectic structure on the space of connections [11]:

$$\Omega = \frac{k}{4\pi} Tr \int_\Sigma \delta_1 A \wedge \delta_2 A. \tag{2.2}$$

This symplectic form is invariant with respect to gauge transformations

$$A^g = gAg^{-1} + dgg^{-1}. \tag{2.3}$$

An easy check shows that the moment map for the gauge group action is proportional to the curvature

$$F = dA + A^2. \tag{2.4}$$

The condition

$$F = 0 \tag{2.5}$$

emerges as a constraint from the Chern Simons action. From this analysis we see, that the phase space of the Chern Simons model is a quotient of the space of flat connections (2.5) over the gauge group action (2.3). In this paper we often refer to this space as moduli space, which should always be understood as a moduli space of flat connections on a Riemann surface.

2.1 Combinatorial Description

Let us introduce a more efficient finite-dimensional description of the moduli space which will be of use throughout the paper. At the same moment we introduce moduli spaces on surfaces with marked points. Let G be a semi-simple connected simply-connected Lie group corresponding to the Lie algebra $\mathcal{G}$, and $\Sigma_{g,m}$ be a surface of genus g with m marked points. Assign a conjugacy class $\mathcal{C}_\nu \in G$ to the ν's marked point. Denote $\pi = \pi_{g,m}$ the fundamental group of the surface $\Sigma_{g,m}$. The group π may be generated by $2g + m$ invertible generators $a_i, b_i, i = 1, \dots, g$, and $l_\nu, \nu = 1, \dots, m$, subject to the relation

$$[b_g, a_g^{-1}] \dots [b_1, a_1^{-1}] l_m \dots l_1 = id \quad . \quad (in\ \pi) \tag{2.6}$$

Here $[x, y]$ stays for the group commutator $xyx^{-1}y^{-1}$.

Definition 1 *The moduli space of flat connections* $\mathbf{m}_{g,m}^{\{\mathcal{C}_\nu\}}$ *on the Riemann surface of genus g with m marked points is defined as*

$$\mathbf{m}_{g,m}^{\{\mathcal{C}_\nu\}} = \{\rho \in Hom(\pi, G), \rho(l_\nu) \in \mathcal{C}_\nu\}/G \quad . \tag{2.7}$$

Here the group G acts on the space of representations of π by conjugations

$$\rho^g(x) = g^{-1}\rho(x)g \quad . \tag{2.8}$$

In order to make contact with the definition which involves flat connections, we represent generators of π as circles on $\Sigma_{g,m}$ intersecting at the base point on the surface. Then a flat connection A induces a representation of the fundamental group via

$$\begin{aligned} M_\nu &= \rho(l_\nu) = Hol(A, l_\nu) \ ; \\ A_i &= \rho(a_i) = Hol(A, a_i) \ ; \\ B_i &= \rho(b_i) = Hol(A, b_i) \ . \end{aligned} \tag{2.9}$$

Here $Hol(A, x)$ is a holonomy of the connection A along the cycle x. Connection A being flat, ρ satisfies the defining relation

$$M_0 = [B_g, A_g^{-1}] \dots [B_1, A_1^{-1}] M_m \dots M_1 = id \quad . \quad (in\ G) \tag{2.10}$$

The case of a surface with marked points requires some further comment. The marked points are deleted from the surface, and the condition of flatness does not hold there. In general we permit the curvature to develop δ-function singularities at the marked points:

$$F = \sum_\nu T_\nu \delta^{(2)}(z - z_\nu) \quad . \tag{2.11}$$

Here T_ν's are elements of the algebra $\mathcal{G}$. However, if one permits arbitrary T_ν's, the space of all connections fails to be symplectic. It regains this property, if we fix conjugacy classes of T_ν in $\mathcal{G}$. This condition may be naturally derived from the Chern-Simons action functional in the same fashion as one obtains the flatness condition for a surface without marked points [12].

Assume that we have fixed some conjugacy classes for T_ν's in the modified flatness condition (2.11). Then via equations (2.9) one can define a subset of representations of the fundamental group of a surface with marked points. It is easy to check that the condition (2.11) implies, that $\rho(l_\nu)$ belong to certain conjugacy classes in the group G, as it is stated in the definition of the moduli space. In fact, one can get the conjugacy class of $\rho(l_\nu)$ by applying the exponential map $exp : \mathcal{G} \to G$ to the conjugacy class of T_ν. In this way we obtain a family of symplectic spaces labeled by two nonnegative integers (genus and the number of marked points), and by conjugacy classes attached to the marked points.

Notations (2.9) prove to be quite useful when constructing functions on the moduli space. Take any function f on the direct product of $2g+m$ copies of the group G, which is invariant with respect to simultaneous conjugations of the arguments,

$$f(g^{-1}g_1g, \dots, g^{-1}g_{(2g+m)}g) = f(g_1, \dots, g_{(2g+m)}) \quad . \tag{2.12}$$

Define a function $f(\rho)$ on the space of representations via

$$f(\rho) = f(M_1, \dots, M_m, A_1, B_1, \dots, A_g, B_g). \tag{2.13}$$

It is easy to see that this function descends to the quotient space. In fact, any analytic function on the moduli space may be obtained in this way.

2.2 Poisson Structure of the Moduli Space

As the quantization program which we are going to develop is close to deformation quantization, we describe the Poisson bracket on the moduli space defined by Atiyah-Bott symplectic structure. This Poisson bracket was described in [13]. However, we use another description of the same object which was especially designed for the needs of deformation quantization [1].

Let us introduce some useful notations. We define left and right invariant differential operators on a Lie group with values in the dual space to its Lie algebra:

$$\begin{aligned} < \nabla_L f(g), X > &= \frac{d}{dt} f(e^{-tX}g)|_{t=0} \quad ; \\ < \nabla_R f(g), X > &= \frac{d}{dt} f(ge^{tX})|_{t=0} \quad . \end{aligned} \tag{2.14}$$

As we work with semi-simple algebras, one can introduce a nondegenerate Killing form $K \in \mathcal{G} \otimes \mathcal{G}$. Let us consider one more element $r \in \mathcal{G} \otimes \mathcal{G}$. It is called classical r-matrix, if it satisfies the classical Yang-Baxter equation in $U(\mathcal{G})^{\otimes 3}$:

$$[r_{12}, r_{13}] + [r_{12}, r_{23}] + [r_{13}, r_{23}] = 0 \ . \tag{2.15}$$

Here we denote $r_{12} = r \otimes 1$, r_{13} and r_{23} are constructed in a similar way. For any solution r of the classical Yang-Baxter equation one can construct another solution r' which is obtained by permutation of two copies of $\mathfrak{g}$. Let us prepare their symmetric and antisymmetric combinations

$$r_s = \frac{1}{2}(r + r') \ , \ \ r_a = \frac{1}{2}(r - r') \ . \tag{2.16}$$

In general, neither r_a nor r_s satisfy the classical Yang-Baxter equation.

In order to describe the Poisson bracket on the moduli space, we use its description via representations of the fundamental group from the previous subsection. As the space of representations is embedded into G^{2g+m}, it is convenient to order the $2(2g+m)$ covariant differential operators in the following way:

$$\begin{aligned} \nabla_{2\nu-1} = \nabla_R^{M_\nu} \ &, \ \ \nabla_{2\nu} = \nabla_L^{M_\nu} \ \text{ for } \ \nu = 1, \ldots, m \ ; \\ \nabla_{m+4i-3} = \nabla_R^{A_i} \ &, \ \ \nabla_{m+4i-1} = \nabla_L^{A_i} \ \text{ for } \ i = 1, \ldots, g \ ; \\ \nabla_{m+4i-2} = \nabla_R^{B_i} \ &, \ \ \nabla_{m+4i-1} = \nabla_L^{B_i} \ \text{ for } \ i = 1, \ldots, g \ . \end{aligned} \tag{2.17}$$

One can rewrite the condition for a function f on G^{2g+m} to be invariant in the sense of (2.12) as

$$\sum_{s=1}^{2(2g+m)} \nabla_s f = 0 \ . \tag{2.18}$$

With these notations we present a description of the Poisson structure on the moduli space.

Theorem 1 (Fock-Rosly [1]) *Let $r \in \mathfrak{g} \otimes \mathfrak{g}$ be a solution of the classical Yang-Baxter equation. Assume that its symmetric part coincides with the Killing bilinear $r_s = K$. Introduce a Poisson bracket on G^{2g+m} by the following formula:*

$$\{f, h\} = \frac{1}{2} \sum_i < r, \nabla_i f \wedge \nabla_i h > + \sum_{i<j} < r, \nabla_i f \wedge \nabla_j h > . \tag{2.19}$$

This Poisson bracket restricts to the space of functions, which are invariant with respect to simultaneous conjugations. Being mapped to the moduli space by means of equation (2.13), the bracket (2.19) coincides with the canonical Poisson bracket defined by the Atiyah-Bott symplectic structure.

The Poisson bracket (2.19) may be easily evaluated for simplest functions on G^{2g+m}. Examples of such functions are given by matrix elements of M_ν, A_i and B_i evaluated in some irreducible representations. We denote a matrix which represents some $X \in G$ in the representation τ^I by X^I. Applying a pair of representations τ^I and τ^J to the classical r-matrix, we produce $r^{IJ} = (\tau^I \otimes \tau^J)(r)$. It is convenient to introduce tensor notations

$X^1 = X \otimes 1, X^2 = 1 \otimes X$. Let us give some examples of Poisson brackets for particular matrix elements:

$$\begin{aligned}
\{\overset{1}{M}{}^I_\nu, \overset{2}{M}{}^J_\mu\} &= r^{IJ}\, \overset{1}{M}{}^I_\nu\, \overset{2}{M}{}^J_\mu - \overset{1}{M}{}^I_\nu r^{IJ}\, \overset{2}{M}{}^J_\mu \\
&\quad - \overset{2}{M}{}^J_\mu r^{IJ}\, \overset{1}{M}{}^I_\nu + \overset{1}{M}{}^I_\nu\, \overset{2}{M}{}^J_\mu r^{IJ} \quad \textit{for } \nu < \mu \ ; \\
\{\overset{1}{M}{}^I_\nu, \overset{2}{M}{}^J_\nu\} &= r^{IJ}\, \overset{1}{M}{}^I_\nu\, \overset{2}{M}{}^J_\nu - \overset{1}{M}{}^I_\nu r^{IJ}\, \overset{2}{M}{}^J_\nu \\
&\quad + \overset{2}{M}{}^J_\nu (r')^{IJ}\, \overset{1}{M}{}^I_\nu - \overset{1}{M}{}^I_\nu\, \overset{2}{M}{}^J_\nu (r')^{IJ} \ ; \qquad (2.20) \\
\{\overset{1}{A}{}^I_i, \overset{2}{B}{}^J_i\} &= r^{IJ}\, \overset{1}{A}{}^I_i\, \overset{2}{B}{}^J_i - \overset{1}{A}{}^I_i r^{IJ}\, \overset{2}{B}{}^J_i + \overset{2}{B}{}^J_i (r')^{IJ}\, \overset{1}{A}{}^I_i + \overset{1}{A}{}^I_i\, \overset{2}{B}{}^J_i r^{IJ} \ .
\end{aligned}$$

As one can see, a Poisson bracket of two matrix elements is quadratic in matrix elements of the same representations. Classical r-matrices r^{IJ} and $(r')^{IJ}$ play the role of structure constants in this Poisson bracket algebra. This point is the most important observation of [1], as it makes possible to proceed with quantization of the algebra of functions on the moduli space.

Matrix elements of M_ν, A_i, B_i which we considered so far do not define functions on the moduli space, as they are not invariant with respect to conjugations. Let us remark that the Poisson bracket (2.19) is not conjugation invariant. Nevertheless it may be consistently restricted to the set of conjugation invariant functions, which means that a Poisson bracket of two such functions is again conjugation invariant. Conjugation invariant functions are produced as linear combinations of elements

$$tr^J \Big(C_1[I_1, \dots, I_{2g+m}|J] M_1^{I_1} \dots M_m^{I_m} A_1^{I_{m+1}} \dots B_g^{I_{2g+m}} C_2[I_1, \dots, I_{2g+m}|J]^* \Big)$$

for arbitrary sets of labels $\{I_\nu\}, J$, and two intertwiners $C_1, C_2 : V^{I_1} \otimes \dots \otimes V^{I_{2g+m}} \mapsto V^J$.

As we have seen in the previous subsection, any invariant function on G^{2g+m} defines a function on the moduli space. In fact, we restrict an invariant function to the subset defined by the conditions (cp. formula 2.10 for notations)

$$M_\nu \in \mathcal{C}_\nu \quad , \quad M_0 = 1 \ . \qquad (2.21)$$

By the theorem of Fock and Rosly this restriction is consistent with the structure of the bracket. This means that constraints (2.21) are preserved by Hamiltonian flows produced by invariant functions. As it is stated in Theorem 2.2, the resulting bracket on the moduli space coincides with the canonical one.

3. Quantization of the Moduli Space

As we mentioned above, the structure constants of Poisson algebra (2.19) are defined by a couple of classical r-matrices r and r'. The key point in the quantization procedure is the fact that we know a family of solutions of the quantum Yang-Baxter equation

$$R_{12}R_{13}R_{23} = R_{23}R_{13}R_{12}, \tag{3.1}$$

labeled by a parameter h so that

$$R^{IJ}|_{h\to 0} = I + hr + \mathcal{O}(h^2) \ . \tag{3.2}$$

We assume that R^{IJ} exist for any I and J, and that they are represented by matrices in the same spaces as r^{IJ}. Our intention is to replace the Poisson algebra with structure constants r, r' by an associative algebra with structure constants R, R'. A natural framework for this construction is provided by the theory of quasi-triangular Hopf algebras.

3.1 Quasi-Triangular Hopf Algebras

The objects we wish to investigate later are associated with a quantum symmetry algebra $\mathcal{G}$. More precisely, $\mathcal{G}$ is a ribbon Hopf-*-algebra, i.e. a $*$-algebra $\mathcal{G}$ with co-unit $\epsilon : \mathcal{G} \mapsto \mathbf{C}$, co-product $\Delta : \mathcal{G} \mapsto \mathcal{G} \otimes \mathcal{G}$, antipode $\mathcal{S} : \mathcal{G} \mapsto \mathcal{G}$, R-matrix $R \in \mathcal{G} \otimes \mathcal{G}$, and the ribbon element $v \in \mathcal{G}$. We do not want to spell out all the standard axioms these objects have to satisfy in order to give a ribbon Hopf algebra (for a complete definition see e.g. [14]). Let us stress, however, that we deal with structures for which the co-product Δ is consistent with the action

$$(\xi \otimes \eta)^* = \eta^* \otimes \xi^* \quad \text{for all} \quad \eta, \xi \in \mathcal{G}$$

of the *-operation $*$ on elements in the tensor product $\mathcal{G} \otimes \mathcal{G}$. This case is of particular interest, since it appears for the quantized universal enveloping algebras $U_q(\mathcal{G})$, when the complex parameter q has values on the unit circle [15].

Given the standard expansion of $R \in \mathcal{G} \otimes \mathcal{G}$, $R = \sum r^1_\sigma \otimes r^2_\sigma$, one constructs the elements

$$u = \sum \mathcal{S}(r^2_\sigma) r^1_\sigma \ . \tag{3.3}$$

Among the properties of u (cp. e.g. [14]) one finds that the product $u\mathcal{S}(u)$ is in the center of $\mathcal{G}$. The ribbon element v is a central square root of $u\mathcal{S}(u)$, which obeys the following relations:

$$v^2 = u\mathcal{S}(u) \quad , \quad \mathcal{S}(v) = v \ , \quad \epsilon(v) = 1 \ , \tag{3.4}$$

$$v^* = v^{-1} \quad , \quad \Delta(v) = (R'R)^{-1}(v \otimes v) \ . \tag{3.5}$$

The elements u and v can be combined to furnish a grouplike unitary element $g = u^{-1}v \in \mathcal{G}$. Examples of ribbon-Hopf-*-algebras are given by the enveloping algebras of all simple Lie algebras [14].

In the following we make several additional assumptions about the ribbon Hopf algebra $\mathcal{G}$. To begin with, we will assume that $\mathcal{G}$ is *semisimple*. More restrictions will be imposed in subsection 3.4 .

For every equivalence class $[J]$ of irreducible *-representations of $\mathcal{G}$ we pick a particular representative τ^J with carrier space V^J. The tensor product $\tau^I \boxtimes \tau^J$ of two representations τ^I, τ^J of the semisimple algebra $\mathcal{G}$ can be decomposed into irreducibles τ^K. This decomposition determines the Clebsch-Gordon maps $C^a[IJ|K] : V^I \otimes V^J \mapsto V^K$,

$$C^a[IJ|K](\tau^I \boxtimes \tau^J)(\xi) = \tau^K(\xi)C^a[IJ|K] \quad . \tag{3.6}$$

The same representations τ^K in general appears with some multiplicity N_K^{IJ}. The superscript $a = 1, \ldots, N_K^{IJ}$ keeps track of these subrepresentations. It is common to call the numbers N_K^{IJ} *fusion rules*. Normalization of these Clebsch Gordon maps is connected with an extra assumption. Notice that the ribbon element v is central, so that the evaluation with irreducible representations τ^I gives complex numbers $v_I = \tau^I(v)$. We suppose that there exists a set of square roots $\kappa_I, \kappa_I^2 = v_I$, such that

$$C^a[IJ|K](R')^{IJ}C^b[IJ|L]^* = \delta_{a,b}\delta_{K,L}\frac{\kappa_I\kappa_J}{\kappa_K} \quad . \tag{3.7}$$

Here $R' = \sum r_\sigma^2 \otimes r_\sigma^1$ and $(R')^{IJ} = (\tau^I \otimes \tau^J)(R')$. The adjoint of the Clebsch Gordon map is meant with respect to the standard scalar product on $V^I \otimes V^J$ induced by the scalar products on V^I, V^J. Let us analyze this relation in more detail. As a consequence of intertwining properties of the Clebsch Gordon maps and the R-element, $\tau^K(\xi)$ commutes with the left hand side of the equation. So by Schurs' lemma, it is equal to the identity e^K times some complex factor $\omega_{ab}(IJ|K)$. After appropriate normalization, $\omega_{ab}(IJ|K) = \delta_{a,b}\omega(IJ|K)$ with a complex phase $\omega(IJ|K)$. Next we exploit the $*$-operation and relation (3.5) to find $\omega_{ab}(IJ|K)^2 = v_I v_J / v_K$. This means that (3.7) can be ensured up to a possible sign $\pm$. Here we assume that this sign is always $+$. This assumption was crucial for the positivity in [2]. It is met by the quantized universal enveloping algebras of all simple Lie algebras, because they are obtained as deformations of Hopf-algebras which clearly satisfy (3.7). As a consequence of the normalization equation (3.7) and the equation $\Delta(e) = e \otimes e$ we obtain the following completeness:

$$\sum_{K,a} \frac{\kappa_K}{\kappa_I\kappa_J}(R')^{IJ}C^a[IJ|K]^*C^a[IJ|K] = e^I \otimes e^J \quad . \tag{3.8}$$

We wish to combine the phases κ_I into one element κ in the center of $\mathcal{G}$, i.e. by definition, κ will denote a central element

$$\kappa \in \mathcal{G} \quad \text{with} \quad \tau^J(\kappa) = \kappa_J \quad . \tag{3.9}$$

Such an element does exist and is unique. It has the property $\kappa^* = \kappa^{-1}$.

The antipode $\mathcal{S}$ of $\mathcal{G}$ furnishes a conjugation in the set of equivalence classes of irreducible representations. We use $[\bar{J}]$ to denote the class conjugate to $[J]$. Some important properties of the fusion rules N_K^{IJ} can be formulated with the help of this conjugation. Among them are the relations

$$N_0^{K\bar{K}} = 1 \ , \ N_K^{IJ} = N_K^{JI} = N_{\bar{I}}^{J\bar{K}} \ . \tag{3.10}$$

The numbers v_I are symmetric under conjugation, i.e. $v_K = v_{\bar{K}}$.

The *q-trace* of an element $X \in End(V^K)$ is defined by

$$tr_q^K(X) = tr^K(X\tau^K(g)), \tag{3.11}$$

where $g = u^{-1}v$ is the grouplike element introduced above. Let us also mention that the q-trace of the identity map $e^K \in End(V^K)$ computes the "quantum dimension" d_K of the representation τ^K [14], i.e.

$$d_K \equiv tr_q^K(e^K) \ . \tag{3.12}$$

The numbers d_K satisfy the equalities $d_I d_J = \sum N_K^{IJ} d_K$ and $d_K = d_{\bar{K}}$. In general, the quantum dimensions d_I differ from the dimensions δ_I of the representation spaces V^I.

3.2 The Graph Algebra $\mathcal{L}_{g,m}$

Equipped with the technique and notations of the preceeding subsection, we introduce a quantized version of the Poisson algebra (2.19) on G^{2g+m}. This is the first step towards quantization of the moduli space.

Definition 2 (Graph algebra $\mathcal{L}_{g,m}$) The graph-algebra $\mathcal{L}_{g,m}$ *is an associative algebra generated by matrix elements of* $M_\nu^I, A_i^I, B_i^I \in End(V^I) \otimes \mathcal{L}_{g,m}$, $\nu = 1, \ldots, m, i = 1, \ldots, g$. *The superscript* I *runs through the set of irreducible representations of a quantum symmetry algebra* $\mathcal{G}$ *with R-element* R. *Elements in* $\mathcal{L}_{g,m}$ *are subject to the following relations*:

$$\overset{1}{M}{}_\nu^I R^{IJ} \overset{2}{M}{}_\nu^J = \sum C^a[IJ|K]^* M_\nu^K C^a[IJ|K] \ , \tag{3.13}$$

$$\overset{1}{A}{}_i^I R^{IJ} \overset{2}{A}{}_i^J = \sum C^a[IJ|K]^* A_i^K C^a[IJ|K] \ , \tag{3.14}$$

$$\overset{1}{B}{}_i^I R^{IJ} \overset{2}{B}{}_i^J = \sum C^a[IJ|K]^* B_i^K C^a[IJ|K] \ , \tag{3.15}$$

$$(R^{-1})^{IJ} \overset{1}{A}{}_i^I R^{IJ} \overset{2}{B}{}_i^J = \overset{2}{B}{}_i^J (R')^{IJ} \overset{1}{A}{}_i^I R^{IJ} \ , \tag{3.16}$$

$$(R^{-1})^{IJ} \overset{1}{M}{}_\nu^I R^{IJ} \overset{2}{M}{}_\mu^J = \overset{2}{M}{}_\mu^J (R^{-1})^{IJ} \overset{1}{M}{}_\nu^I R^{IJ} \quad \text{for} \ \ \nu < \mu \ . \tag{3.17}$$

$$(R^{-1})^{IJ} \overset{1}{M}{}_\nu^I R^{IJ} \overset{2}{A}{}_j^J = \overset{2}{A}{}_j^J (R^{-1})^{IJ} \overset{1}{M}{}_\nu^I R^{IJ} \quad \text{for all} \ \ \nu, j \ , \tag{3.18}$$

$$(R^{-1})^{IJ} \overset{1}{M}{}_\nu^I R^{IJ} \overset{2}{B}{}_j^J = \overset{2}{B}{}_j^J (R^{-1})^{IJ} \overset{1}{M}{}_\nu^I R^{IJ} \quad \text{for all} \ \ \nu, j \ , \tag{3.19}$$

$$(R^{-1})^{IJ} \overset{1}{A}{}_i^I R^{IJ} \overset{2}{A}{}_j^J = \overset{2}{A}{}_j^J (R^{-1})^{IJ} \overset{1}{A}{}_i^I R^{IJ} \quad \text{for} \quad i<j \ , \tag{3.20}$$

$$(R^{-1})^{IJ} \overset{1}{B}{}_i^I R^{IJ} \overset{2}{B}{}_j^J = \overset{2}{B}{}_j^J (R^{-1})^{IJ} \overset{1}{B}{}_i^I R^{IJ} \quad \text{for} \quad i<j \ , \tag{3.21}$$

$$(R^{-1})^{IJ} \overset{1}{A}{}_i^I R^{IJ} \overset{2}{B}{}_j^J = \overset{2}{B}{}_j^J (R^{-1})^{IJ} \overset{1}{A}{}_i^I R^{IJ} \quad \text{for} \quad i<j \ , \tag{3.22}$$

$$(R^{-1})^{IJ} \overset{1}{B}{}_i^I R^{IJ} \overset{2}{A}{}_j^J = \overset{2}{A}{}_j^J (R^{-1})^{IJ} \overset{1}{B}{}_i^I R^{IJ} \quad \text{for} \quad i<j \ . \tag{3.23}$$

Now we motivate and discuss some pieces of this long definition.

First of all, the most part of defining relations are quantized counterparts of Poisson brackets (2.19). Let us pick up as an example the equation

$$(R^{-1})^{IJ} \overset{1}{A}{}_i^I R^{IJ} \overset{2}{B}{}_i^J = \overset{2}{B}{}_i^J (R')^{IJ} \overset{1}{A}{}_i^I R^{IJ} \ . \tag{3.24}$$

In order to recover the Poisson algebra, one has to use the standard ansatz of quantum mechanics (axiomatized by the deformation quantization theory).

Definition 3 *Let X and Y be elements of an associative algebra defined over formal power series in h. Assume that this algebra becomes Abelian for $h=0$, and*

$$XY - YX = hZ_1 + h^2 Z_2 + \dots \ . \tag{3.25}$$

Then the Poisson bracket of X and Y is defined as

$$\{X,Y\} = Z_1 \ . \tag{3.26}$$

Applying this definition to equation (3.24), we expand quantum R-matrices into the power series in h. As a coefficient at h^0 we discover the commutator of $\overset{1}{A}{}_i^I$ and $\overset{2}{B}{}_i^J$, which fits to the l.h.s. of (3.25). To evaluate the first order in h, we use formula (3.2) and recover the Poisson bracket as presented in the r.h.s. of the last equation in (2.20).

What seems to be missing in this picture are counterparts of mutual Poisson brackets of matrix elements of the same holonomy. Let us temporarily reintroduce these missing relations:

$$\begin{aligned}
(R^{-1})^{IJ} \overset{1}{A}{}_i^I R^{IJ} \overset{2}{A}{}_i^J &= \overset{2}{A}{}_i^J (R')^{IJ} \overset{1}{A}{}_i^I (R'^{-1})^{IJ} \ , \\
(R^{-1})^{IJ} \overset{1}{B}{}_i^I R^{IJ} \overset{2}{B}{}_i^J &= \overset{2}{B}{}_i^J (R')^{IJ} \overset{1}{B}{}_i^I (R'^{-1})^{IJ} \ , \\
(R^{-1})^{IJ} \overset{1}{M}{}_\nu^I R^{IJ} \overset{2}{M}{}_\nu^J &= \overset{2}{M}{}_\nu^J (R')^{IJ} \overset{1}{M}{}_\nu^I (R'^{-1})^{IJ} \ .
\end{aligned} \tag{3.27}$$

Together with other quadratic relations of the last definition, equations (3.27) provide a quantization of the Poisson bracket (2.19). Let us mention, that in fact formula (2.19) encodes the same number of equations as (3.16-3.23) together with (3.27). Quantum exchange relations look somewhat more complicated only for notational reasons.

The key observation, which one can make looking at the quadratic relations (3.16-3.23), (3.27), is the presence of the Hopf algebra symmetry in the graph algebra. More explicitly, let ξ be an element of $\mathcal{G}$, and

$$\Delta(\xi) = \sum \xi^1_\sigma \otimes \xi^2_\sigma. \tag{3.28}$$

The action of ξ on the generators of $\mathcal{L}_{g,m}$ is defined as follows:

$$\begin{aligned} \xi(M^I_\nu) &= \sum_\sigma \tau^I(S(\xi^1_\sigma))M^I_\nu \tau^I(\xi^2_\sigma) \ , \\ \xi(A^I_i) &= \sum_\sigma \tau^I(S(\xi^1_\sigma))A^I_i \tau^I(\xi^2_\sigma) \ , \\ \xi(B^I_i) &= \sum_\sigma \tau^I(S(\xi^1_\sigma))B^I_i \tau^I(\xi^2_\sigma) \ . \end{aligned} \tag{3.29}$$

One can continue this action to the whole of $\mathcal{L}_{g,m}$, using the property of generalized derivations [16]

$$\xi(XY) = \sum_\sigma \xi^1_\sigma(X)\xi^2_\sigma(Y) \ . \tag{3.30}$$

Here X and Y are arbitrary elements of $\mathcal{L}_{g,m}$. Formulas (3.29) and (3.30) provide a proper generalization for the quantized case of simultaneous conjugations (2.12). Quadratic exchange relations, which define the algebra $\mathcal{L}_{g,m}$, are invariant with respect to the action of the quantum symmetry in the following sense. For each of them

$$\xi(l.h.s) = \xi(r.h.s) \tag{3.31}$$

for any $\xi \in \mathcal{G}$.

There is no surprise that we discover the quantum invariance in the quadratic exchange relations. Indeed, they are defined by the same set of R-matrices as the quasi-triangular Hopf symmetry algebra.

In fact, we can make the quantum description closer to the classical one, if we trade the action of $\mathcal{G}$ for the coaction of the dual Hopf algebra $\mathcal{G}^*$. The latter is generated by the matrix elements of $g^I \in End(V^I) \otimes \mathcal{G}^*$. Among the others they satisfy quadratic exchange relations

$$R^{IJ} g^I g^J = g^J g^I R^{IJ}. \tag{3.32}$$

These are famous defining relations for the algebra of functions on a quantum group [17].

It is easy to check that the mapping $\mathbf{g} : \mathcal{L}_{g,m} \to \mathcal{G}^* \otimes \mathcal{L}_{g,m}$ defined as

$$\begin{aligned} M^I_\nu &\to (g^I)^{-1} M^I_\nu g^I \ , \\ A^I_i &\to (g^I)^{-1} A^I_i g^I \ , \\ B^I_i &\to (g^I)^{-1} B^I_i g^I , \end{aligned} \tag{3.33}$$

preserves exchange relations. The formalism (3.33) is more transparent. In particular, it has been extensively used in [8] for description of lattice gauge models with quantum gauge group. However, in view of the generalizations for the quasi-Hopf algebras, we stick to the more formal definition of the quantum invariance, which involves the action of $\mathcal{G}$ rather than the coaction of $\mathcal{G}^*$. Let us stress that in the Hopf algebra case these two approaches are completely equivalent.

Now we can explain the origin of the first three relations in the definition of the graph algebra. Before quantization holonomies M_ν, A_i, and B_i take values in the group G. This implies that matrix elements of the same holonomy in different representations are not algebraically independent. More explicitly, a product of two matrix elements of some holonomy matrix in two different representations may be decomposed into a sum of certain matrix elements by means of Clebsch-Gordon maps:

$$\overset{1}{M}{}^I_\nu \overset{2}{M}{}^J_\nu = \sum C_0^a[IJ|K]^* M_\nu^K C_0^a[IJ|K] \ . \tag{3.34}$$

Here the Clebsch-Gordon maps with the subscript 0 refer to the undeformed Lie group G.

Relations of the type (3.34) do not hold in the quantized algebra. They would contradict e.g. the quadratic exchange relations (3.27). The first three relations in the definition of the graph algebra provide a proper substitute for (3.34). They are chosen to be consistent with the symmetry Hopf algebra action (3.29), (3.30) on the graph algebra. Equations (3.27) follow from the multiplication laws (3.13-3.15). That is why we do not include (3.27) into the basic definition.

Let us formulate the outcome of the preceeding discussion.

Lemma 3.1. *For generic values of the deformation parameter q, the graph algebra $\mathcal{L}_{g,m}$ provides a deformation quantization of the algebra of analytic functions on G^{2g+m}, consistent with the Fock-Rosly Poisson bracket (2.19).*

The deformation parameter q in the formulation of Lemma 1 is the parameter, which enters into R-matrices and Clebsch-Gordon maps which define structure constants of $\mathcal{L}_{g,m}$. The algebra of analytic functions on the compact group G^{2g+m} is generated by matrix elements of irreducible finite dimensional representations of G, applied to each term in the Cartesian product.

Lemma 1 motivates our interest to the algebra $\mathcal{L}_{g,m}$ for both generic vaules of q and roots of unity.

3.3 Integration and $*$-Properties of Graph Algebras

This subsection is devoted to the $*$-operation and the integration measure on the graph algebras. These two objects are useful technical tools in the analysis of the representation theory. Also, the $*$-operation is important for the physical interpretation, as an algebra of observables of a quantum system is always a $*$-algebra.

Let us recall, that we assume consistency of the co-product for the Hopf symmetry algebra with the special kind of $*$-operation, which reverses the order in the tensor product (see subsection 3.1). As a consequence of this choice the algebra $\mathcal{L}_{g,m}$ is *not* equipped with a natural $*$-operation. However, we can save the situation using the following trick.

The algebra $\mathcal{G}$ acting on $\mathcal{L}_{g,m}$, we can define a semi-direct product $\mathcal{S}_{g,m}$ of these two objects. It is generated by the elements $\xi \in \mathcal{G}$ and by the elements of $\mathcal{L}_{g,m}$. In order to describe the commutation relations of the symmetry generators and the matrix elements of quantized holonomies, it is convenient to introduce the generating matrices $\mu^I(\xi) \in End(V^I) \otimes \mathcal{G}$ for the symmetry algebra:

$$\mu^I(\xi) = (\tau^I \otimes id)\Delta(\xi) \quad . \tag{3.35}$$

The cross relation between ξ and quantum holonomies in $\mathcal{S}_{g,m}$ look like

$$\mu^J(\xi)M^J_\nu = M^J_\nu \mu^J(\xi) \ , \tag{3.36}$$

$$\mu^J(\xi)A^J_i = A^J_i \mu^J(\xi) \quad , \quad \mu^J(\xi)B^J_i = B^J_i \mu^J(\xi) \ . \tag{3.37}$$

The semi-direct product $\mathcal{S}_{g,m}$ is already a $*$-algebra. The $*$-operation on $\mathcal{G}$ coincides with the genuine $*$-operation of the symmetry algebra. It is continued to the quantum holonomies as [2]:

$$\begin{aligned}(M^I_\nu)^* &= \sigma_\kappa(R^I(M^I_\nu)^{-1}(R^{-1})^I) \ , \\ (A^I_i)^* &= \sigma_\kappa(R^I(A^I_i)^{-1}(R^{-1})^I) \ , \\ (B^I_i)^* &= \sigma_\kappa(R^I(B^I_i)^{-1}(R^{-1})^I) \ .\end{aligned} \tag{3.38}$$

Here we introduced $(M^I_\nu)^{-1}, (A^I_i)^{-1}, (B^I_i)^{-1} \in End(V^I) \otimes \mathcal{S}_{g,m}$ being the unique solutions of

$$M^I_\nu(M^I_\nu)^{-1} = e^I = (M^I_\nu)^{-1}M^I_\nu,$$
$$A^I_i(A^I_i)^{-1} = e^I = (A^I_i)^{-1}A^I_i,$$
$$B^I_i(B^I_i)^{-1} = e^I = (B^I_i)^{-1}B^I_i.$$

The symbol σ_κ stands for the automorphism of $\mathcal{S}_{g,m}$ obtained by conjugation with the unitary element $\kappa \in \mathcal{G}$,

$$\sigma_\kappa(F) = \kappa^{-1}F\kappa$$

for any F.

Another important object which can be introduced for $\mathcal{S}_{g,m}$ is an invariant integration measure. We define a linear functional $\omega : \mathcal{S}_{g,m} \to \mathbf{C}$ as

$$\omega(\ M_1^{I_1} \dots B_g^{I_{m+2g}}\xi\ ) = \epsilon(\xi) \prod_{s=1}^{m+2g} \delta_{I_s,0} \ . \tag{3.39}$$

Formula (3.39) is defined on the set of monomials which span the algebra $\mathcal{S}_{g,m}$. So, ω is continued to the whole algebra as a linear functional. The integral (3.39) may be restricted to the algebra $\mathcal{L}_{g,m}$. There it furnishes the quantum analog of the multidimensional Haar measure on G^{m+2g}. In particular, one can formulate the invariance of ω as

$$\omega(\xi(X)) = \epsilon(\xi)\,\omega(X) \tag{3.40}$$

for any $\xi \in \mathcal{G}$ and $X \in \mathcal{L}_{g,m}$.

Usually an interplay between the $*$-operation and the integration is provided by the positivity property of the Hermitian scalar product defined by the integration functional

$$(X, Y) = \omega(X^* Y) \ .$$

However, this property never holds on $\mathcal{S}_{g,m}$, as ω always has a big kernel in $\mathcal{G}$. On the other hand, one cannot formulate positivity on $\mathcal{L}_{g,m}$, as this is not a $*$-algebra. So, instead of the usual positivity we formulate the following substitute.

Theorem 2 (Positivity [2]) *Assume that the quantum dimensions of all irreducible representations of the Hopf symmetry algebra $\mathcal{G}$ are strictly positive,*

$$d_I > 0 \ \mathit{for\ every}\ I \ . \tag{3.41}$$

Then the restriction of the integral ω to $\mathcal{L}_{g,m}$ is positive in the following sense:

$$\begin{aligned} &\omega(X^*X) \geq 0 \ \text{ and} \\ &\omega(X^*X) = 0 \Rightarrow X = 0 \quad \text{for all } X \in \mathcal{L}_{g,m}. \end{aligned} \tag{3.42}$$

One can find a proof of this theorem in [2].

Let us remark that the element X^* does not belong to $\mathcal{L}_{g,m}$. So, we need the bigger algebra $\mathcal{S}_{g,m}$ in order to make sense of (3.42). In fact, this kind of positivity on the level of graph algebras will be sufficient to provide positivity of the integration measure on the quantized moduli algebras.

3.4 The Moduli Algebra

The construction of the moduli space starting from G^{2g+m} involves two additional steps. First, one has to restrict the consideration to the subspace of invariant functions. Next, one imposes the conditions (2.21). We should find quantum counterparts of both operations.

It is straightforward to generalize the first step.

Definition 4 (Algebras $\mathcal{A}_{g,m}$) *$\mathcal{A}_{g,m}$ is defined as the subalgebra of elements $A \in \mathcal{L}_{g,m}$, which are invariant with respect to the natural action of $\mathcal{G}$, i.e. $\xi(A) = A\epsilon(\xi)$ for all $A \in \mathcal{A}_{g,m}$ and $\xi \in \mathcal{G}$.*

Since elements $\xi \in \mathcal{G}$ act trivially on $\mathcal{A}_{g,m}$, the semi-direct product of $\mathcal{G}$ and $\mathcal{A}_{g,m}$ coincides with the usual Cartesian product $\mathcal{G} \times \mathcal{A}_{g,m}$, and hence the $*$-operation on $\mathcal{S}_{g,m}$ furnishes a $*$-operation on $\mathcal{A}_{g,m}$. Then the modified positivity, which we defined on $\mathcal{L}_{g,m}$, ensures the usual positivity property of ω on $\mathcal{A}_{g,m}$. Let us also remark that elements of $\mathcal{A}_{g,m}$ are linear combinations of expressions of the form:

$$tr^J_q \left(C^q_1[I_1, \ldots, I_{2g+m}|J] M^{I_1}_1 \ldots M^{I_m}_m A^{I_{m+1}}_1 \ldots B^{I_{2g+m}}_g C^q_2[I_1, \ldots, I_{2g+m}|J]^* \right) .$$

Here tr_q is the q-trace, C^q_1, C^q_2 are intertwining operators for the Hopf algebra action. For generic values of q one can establish an isomorphism of the linear spaces $\mathcal{A}_{g,m}$ and $\mathcal{A}^0_{g,m}$. The latter is the space of conjugation invariant analytic functions on G^{2g+m}. Obviously, this isomorphism cannot be lifted to the level of algebraic structures, as the space of functions is Abelian, whereas $\mathcal{A}_{g,m}$ is not.

To perform the second step of the reduction to the moduli space, we single out a particular graph algebra corresponding to one marked point on a Riemann surface.

Definition 5 (Loop algebra $\mathcal{L}$) *The loop algebra $\mathcal{L}$ is an associative algebra isomorphic to the graph algebra $\mathcal{L}_{0,1}$. It is generated by matrix elements of the monodromies $M^I \in End(V^I) \otimes \mathcal{L}$.*

Closely related to $\mathcal{L}$ is the abelian *fusion* (or *Verlinde) algebra.*

Definition 6 (Fusion (or Verlinde) algebra) *Let $\mathcal{G}$ be a semi-simple (quasi-) Hopf $*$-algebra. Its fusion (or Verlinde) algebra $\mathcal{V}$ is an abelian $*$-algebra spanned by the set of generators c^I. Here I runs through the set of all irreducible representations of $\mathcal{G}$. The multiplication law and the $*$-operation in $\mathcal{V}$ are defined as follows:*

$$c^I c^J = \sum N^{IJ}_K c^K, \text{ and } (c^I)^* = c^{\bar{I}} . \tag{3.43}$$

One can employ the q-traces tr^I_q to construct central elements $c^I \in \mathcal{L}$ from the monodromies M^I as

$$c^I = \kappa_I tr^I_q(M^I) . \tag{3.44}$$

In [3] we have demonstrated that the elements (3.44) generate the Verlinde algebra.

It is important that under certain conditions the representations of the loop algebra $\mathcal{L}$ and of the corresponding Verlinde algebra $\mathcal{V}$ may be labeled by the same set of labels as the representations of the symmetry algebra $\mathcal{G}$. Here we describe the representations of the Verlinde algebra by explicit formulas.

Let us introduce a (possibly infinite) matrix

$$s^{IJ} = (tr^I_q \otimes tr^J_q)(R'R) \tag{3.45}$$

with rows and columns labeled by the representations of $\mathcal{G}$. Equations

$$\vartheta^J(c^I) = \frac{s^{IJ}}{d^J} \tag{3.46}$$

define the set of representations of the Verlinde algebra. In a quite general situation this set of representations is complete.

It is convenient to introduce a special notation for the relations

$$\Phi^J = \{c^I = \vartheta^J(c^I)\}, \tag{3.47}$$

which restricts the generators c^I to a certain representation. Imposing these central relations we may get ideals in both $\mathcal{V}$ and $\mathcal{L}$.

Returning to arbitrary values of g and m we introduce $m+1$ embeddings of $\mathcal{L}$ into $\mathcal{L}_{g,m}$, defined by

$$\begin{aligned} e_\nu(M) &= M_\nu \ \text{ for } \nu = 1, \dots, m \ ; \\ e_0(M) &= [B_g, A_g^{-1}] \dots [B_1, A_1^{-1}] M_m \dots M_1 \ . \end{aligned} \tag{3.48}$$

If necessary, these embeddings may be lifted to the corresponding semi-direct products with the symmetry algebra.

The embeddings (3.48) provide a set of elements in $\mathcal{A}_{g,m}$

$$c_\nu^I = \kappa_I tr_q^I(M_\nu^I) \ , \ \nu = 0, \dots, m \ . \tag{3.49}$$

It was shown in [3] that all of them belong to the centre of $\mathcal{A}_{g,m}$. In particular, they commute with each other. Now we are ready to define the moduli algebra.

Definition 7 (Moduli algebra) *Let $\mathcal{G}$ be a quasi-triangular Hopf symmetry algebra, $\Sigma_{g,m}$ a closed oriented 2-dimensional surface of genus g with m marked points, and $I_1, \dots, I_m$ a set of m irreducible representations of $\mathcal{G}$ assigned to the marked points. The moduli algebra $\mathcal{M}_{g,m}^{\{I_\nu\}}$ is defined by these data as a quotient of the invariant algebra $\mathcal{A}_{g,m}$,*

$$\mathcal{M}_{g,m}^{\{I_\nu\}} = \mathcal{A}_{g,m} / \{\Phi^0(M_0), \Phi^{I_\nu}(M_\nu), \nu = 1, \dots, m\}. \tag{3.50}$$

Here 0 labels the trivial representation of $\mathcal{G}$. The moduli algebra $\mathcal{M}_{g,m}^{I_\nu}$ inherits the $*$-operation and the positive integration functional ω from $\mathcal{A}_{g,m}$.

In fact, relations Φ^I are proper quantum counterparts of fixing the eigenvalues of the corresponding quantum holonomy. In particular, the set of relations Φ^0 is equivalent to $M^I = \kappa_I^{-1} e^I$ for any I. The scalar factor κ_I gives a "quantum correction" to the classical flatness condition $M^I = e^I$.

It is plausible, that the moduli algebra is related to the algebra of functions on the moduli space of flat connections in the same fashion, as the algebra $\mathcal{L}_{g,m}$ is related to the algebra of analytic functions on G^{2g+m}.

Before formulating a conjecture, let us introduce some notations. If a Hopf or quasi-Hopf algebra $\mathcal{G}$ is obtained as (possibly truncated) universal enveloping algebra $U_q(\mathcal{g})$, its finite dimensional irreducible representations of nonzero quantum dimension are deformations of certain finite dimensional representations of $U(\mathcal{G})$. One way to see it is to use the Lusztig basis in the universal enveloping algebras. In this way to each representation I_ν, which enters the definition of $\mathcal{M}$, corresponds a representation of $U(\mathcal{G})$ of the same dimension, which we label by the same letter I_ν.

The representation I_ν has a highest weight, which may be viewed as an element of the dual space $\mathcal{G}^*$ to the Lie algebra $\mathcal{G}$. We can assign to this element of $\mathcal{G}^*$ a coadjoint orbit which passes through it. In fact, the orbit, which we obtain in this way, corresponds to the representation I_ν in the orbit-representation correspondance provided by geometric quantization.

Here we assume that the Lie algebra $\mathcal{G}$ is semi-simple. Than the space $\mathcal{G}^*$ may be identified with $\mathcal{G}$ by means of the Killing form. Under this identification, coadjoint orbits are identified with the orbits of adjoint action in $\mathcal{G}$. Summarizing, to each irreducible representation of $U_q(\mathcal{G})$ in the definition of the moduli algebra we assigned an adjoint orbit (or conjugacy class) in $\mathcal{G}$. Let us exponentiate this cojugacy class by means of the exponential map from Lie algebra to the Lie group. We denote the resulting conjugacy class in the group G by $exp(I_\nu)$.

Conjecture 1 *For generic value of the deformation parameter q the moduli algebra $\mathcal{M}_{g,m}^{\{I_\nu\}}$ provides a deformation quantization of the algebra of analytic functions on the moduli space of flat connection $\mathbf{m}_{g,m}^{\{C_\nu\}}$ for $C_\nu = exp(I_\nu)$.*

Here the space of analytic functions on the moduli space is defined as follows. These are functions obtained from conjugation invariant polynomials of matrix elements of finite dimensional representations of G^{2g+m}. In this way we avoid the discussion of possible singularities of the moduli space.

Though we formulate the above statement as conjecture, it is probably not very hard to prove. Let us also notice that, as the moduli space has a symplectic structure and the second Betti number is equal to 1, the deformation quantization is unique up to isomorphism and up to renormalization of the deformation parameter. Strictly speaking this applies only to smooth moduli spaces. So, assuming that Conjecture 1 holds true, we suggest an explicit construction of the unique deformation quantization of the moduli space of flat connections on a Riemann surface.

3.5 The Finite-Dimensional Case

As we already mentioned, we are mostly concerned with the case of q being a root of unity. Then one cannot view the moduli algebra as a deformation of the algebra of functions on the moduli space. Instead, one can reverse the logic and reinterpret the definition of the graph algebra. The latter is defined by choosing a Riemann surface with marked points, a ribbon Hopf $*$-algebra, and a set of representations of this algebra, one representation for each marked point. The main ingredient in these data is the ribbon Hopf algebra. Instead of looking what happens to the the moduli algebra at roots of unity, we look at the symmetry Hopf algebra, define it at roots of unity, and thus induce a new definition of the moduli algebra. Here we follow this strategy. However, the relation between the moduli algebras at generic q and at roots of unity still needs to be clarified.

The first important observation is that at roots of unity quantum universal enveloping algebras have a big centre [18]. A natural object is a quotient over certain central relations which is already finite dimensional [19]. This is a motivation to consider a symmetry Hopf algebra with only a finite number of irreducible representations.

It is instructive to consider the special case of $\mathcal{G} = U_q(sl(2))$ at this point. This algebra is generated by X_+, X_-, and an invertible generator K. For $q^p = 1$, the monomials $X_\pm^p, K^p$ are in the centre of $U_q(sl(2))$. One can form a quotient of this algebra over the ideal generated by relations

$$X_\pm^p = 0 \ , \ K^{2p} = 1. \tag{3.51}$$

The comultiplication lifts to the factor-algebra. We denote the resulting finite-dimensional Hopf algebra by $U_q^F(sl(2))$ We do not factorize over relation $K^p = 1$, in order to preserve half-integer spins in the representation theory.

It is easy to show that the irreducible representations of $U_q^F(sl(2))$ are highest weight representations of spins $j = -p/2, \ldots, (p-1)/2$. Here the representations of interger spin are finite dimensional, of the dimension $p + (2j+1)$. In this series there are representations of spin $j = -p/2$ and of spin $j = (p-1)/2$, which have quantum dimensions equal to zero. Let us call all other irreducible representations *physical.*

To ensure the positivity of ω (see Subsection 3.3) we require the quantum dimensions of all irreducible physical representations to be strictly positive:

$$d_I > 0 \quad \text{for all} \quad I \ . \tag{3.52}$$

In fact, this requirement may be satisfied only at roots of unity. Let us remark that there are still indecomposable representations with vanishing quantum dimensions. We will deal with them later in this subsection.

As the number of representations is finite, one can introduce a normalization constant

$$\mathcal{N} \equiv (\sum_K d_K^2)^{-1/2} < \infty \ . \tag{3.53}$$

It is useful to have a matrix S^{IJ} which differs from s^{IJ} by this scalar factor:

$$S_{IJ} \ \equiv \ \mathcal{N}(tr_q^I \otimes tr_q^J)(R'R) \ . \tag{3.54}$$

We assume that *the matrix S is invertible.* A number of standard properties of S can be derived from the invertibility (and properties of the ribbon Hopf-$*$-algebra). We list them here without further discussion. Proofs can be found e.g. in [20].

$$S_{IJ} = S_{JI} \quad , \quad S_{0J} = \mathcal{N} d_J \ ,$$
$$\sum_J S_{IJ}\overline{S_{KJ}} = \delta_{IK} \quad , \quad \sum_J S_{IJ} S_{JK} = C_{IK} \ , \tag{3.55}$$
$$\sum_K N_K^{IJ} S_{KL} \ = \ S_{JL} S_{IL} (\mathcal{N} d_L)^{-1},$$

with $C_{IJ} = N_0^{IJ}$. For the relations in the second line, the existence of an inverse of S is obviously necessary. Invertibility of S is also among the defining features of a modular Hopf-algebra in [21]. The last equation in the set (3.55) is usually referred to as "diagonalization" of fusion rules [22].

In the finite-dimensional situation on can consider certain linear combinations χ^K of the c^I,

$$\chi^K = \mathcal{N} d_K S_{KI} c^{\bar{I}} \ .$$

Here $\mathcal{N} = (\sum d_I^2)^{-1/2}$, and S_{KI} are components of the S-matrix (3.54). The projectors χ^K are central orthogonal projectors, i.e. $(\chi^K)^* = \chi^K$, and $\chi^K \chi^L = \delta_{K,L} \chi^K$ (see [3]). They represent characteristic functions for the 1-dimensional ideals in the Verlinde algebra corresponding to the representations considered in the previous subsection.

One serious technical problem, which arises when we consider quantum universal enveloping algebras at roots of unity, is the fact that they loose semi-simplicity. In particular, the finite-dimensional Hopf algebras under discussion are not semi-simple. This problem may be cured in two different ways. One way is to work with non semi-simple algebras. Then the resulting moduli algebra is expected to have a big ideal formed by functions which include matrix elements of indecomposable representations. If one wants to introduce the moduli algebra as a $*$-algebra, this ideal should be factored out. We refer to this operation as *truncation.*

Another way is to force the universal enveloping algebra to be semi-simple [15]. Technically, one moves to the class of quasi-Hopf algebras and relaxes the axiom

$$\Delta(e) = e \otimes e \ . \tag{3.56}$$

This defines a class of weak quasi-Hopf algebras. In fact, other axioms of the Hopf algebra theory should be modified as well (for details see [15]).

Let us illustrate this concept on the example of $U_q^F(sl(2))$.

Definition 8 *The truncated universal enveloping algebra $U_q^T(sl(2))$ is defined as the quotient of $U_q^T(sl(2))$ over the ideal formed by all elements which vanish in all physical representations of nonnegative spin.*

The resulting algebra $U_q^T(sl(2))$ is semi-simple by the construction. The physical representations of spins $j = 0, \ldots, (p-1)/2$ are the only irreducible representations of this new object. This construction is justified by the following observation [15].

Theorem 8 *The algebra $U_q^T(sl(2))$ carries a weak quasi-Hopf structure induced by the Hopf structure of $U_q^F(sl(2))$.*

In fact, this is the truncated algebra $U_q^T(sl(2))$ and its higher rank counterparts, which we wish to use to construct the moduli algebra at roots of unity.

Admitting weak quasi-Hopf algebras we have to prove that moduli algebras may be defined by these data. This has been done in [2], [3]. The qualitative difference is that the graph algebra now is only quasi-associative. Indeed, it is designed as an algebra of the objects covariant with respect to the symmetry action. As the co-multiplication of the symmetry algebra is quasi-co-associative, the multiplication of tensors is forced to be quasi-associative. However, the moduli algebra is always an associative algebra.

All our results are valid for truncated weak quasi-Hopf symmetry algebras. This includes the most interesting cases of $U_q(\mathcal{G})$ at roots of unity, which correspond to quantization of the Chern-Simons theory for integer values of k and for the compact Lie group G.

4. Summary of Results

4.1 Representation Theory of the Moduli Algebra

The moduli algebra $\mathcal{M}_{g,m}^{\{I_\nu\}}$ corresponding to the truncated universal enveloping algebra $U_q^T(\mathcal{G})$ is supposed to coincide with the algebra of observables of the Chern-Simons theory. Naturally, its $*$-representations may be considered as candidates for the role of the Hilbert space in this model.

Assuming that we construct the moduli algebra starting from a semi-simple quasi-triangular (quasi)-Hopf algebra with only finite number of representations, we arrive at the following result.

Theorem 4 (Representations of the moduli algebra [4]) *For any set of representations $I_1, \ldots, I_m$ assigned to the marked points, there exists a unique irreducible $*$-representation of the moduli algebra $\mathcal{M}_{g,m}^{\{I_\nu\}}$, which acts in the space*

$$\begin{aligned} W_g^0(I_1, \ldots, I_m) &= Inv(V^{I_1} \otimes \ldots V^{I_m} \otimes \Re^{\otimes g}), \qquad (4.1) \\ \text{where} \qquad & \Re = \oplus_I V^I \otimes V^{\bar{I}} \ . \end{aligned}$$

Here Inv stands for invariant subspace with respect to the natural action of the symmetry algebra.

It is remarkable that the moduli algebra has a unique representation. Thus it is identified with the full matrix algebra with the natural $*$-operation and may be regarded as an algebra of observables of some quantum mechanical system with a finite-dimensional space of states. Apparently, this is the case of the Chern-Simons theory in the Hamiltonian formulation.

The space $\Re$, which enters the definition of $W_g^0(I_1, \ldots, I_m)$, is an analog of the regular representation of a finite or a compact Lie group.

Let us remark that there is an important difference between the moduli algebras of zero and nonzero genus. The former involve only R-matrices in the expressions for the matrix elements of their representations. However, if the genus of the surface is nontrivial, the knowledge of Clebsch-Gordon maps is required.

4.2 The Action of the Mapping Class Group on the Moduli Algebra

There is an important structure on the moduli space which we did not touch before. The moduli space of flat connections on a surface $\Sigma_{g,m}$ carries the action of the pure mapping class group $PM(g,m)$ of the surface. The pure mapping class group is a subgroup of the mapping class group $M(g,m)$, which preserves the order of marked points. $PM(g,m)$ acts by automorphisms of the fundamental group of the surface. This action lifts to $Hom(\pi, G)$ as

$$\rho^{\eta}(x) = \rho(x^{\eta}) \quad \text{for all} \quad x \in \pi_1(\Sigma_{g,m}) \ . \tag{4.2}$$

Here ρ is a representation of π in G, x is an element of the fundamental group, and η is an element of the mapping class group. As η defines an automorphism of π, this action descends to the moduli space.

In the course of quantization symmetries are usually very important, as they give us guidelines which features of the classical theory should be preserved by quantization. The action of the mapping class group on the moduli space of flat connections preserves the symplectic structure. So, we should expect that this action lifts to the quantized algebra of functions on the moduli space. This is indeed the case. We need some more notations to describe this action.

The pure mapping class group may be generated by so called Dehn twists. A Dehn twist corresponds to a circle on a Riemann surface. The mapping class group transformation includes cutting the surface along this fixed circle, relative rotation of the boundaries of the cut by the angle of 2π, and gluing the sides of the cut back together. Thereby one defines a smooth mapping of the surface into itself, which does not belong to the connected component of the identical mapping.

As a first step we define a Verlinde subalgebra in the moduli algebra for each circle on a Riemann surface. A circle on a surface (or, more exactly, its homotopy type) defines a conjugacy class in the fundamental group. Let us take any element $\mathcal{C}$ of this conjugacy class and represent it as a word in the generators of the standard basis l_ν, a_i, b_i.

We wish to define a holonomy $M^I(\mathcal{C})$ corresponding to the element $\mathcal{C}$ of the fundamental group. Suppose that $\mathcal{C} = \mathcal{C}_1\mathcal{C}_2$. Then we set

$$M^I(\mathcal{C}) \equiv \kappa_I^{w(\mathcal{C}_1,\mathcal{C}_2)} M^I(\mathcal{C}_1) M^I(\mathcal{C}_2) \ . \tag{4.3}$$

Here $w(\mathcal{C}_1, \mathcal{C}_2) = \pm 1$ is determined from the pair $(\mathcal{C}_1, \mathcal{C}_2)$ as follows. Suppose that we have presentations of $\mathcal{C}_1$ and $\mathcal{C}_2$ in terms of generators $l_\nu^\pm, a_i^\pm, b_i^\pm$. We assume that these presentations are reduced, so that neighboring elements are never inverse to each other. From these presentations we extract two generators $c_i \in \{l_\nu^\pm, a_i^\pm, b_i^\pm\}$, such that $\mathcal{C}_1 = \mathcal{C}_1' c_1$ and $\mathcal{C}_2 = c_2 \mathcal{C}_2'$. Without restriction we can assume that $c_1 \neq c_2^{-1}$. The weight w that we want to describe satisfies $w(\mathcal{C}_1, \mathcal{C}_2) = w(c_1, c_2)$. To define the latter we need to introduce two maps t, s from the set of generators $\{l_\nu^\pm, a_I^\pm, b_i^\pm\}$ to the integers $1, \dots, 2m + 4g$:

$$t(l_\nu) = 2\nu = s(l_\nu) + 1 \quad , \quad t(a_i) = 2m + 4i = s(a_i) + 2 \ ,$$
$$t(b_i) = \ 2m + 4i - 1 \ = s(b_i) + 2,$$

and $t(c^-) = s(c)$ for every $c \in \{l_\nu^\pm, a_I^\pm, b_i^\pm\}$. Now we can complete the description of the weight w.

$$w(c_1, c_2) \equiv \begin{cases} +1 & \text{if} \quad t(c_1) < s(c_2) \\ -1 & \text{if} \quad t(c_1) > s(c_2) \end{cases} \tag{4.4}$$

for all $c_1, c_2 \in \{l_\nu^\pm, a_I^\pm, b_i^\pm\}$. Our choice of the weight factor in the definition (4.3) is designed such that all monodromies $M^I(\mathcal{C})$ satisfy relations of the form (2.11).

As an example let us pick up an element

$$\mathcal{C} = l_1 b_1 a_1 \tag{4.5}$$

Then

$$M^I(\mathcal{C}) = \kappa_I^{-2} M_1^I B_1^I A_1^I \tag{4.6}$$

In the standard way we define the generators of the Verlinde algebra

$$c^I(\mathcal{C}) = \kappa_I tr_q M^I(\mathcal{C}) \tag{4.7}$$

corresponding to the circle $\mathcal{C}$. One can easily repeat this procedure for an arbitrary circle on the surface. The choice of the representative in the conjugacy class appears to be irrelevant for the definition of c^I, due to the properties of the q-trace. It is convenient to introduce a notation $\mathcal{V}(\mathcal{C})$ for a Verlinde algebra corresponding to a circle $\mathcal{C}$.

The last ingredient which we need is a particular element in the Verlinde algebra $\mathcal{V}(\mathcal{C})$, defined as

$$\hat{h}(\mathcal{C}) = \sum_I v_I^{-1} \chi^I(\mathcal{C}) \ . \tag{4.8}$$

Here projectors χ^I are constructed of c^I as in the previous section.

We collect the main results concerning the action of the mapping class group in the following theorem.

Theorem 5 (Action of the mapping class group [4]) *The moduli algebra $\mathcal{M}_{g,m}^{\{I_\nu\}}$ corresponding to a surface of genus g with m marked points carries* *a natural action of the pure mapping class group $PM(g,m)$, which preserves the order of the marked points. $PM(g,m)$ acts by inner automorphisms of the moduli algebra. The unitary generator of a particular Dehn twist defined by a circle x on the surface is given by $\hat{h}(x)$. This gives rise to a projective representation of the mapping class group, which is unitary equivalent to the one described in [14].*

4.3 Comparison to Topological Field Theory

The different algebras $\mathcal{M}_{g,m}^{\{K_\nu\}}$ are related with each other. In fact one can construct various inclusions which are parametrized by the choice of a circle on the punctured surface. Geometrically, the inclusions corresponds to a pinching of the surface. We will explain this only for some specific pinching-circles, but the idea is more general. In this way we make contact with Topological Field Theory defined by Atiyah axioms. The peculiar point is that in our case linear spaces assigned to geometric objects carry an algebraic structure.

To state the *pinching theorems* we need to introduce some new notations. Suppose that $\mathcal{X}$ is a subalgebra of an algebra $\mathcal{Y}$. The (relative) commutant of $\mathcal{X} \subset \mathcal{Y}$ will be denoted by $\mathcal{C}(\mathcal{X}, \mathcal{Y})$.

Consider the moduli algebra $\mathcal{M}_m^{K_1,\dots,K_m}$ corresponding to a sphere with m marked points. Pick up a cycle $l = l_m l_{m-1}$ (i.e. the product of the two elementary loops l_m and l_{m-1}), and construct the fusion algebra $\mathcal{V}(l)$ assigned to l, i.e. the algebra generated by

$$c^I(l) = tr_q^I(M_m^I M_{m-1}^I) \quad .$$

We wish to investigate the commutant of $\mathcal{V}(l)$ in $\mathcal{M}_m^{K_1,\dots,K_m}$. The result is given by the following theorem.

Theorem 6 (First pinching theorem [4]) *The commutant of $\mathcal{V}(l)$ in $\mathcal{M}_m^{K_1,\dots,K_m}$ splits into the direct sum of products of moduli algebras corresponding to $m-1$ and 3 marked points,*

$$\mathcal{C}(\mathcal{V}(l), \mathcal{M}_m^{K_1,\dots,K_m}) \cong \bigoplus_K \mathcal{M}_{m-1}^{K_1,\dots,K_{m-2},K} \otimes \mathcal{M}_3^{\bar{K},K_{m-1},K_m} . \tag{4.9}$$

Here the sum runs over all classes of irreducible representations of the symmetry Hopf algebra $\mathcal{G}$.

In Topological Field Theory the evaluation of the commutant should be interpreted as a fusion of two marked points into one. One can imagine that we create a long neck which separates these two points from the rest of the surface. When we cut the neck, the surface splits into two pieces. The "main part" carries the rest of marked points and a new one created by the cut. The other piece has only three punctures, two of them are those that we wish to fuse, and the new one appears because of the cut. Iteration of this procedure results in a product of 3-punctured spheres.

We would like to extend our description of relations between moduli algebras to higher genus surfaces. Now we have a new option: we can also shrink the surface along a circle that wraps around some handle. To be specific, we choose the fusion algebra $\mathcal{V}(b_1)$ constructed from the elements B_1^I and evaluate its commutant. Recall that $\mathcal{V}(b_1)$ is generated by

$$c^I = \kappa_I tr_q^I(B_1^I) \quad .$$

Theorem 7 (Second Pinching Theorem [4]) *The commutant of $\mathcal{V}(b_1)$ in the moduli algebra $\mathcal{M}_{g,m}^{K_1,\dots,K_m}$ splits into the direct sum of moduli algebras of genus $g-1$ with $m+2$ marked points,*

$$\mathcal{C}(\mathcal{V}(b_1), \mathcal{M}_{g,m}^{K_1,\dots,K_m}) \cong \bigoplus_K \mathcal{M}_{g-1,m+2}^{K_1,\dots,K_m,K,\bar{K}} . \tag{4.10}$$

Here the sum runs over all classes of irreducible representations of the symmetry Hopf algebra.

In the language of Topological Field Theory, evaluation of the commutant corresponds to shrinking the cycle b_1, so that we get a surface of lower genus. It has two marked points at the place where the handle is pinched – one on either side of the cut. Shrinking all the b-cycles one after another, one produces spheres with $m+2g$ marked points.

4.4 Comparison to Other Approaches

We can compare the Combinatorial quantization approach to two other quantization schemes.

Let us recall that the Hilbert space of the Chern-Simons theory has been identified with the space of conformal blocks in the WZW Conformal Field Theory corresponding to the same group G and with the same value of the coupling constant k [23]. The guess about the relation of the CS and WZW systems has been explained in the framework of Geometric Quantization [24]. Technically speaking the space of conformal blocks may be characterized as a space of solutions of certain linear differential equations. In the simplest case of the 2-dimensional surface being a sphere with marked points, these equations were suggested in [25] and usually called Knizhnik-Zamolodchikov (KZ) equations.

One can notice that the dimension of the representation space (4.1) coincides with the Verlinde formula for the dimension of the space of conformal blocks. The natural isomorphism between these two spaces emerges when one considers certain asymptotics of the KZ-equations [26]. This is one of the aspects of the relation between quantum groups and Kac-Moody algebras discovered in [27]. The particular case of roots of unity has been worked out in [28]. We expect that the general results of [27, 28] imply the identification of the geometric quantization and quantum group pictures for the space of conformal blocks. However, the current status of this construction is not clear to us.

Another approach which we follow quite closely is deformation quantization. The r-matrix presentation of the Poisson brackets on the moduli space is especially designed to make deformation quantization easy. We conjecture (see subsection 2.4), that for generic values of the deformation parameter q the moduli algebra $\mathcal{M}_{g,m}^{\{I_\nu\}}$ provides a deformation quantization of the algebra of functions on the moduli space.

Acknowledgement
We would like to thank the organizers of the Schladming Winter School for the opportunity to present this material. We are grateful to H.Grosse, with whom we were working together on this subject.

Special thanks are due to V. Fock, A. Rosly, and D. Boulatov for patient explanations of their works. We would like to thank A. Bytsko, V. Drinfel'd, L. Faddeev, J. Fröhlich, K. Gawedzki, A. Gerasimov, A. Malkin, A. Niemi, M. Semenov-Tian-Shansky, E. Sklyanin for the interest in this work and a lot of useful discussions.

References

1. V.V. Fock, A.A. Rosly, *Poisson structures on moduli of flat connections on Riemann surfaces and r-matrices*, preprint ITEP 72-92, June 1992, Moscow
2. A. Yu. Alekseev, H. Grosse, V. Schomerus, *Combinatorial quantization of the Hamiltonian Chern Simons theory I*, HUTMP 94-B336, HEP-TH/9403066, Commun. Math. Phys., to appear
3. A. Yu. Alekseev, H. Grosse, V. Schomerus, *Combinatorial quantization of the Hamiltonian Chern Simons theory II*, HUTMP 94-B336, HEP-TH/9408097, Commun. Math. Phys., to appear
4. A.Yu. Alekseev, V. Schomerus, *Representation Theory of Chern-Simons Observables*, Q-ALG/9503016.
5. A.Yu.Alekseev, *Integrability in the Hamiltonian Chern-Simons theory*, Uppsala preprint HEP-TH/9311074, St.-Petersburg Math. J., **vol. 6**, no. 2 (1994) 53
6. D.V. Boulatov, *q-deformed lattice gauge theory and three manifold invariants*, Int. J. Mod. Phys. **A8**,(1993), 3139
7. A.Yu.Alekseev, L.D.Faddeev, M.A.Semenov-Tian-Shansky *Hidden Quantum groups inside Kac-Moody algebras*, Commun. Math. Phys. **149**, no.2 (1992) p.335.
8. E. Buffenoir, Ph. Roche, *Two dimensional lattice gauge theory based on a quantum group*, preprint CPTH A 302-05/94 and HEP-TH/9405126
9. J. Andersen, J. Mattis, N. Reshetikhin, to be published
10. S.A. Frolov, *Hamiltonian lattice gauge models and the Heisenberg double*, Munich preprint LMU-TPW 95-3, HEP-TH/9502121.
11. M. Atiyah, R. Bott, *The Yang-Mills equation over Riemann surfaces*, Phil. Trans. of the Royal Soc. of London, Ser A **308** (1982) 523
12. S. Elitzur, G. Moore, A. Schwimmer, N. Seiberg, *Remarks on the canonical quantization of the Chern-Simons-Witten theory*, Nucl. Phys. **326**, (1989) 108

13. W. Goldman, *Invariant functions on Lie groups and Hamiltonian flows of surface group representations*, Invent. Math. **85** (1986) 263;
W.Goldman, *The symplectic nature of fundamental groups of surfaces*, Adv. Math. **54** (1984) 200
14. N. Reshetikhin, V. G. Turaev, *Ribbon graphs and their invariants derived from quantum groups*, Commun. Math. Phys. **127**, (1990) 1
15. G. Mack, V. Schomerus, *Quasi Hopf quantum symmetry in quantum theory*, Nucl. Phys. **B370** (1992) 185
16. A. Sudbury, *Non-commuting coordinates and differential operators* in: Quantum groups, T. Curtright et al. (eds), World Scientific, Singapore 1991
17. L.D. Faddeev, N. Yu. Reshetikhin, L.A. Takhtadzhyan, *Quantization of Lie Groups and Lie Algebras*, Algebra and Analysis **1** (1989), 1 and Leningrad Math. J. Vol. 1 (1990), No. 1
18. DeConcini, V. Kac, *Representations of quantum groups at roots of 1*, in *Progress in Mathematics* vol. **92**, Birkhäuser 1990
19. G.Lusztig, *Finite dimensional Hopf algebras arising from quantized universal enveloping algebras* J. Am. Math. Soc. **vol. 3** 1 (1990) 257
20. J. Fröhlich, F. Gabbiani, *Braid statistics in local quantum theory*, Rev. Math. Phys. **2** (1991) 251
21. N. Reshetikhin, V. G. Turaev, *Invariants of 3-manifolds via link polynomials and quantum groups*, Invent. Math. **103** (1991) 547
22. E. Verlinde, *Fusion rules and modular transformations in 2D conformal field theory*, Nucl. Phys. **B 300** (1988) 360
23. E.Witten, *Quantum field theory and the Jones polynomial*, Commun. Math. Phys. **121**, (1989) 351
24. S. Axelrod, S. Della Pietra, E. Witten, *Geometric quantization of Chern-Simons gauge theory*, J. Diff. Geom. **33** (1991) 787
25. V. Knizhnik, A. B. Zamolodchikov, *Current algebra and Wess-Zumino model in two dimensions*, Nucl. Phys. **B247** (1984) 83
26. V.G. Drinfel'd, *Quasi Hopf algebras and Knizhnik Zamolodchikov equations*, in: Problems of modern quantum field theory, Proceedings Alushta 1989, Research reports in physics, Springer Verlag Heidelberg 1989
V.G. Drinfel'd, *Quasi-Hopf algebras*, Leningrad. Math. J. Vol. **1** (1990), No. 6
27. D. Kazhdan, G. Lusztig, *Tensor structures arising from affine Lie algebras I/II*, J. Am. Math. Soc. **vol. 6** 4 (1993) 905
28. M.Finkelberg, *Fusion Categories*, Harvard University thesis, May 1993

Comments About Higgs Fields, Noncommutative Geometry, and the Standard Model

G. Cammarata and R. Coquereaux

Centre de Physique Théorique - CNRS - Luminy, Case 907, F-13288 Marseille Cedex 9 - France

1. Introduction

This paper is written for those who are not willing to become experts in the field of noncommutative geometry, but nevertheless want to understand the link between this approach and the usual formulation of the standard model of electro-weak interactions. The paper tries to give simple answers to the following questions:

- What has been done in this field so far?
- What are the several approaches?
- What are the results that are going to stay, and what are those which are, for the moment, conjectural?
- Is there any relation between this and the formalism of superconnections (and with super-algebras)?

We also make a number of comments that may help the reader to see what is going on in this field.

The construction of the full standard model (with usual quarks and leptons, but also with right neutrinos) is carried out by following the simplest possible route (at least the simplest, from the point of view of the present authors) and using an appropriate generalization of the notion of connection. The present paper can be considered as a sequel of [4], but can also be read independently; it should not be considered as an expository lecture on the approach initiated by [1].

2. The Meaning of Noncommutative Geometry

From the point of view of physics, one can summarize the situation very simply by saying that "commutative geometry" is the collection of mathematical tools describing classical physics, whereas "non commutative geometry" is the collection of mathematical tools describing quantum physics.

Commutative geometry (or better "commutative mathematics") deals with mathematical properties of spaces (measurable, topological, differentiable, Riemannian, homogeneous...). For the physicist, these "spaces" provide a mathematical model for the system under study, and all the properties of interest can be expressed in terms of an appropriate class of (numerical) functions defined on such spaces. It is a fact – physically obvious but also mathematically rooted – that properties of "spaces" are entirely encoded in terms of properties of algebras of numerical functions (coordinates for example) or of objects themselves defined from numerical functions (forms, tensors etc.). The name "commutative mathematics" comes from the fact that a set of numerical functions defined on a space is a commutative (and associative) algebra for pointwise multiplication and addition of functions.

Non commutative geometry (or better "non commutative mathematics") deals with mathematical properties of algebras which are not necessarily commutative, and generalizes – or tries to generalize – the constructions already known for commutative algebras (i.e. spaces) to non commutative situations (i.e. to operators).

This shows that one should maybe not always speak of "commutative geometry" or of "non commutative geometry", but of "commutative mathematics" or of "non commutative mathematics". What we have in mind in the present paper is not the use of non commutative mathematics in physics (because this could include, among many other things, the mathematics of quantum statistical mechanics), but non commutative differential geometry.

From an epistemological point of view, and once the concepts of commutative geometry and/or non commutative geometry have been mathematically studied, one should probably revert the first general statement of the present paragraph and define classical physics itself as a human activity characterized by the wish of understanding what we call "Nature" in terms of commutative mathematics, and define quantum physics in the same way but where the models are now expressed in terms of non commutative mathematics. One could even go further and declare that only the choice of the mathematical model (or models) gives a meaning (meanings) to the whole thing (Nature), and that there is no such thing as "reality" per se..., but we now abandon these philosophical considerations to return to the differential calculus, commutative or not.

3. Non Commutative Versus Commutative Differential Calculus

A branch of non commutative differential geometry is non commutative differential calculus. The aim is to be able to consider objects like df or ∇f, i.e. differentials or covariant differentials, and to perform computations with them, assuming that f is no longer a function but an operator acting in some Hilbert space. Such a calculus has been developed in the recent years. There exist several kinds of non commutative differential calculi (for instance [2, 3, 4, 5, 9, 10]) and we do not intend here to describe them all. As a matter of fact, we shall describe none of them. Indeed, it turns out that a very simple by-product of this (these) generalization(s) gives us the necessary tools to understand Higgs fields as generalization of connections (Yang-Mills fields). In some cases, for instance the $U(1) \times U(1)$ model studied in [4], this by-product actually belongs to the realm of commutative geometry, because it involves only commutative algebras of functions on "spaces"! The point is that it was discovered historically [1] only after the new developments of non commutative calculus. It is maybe a little bit misleading to call "non commutative" some of these considerations, first because, in simple cases, they are not so, and next because they give the reader the feeling that he should first master all (or most of) the niceties of non commutative differential calculus to understand the constructions. Of course, this is a matter of taste, and some people could as well argue that one should always understand the general before going to the particular...

4. Commutative Non Local Differential Calculus

In the previous paragraph, we said that the constructions, that are at the root of our understanding of Higgs field as generalized connections, do not really belong to the realm of non commutative differential geometry (they are a by-product). They however correspond to some commutative – but non local – geometry. Let us see why.

Consider a discrete set $\{L, R\}$ with two elements that we call L and R. Call x the coordinate function $x(L) \doteq 1, x(R) \doteq 0$, and y the coordinate function $y(L) \doteq 0, y(R) \doteq 1$. Notice that $xy = yx = 0$, $x^2 = x, y^2 = y$ and $x + y = 1$, where 1 is the unit function $1(L) = 1, 1(R) = 1$. An arbitrary element of the associative (and commutative) algebra $\mathcal{A}$ generated by x and y can be written $\lambda x + \mu y$ (where λ and μ are two complex numbers) and can be represented as a diagonal matrix $\begin{pmatrix} \lambda & 0 \\ 0 & \mu \end{pmatrix}$. One can write $\mathcal{A} = \mathbb{C}x \oplus \mathbb{C}y$ and is isomorphic with $\mathbb{C} \oplus \mathbb{C}$. We now introduce a differential δ satisfying $\delta^2 = 0$, $\delta 1 = 0$, and the usual Leibnitz rule, along with formal symbols δx and δy. It is clear that Ω^1, the space of differentials of degree 1, is generated by the two independent quantities $x\delta x$ and $y\delta y$. Indeed, the relation $x + y = 1$ implies $\delta x + \delta y = 0$, the relations $x^2 = x$ and $y^2 = y$ imply $(\delta x)x + x(\delta x) = (\delta x)$, therefore $(\delta x)x = (1 - x)\delta x$ and $(\delta y)y = (1 - y)\delta y$. This implies also, for example, $\delta x = 1\delta x = x\delta x + y\delta x$, $x\delta x = -x\delta y$, $y\delta x = (1 - x)\delta x$, $(\delta x)x = y\delta x = -y\delta y$ etc . More generally, let us call Ω^p the space of differentials of degree p; the above relations imply that a basis of this vector space is given by $\{x\delta x\delta x \dots \delta x, y\delta y\delta y \dots \delta y\}$. Call $\Omega^0 = \mathcal{A}$ and $\Omega = \bigoplus_p \Omega^p$. This space Ω is an algebra: We can multiply forms freely, but one of course has to take into account the Leibnitz rule, for instance $x(\delta x)x(\delta x) = x(1-x)(\delta x)^2$. Since each Ω^p is two dimensional, we can easily represent it in terms of matrices. More precisely, we can represent the element $\lambda x(\delta x)^{2p} + \mu y(\delta y)^{2p}$ as the diagonal matrix $\begin{pmatrix} \lambda & 0 \\ 0 & \mu \end{pmatrix}$, and the the element $\alpha x(\delta x)^{2p+1} + \beta y(\delta y)^{2p+1}$ as the off diagonal matrix $\begin{pmatrix} 0 & i\alpha \\ i\beta & 0 \end{pmatrix}$. In other words, we represent even forms by even (i.e. diagonal) matrices, and odd forms by odd (i.e. off diagonal) matrices; doing so is not only natural but compulsory if we want the multiplication of matrices to be compatible with the multiplication in Ω. Indeed, the relations

$$\begin{aligned} x(\delta x)^{2p}x &= x(\delta x)^{2p} \\ x(\delta x)^{2p}y &= 0 \\ x(\delta x)^{2p+1}x &= 0 \\ x(\delta x)^{2p+1}y &= x(\delta x)^{2p+1} \end{aligned}$$

imply that the above representation using 2×2 matrices is indeed a homomorphism of $\mathbb{Z}_2$−graded algebras from the algebra of universal forms Ω (graded by the parity of p) to the algebra of 2×2 complex matrices (with $\mathbb{Z}_2$−grading associated with the decomposition of a matrix into diagonal and non diagonal components). The presence of a factor i in the off diagonal matrices representing odd elements (see above expressions) is necessary for the matrix product to be compatible with the product in Ω. Notice that the algebra Ω is infinite dimensional (since p ranges from 0 to infinity), but if we represent the whole of Ω in terms of 2×2 matrices acting on a fixed 2-dimensional vector space, the p grading is lost and only the $\mathbb{Z}_2$-grading is left. The differential δ obeys the usual Leibnitz rule when it acts on elements of $\mathcal{A}$, but a graded Leibnitz rule when it acts on elements of Ω, namely $\delta(\omega_1\omega_2) = \delta(\omega_1)\omega_2 + (-1)^{\partial\omega_1}\omega_1\delta(\omega_2)$, where $\partial\omega_1$ denotes 0 or 1 depending if ω_1 is even or odd.

A one-form (this will be interpreted as a Higgs field and can be seen to define a generalized connection) is an element of Ω^1. Take $A \doteq (\varphi\, x\delta x + \overline{\varphi}\, y\delta y)$. The matrix representation of A reads therefore

$$A = \begin{pmatrix} 0 & i\varphi \\ i\overline{\varphi} & 0 \end{pmatrix}.$$

The corresponding curvature is then $F = \delta A + A^2$, but $A^2 = -\varphi\overline{\varphi}x^2\,\delta x\delta x - \overline{\varphi}\varphi\, y^2\delta y\delta y = -\varphi\overline{\varphi}x\,\delta x\delta x - \overline{\varphi}\varphi\, y\delta y\delta y$, and $\delta A = \varphi\delta x\delta x + \overline{\varphi}\delta y\delta y = (\overline{\varphi} + \varphi)(x\delta x\delta x + y\delta y\delta y)$, so that the curvature can also be written

$$F = (\varphi + \overline{\varphi} - \varphi\overline{\varphi})(x\delta x\delta x + y\delta y\delta y) = \begin{pmatrix} \varphi + \overline{\varphi} - \varphi\overline{\varphi} & 0 \\ 0 & \varphi + \overline{\varphi} - \varphi\overline{\varphi} \end{pmatrix}.$$

We now choose a hermitian product on Ω by declaring the basis $x(\delta x)^p, y(\delta y)^q$ to be orthonormal. Then $|F|^2 = \overline{F}F = (\varphi + \overline{\varphi} - \varphi\overline{\varphi})^2$. One recognizes here a (shifted) Higgs potential. The previous calculation (expressed in the language of K-cycles) is already discussed in [1, 2] and can be recognized in [4] where it is written in the language of 2×2 matrices. The previous construction could of course be generalized. For instance, we could take three points rather than two. It is easy to show that in such a case Ω^1 is of dimension 6 and Ω^2 of dimension 12. If we take q points, the dimension of Ω^p is $q(q-1)^p$. More generally, if we take infinitely many points – take for instance points belonging to a manifold X –, it is easy to see that elements of Ω^1 can be defined as functions $A(x,y)$ of two variables on X such that $A(x,x)=0$, and that elements of Ω^2 can be defined as functions $F(x,y,z)$ of three variables on X such that $F(x,y,y)=F(x,x,y)=0$.

In the case of the geometry on the discrete set $\{L,R\}$ – that is our main example in the present paper – we recover the fact, that an element A of Ω^1 considered as a function of two variables should satisfy the constraints $A(L,L)=A(R,R)=0$ and can therefore be written as off-diagonal 2×2 matrices indexed by L and R. An element F of Ω^2 considered as a function of three variables should satisfy the constraints $F(L,L,R)=F(R,R,L)=F(L,R,R)=F(R,L,L)=F(R,R,R)=F(L,L,L)=0$, so that non zero components are $F(L,R,L)$ and $F(R,L,R)$. The fact that $dim(\Omega^p)=2$ for all p explains that we can use a representation of fixed dimension (namely 2×2 matrices) for all values of p, but one should maybe remember that it would not be so if we were considering a geometry on more than 2 points.

Notice that we are here doing commutative differential calculus (because the associative algebra of functions on a set of 2 elements is just the commutative algebra of diagonal 2 by 2 matrices with real or complex entries), but that we are doing a non local differential calculus, because the distance between the two points labelled L and R can not be made infinitesimally small. The reader will have recognized that one can interpret the above results in terms of Higgs fields. This is the subject of our next section.

5. What are the Higgses ? The Yukawa Interaction Term

Higgs fields (φ) allow left and right fermions (ψ) to communicate. In four dimensional Minkowski space, this is clear from the trilinear Yukawa couplings such as $\overline{\psi}_L \varphi \psi_R + h.c.$, that appear in the Lagrangian density of the standard model. This should be contrasted with terms like $\overline{\psi}_L \gamma^\mu A_\mu \psi_L + h.c.$, or $\overline{\psi}_R \gamma^\mu B_\mu \psi_R + h.c.$, where A_μ or B_μ denote usual Yang-Mills gauge fields. If we had no Higgs fields, of course we would have no mass term, but also no possible communication (interaction) between right and left. There would be no justification for choosing a single connected manifold to modelize our universe. We would have a Minkowski space-time for the right movers and a Minkowski space-time for the left movers. Existence of chirality in four dimensions leads therefore to the conclusion that we live in two parallel universes, one labelled by L and the other by R. Usual connections – Yang-Mills fields – connect (infinitesimally) L and L together, and R and R together, whereas Higgs fields are non local connections that connect L and R and allow us to identify the two copies of our universe.

As explained in all books of particle physics, the scalar interaction (Yukawa) of quarks is a priori of the form

$$\mathcal{L} = (\overline{U}_L\overline{D}_L)\begin{pmatrix}\varphi_+\\ \varphi_0\end{pmatrix}\mathcal{M}_D D_R + (\overline{U}_L\overline{D}_L)\begin{pmatrix}\overline{\varphi}_0\\ -\varphi_-\end{pmatrix}\mathcal{M}_U U_R + h.c.,$$

where all quark fields of charge 2/3 are collected into the multi-spinor field $U = (u\,c\,t)^t$, and similarly for quarks of charge $-1/3$ with $D = (d\,s\,b)^t$. The 3×3 complex matrices $\mathcal{M}_D$ and $\mathcal{M}_U$ encode all the dimensionless Yukawa coupling constants (here spinor and scalar fields have their usual dimensions, namely 3/2 and 1 in units of mass). If we expand the previous expression, we find

$$\begin{aligned}\mathcal{L} = \;& (\overline{U}_L\mathcal{M}_U\overline{\varphi}_0U_R + \overline{U}_R\mathcal{M}_U^\dagger\varphi_0U_L) + (\overline{D}_L\mathcal{M}_D\varphi_0D_R + \overline{D}_R\mathcal{M}_D^\dagger\overline{\varphi}_0D_L) +\\ & (\overline{U}_L\mathcal{M}_D\varphi_+D_R + \overline{D}_R\mathcal{M}_D^\dagger\varphi_-U_L) - (\overline{U}_R\mathcal{M}_U^\dagger\varphi_+D_L + \overline{D}_L\mathcal{M}_U\varphi_-U_R).\end{aligned}$$

Let us now collect *all* left-handed quark fields (of charge 2/3 and $-1/3$) of the standard model into a single spinor $\Psi_L = (U_LD_L)$, and *all* right-handed quark fields into a single spinor $\Psi_R = (U_RD_R)$. Here $U = (u\,c\,t)$ and $D = (d\,s\,b)$. The above Yukawa interaction term reads

$$\mathcal{L} = (\overline{\Psi}_L \quad \overline{\Psi}_R)\begin{pmatrix}0 & \Phi\\ \Phi^\dagger & 0\end{pmatrix}\begin{pmatrix}\Psi_L\\ \Psi_R\end{pmatrix},$$

where

$$\Phi = \begin{pmatrix}\mathcal{M}_U\overline{\varphi}_0 & \mathcal{M}_D\varphi_+\\ -\mathcal{M}_U\varphi_- & \mathcal{M}_D\varphi_0\end{pmatrix}.$$

The mass term is obtained by shifting φ_0 and $\overline{\varphi}_0$ by a real constant with dimension of a mass that we call $\nu/\sqrt{2}$. The mass term itself is therefore described by the mass matrix

$$\mathcal{M} = \nu/\sqrt{2}\begin{pmatrix}\mathcal{M}_U & 0\\ 0 & \mathcal{M}_D\end{pmatrix}.$$

Writing $\Psi = (\Psi_L, \Psi_R)^t$, the whole fermionic Lagrangian for quarks reads $\overline{\Psi}(\mathcal{D} + \mathcal{A})\Psi$ with

$$\mathcal{D} + \mathcal{A} = \begin{pmatrix}i\gamma^\mu\partial_\mu & -\mathcal{M}\\ -\mathcal{M}^\dagger & i\gamma^\mu\partial_\mu\end{pmatrix} + \begin{pmatrix}\gamma^\mu L_\mu & \Phi\\ \Phi^\dagger & \gamma^\mu R_\mu\end{pmatrix},$$

where L_μ and R_μ collectively refer to those components of the gauge fields coupled to the left and right handed sectors, and Φ collectively refers as before to Higgs fields couplings. The scalar interaction (Yukawa) of leptons is exactly of the same type. The only possible difference is that, in the minimal standard model, one does not usually add right neutrinos. We shall actually add such right neutrinos: They will not be coupled to the gauge fields, of course, but they will give a mass to the different kinds of Dirac neutrinos and will be also coupled between themselves – via mass matrices – and to the Higgses (and also therefore, in the unitary gauge, to the longitudinal part of the gauge bosons). Notice that we do not consider Majorana neutrinos. Introducing right neutrinos, not only allows us to use the same formalism for quarks and leptons (the only difference in the Yukawa interaction term is the replacement of matrices $\mathcal{M}_D$ and $\mathcal{M}_U$ by $\mathcal{M}_E$ and $\mathcal{M}_\nu$, respectively), but also, as we shall see later, simplifies our analysis. The Yukawa interaction for leptons is

$$\mathcal{L} = (\overline{\nu}_L \overline{E}_L) \begin{pmatrix} \varphi_+ \\ \varphi_0 \end{pmatrix} \mathcal{M}_E E_R + (\overline{\nu}_L \overline{E}_L) \begin{pmatrix} \overline{\varphi}_0 \\ -\varphi_- \end{pmatrix} \mathcal{M}_\nu \nu_R + h.c.,$$

where all leptons fields of charge -1 are collected into the multi-spinor field $E = (e\mu\tau)^t$, and similarly for the neutrinos $\nu = (\nu_e \nu_\mu \nu_\tau)^t$. The 3×3 complex matrices $\mathcal{M}_E$ and $\mathcal{M}_\nu$ encode all the Yukawa coupling constants. The whole fermionic Lagrangian for leptons reads as before, but with

$$\Phi = \begin{pmatrix} \mathcal{M}_\nu \overline{\varphi}_0 & \mathcal{M}_E \varphi_+ \\ -\mathcal{M}_\nu \varphi_- & \mathcal{M}_E \varphi_0 \end{pmatrix}.$$

In the standard model one should consider simultaneously not only the three generations of leptons, but also three copies (for color) of the three generations of quarks. Taking into account – as above – the presence of three kinds of right neutrinos, we get an interaction term $\overline{\Psi}(\mathcal{D} + \mathcal{A})\Psi$, with $\Psi = (\Psi_L \Psi_R)$, and where both Ψ_L and Ψ_R are multi-spinor fields – they are column vectors with 24 components (since $24 = 3+3+3(3+3)$), each component being itself a Weyl fermion.

In the spirit of noncommutative geometry, one should think of $\mathcal{D} + \mathcal{A}$ as a generalization of the Dirac operator (it incorporates masses and Yukawa couplings) coupled to an algebraic connection. It should be called the Dirac - Yukawa operator. The first piece in this expression is a generalized differential operator, since the mass matrix $\mathcal{M}$ appears as the inverse of a quantity encoding a discrete set of fundamental lengths. The second piece $\mathcal{A}$ is a generalized connection: it incorporates both Yang-Mills and Higgs fields.

6. The Bosonic Lagrangian

The theory of – usual – connections explains, why $F = dA + A^2$ is the natural object (curvature) associated with a Yang-Mills field, the root of the explanation being that the square of the corresponding covariant differential is a linear object whose expression is precisely given by the above formula. In the same way, and as explained (section 4) in a very simple case, the theory of generalized connections shows that $\varphi + \overline{\varphi} + \varphi\overline{\varphi}$ is the natural object (curvature) associated with the Higgs field φ introduced in the section 4 and defined as a non local connection on the discrete set $\{L, R\}$.

Now we do not have a discrete set, but a space $X = M_L \cup M_R$ that is the union of space-time for left-handed movers and space-time for right-handed movers, in other words, we have the product of Minkowski space by a discrete set of two elements called L and R. The generalized curvature $\mathcal{F}$ associated with the generalized connection $\mathcal{A}$ introduced in the previous paragraph is

$$\mathcal{F} = \begin{pmatrix} \mathcal{F}^{LL} & \mathcal{F}^{LR} \\ \mathcal{F}^{RL} & \mathcal{F}^{RR} \end{pmatrix},$$

with

$$\begin{aligned} \mathcal{F}^{LL} &= F^{LL} - ((\Phi + \Phi^\dagger)\nu/\sqrt{2} + \Phi\Phi^\dagger) \\ \mathcal{F}^{RR} &= F^{RR} - ((\Phi + \Phi^\dagger)\nu/\sqrt{2} + \Phi^\dagger\Phi) \\ \mathcal{F}^{LR} &= \nabla\Phi + i(L - R)\nu/\sqrt{2} \\ \mathcal{F}^{RL} &= \nabla\Phi^\dagger - i(L - R)\nu/\sqrt{2} \end{aligned}$$

and

$$\begin{aligned} \nabla\Phi &= \gamma^\mu(\partial_\mu\Phi + i(L_\mu\Phi - \Phi R_\mu)) \\ L &= \gamma^\mu L_\mu \\ F^{LL} &= \tfrac{1}{2}\gamma^\mu\gamma^\nu F^L_{\mu\nu} \\ R &= \gamma^\mu R_\mu \\ F^{RR} &= \tfrac{1}{2}\gamma^\mu\gamma^\nu F^R_{\mu\nu}. \end{aligned}$$

The symbols F^L and F^R denote the usual curvatures of Yang-Mills fields associated with hermitian fields L and R. The expression of matrix elements of $\mathcal{F}$ given before is a non trivial consequence of the formalism of non commutative geometry (or of a non local commutative differential calculus!) and can here be taken as a definition. These expressions can indeed be computed from the theory of general connections (commutative or not). The components of the curvature were obtained first by [1]. Up to different normalization factors and the presence of spurious fields, their expression agrees with the one given just above. This analysis was later improved in [2] (replacement of the so-called algebra Ω by Ω_D). A detailed exposition of the formalism of [2] using K-cycles and Dixmier trace can now be found in several places [3, 16, 17]. The matrix elements of $\mathcal{F}$ given above were obtained by [4, 7] in a simple way (and using the above notations). Our method is briefly recalled in one of the "comments" of section 8.

Notice that the above expressions for $\mathcal{F}$ have a dimension of a mass squared and that, as a consequence, an arbitrary mass scale $\nu/\sqrt{2}$ appears in the formula. Explicitely, the term $(\Phi + \Phi^\dagger)\nu/\sqrt{2} + \Phi\Phi^\dagger$ and its adjoint can be computed from the expressions of Φ given previously, both in the quark and leptonic sectors.

Up to a normalization factors (we shall come back later to this physically important problem) one recognizes that the trace of $\overline{\mathcal{F}}\mathcal{F}$ is nothing else than the Lagrangian describing the bosonic sector of the standard model: One obtains directly the expression that usually comes after a shift by $\nu/\sqrt{2}$ in the Higgs fields φ_0 and $\overline{\varphi}_0$ (see [4] for a discussion of this point).

In a sense, the discussion could stop at this point. Indeed, we have seen in section 5 how to re-write the Dirac-Yukawa interaction term of fermions, and in this section how to recover the whole bosonic sector of the standard model by treating Yang-Mills fields together with Higgs fields as different components of a generalized connection. However, there are several claims made in the literature about possible constraints on the parameters of the Lagrangian, that one could obtain thanks to a formalism of non commutative geometry. Because we want to clarify this point (at least in the present formalism), we shall continue the discussion a little further.

The whole discussion comes actually from our understanding of the notation $Tr\overline{\mathcal{F}}\mathcal{F}$, that should denote a real number. On the one hand, if we decide to introduce, by hand, as many arbitrary constants in the expansion of this quantity (that gives rise to the full bosonic Lagrangian of the standard model) as gauge invariance allows, we recover exactly the standard model with the same $Tr\overline{\mathcal{F}}\mathcal{F}$ in order to normalize simultaneously all the gauge fields and Higgs fields, and we obtain non trivial relations. The interest of the formalism of non commutative differential geometry is not, for us, tied up with the existence of such relations; it may be, however, that such relations turn out to acquire, some day, a better status. For this reason, and also because the reader may be interested, we shall devote the end of this section to discuss them.

After global multiplication by $1/g^2$, we can rescale gauge fields as usual by $A \longrightarrow gA$, and also the Higgs fields by $\Phi \longrightarrow g\Phi$. Under identification with the usual Lagrangian one obtains immediately $g^2 = \lambda$; this relation is quite natural from a point of view that identifies gauge fields and Yang-Mills fields as different components of a generalized connection. In that case, the first general relation giving M_H is not modified, but the second relation becomes $M_W = M_H/2$. Moreover, as we shall see below, the value of θ also gets constrained.

Rather than writing again in full the well known bosonic Lagrangian of the standard model, we shall examine several of the terms, as they appear here. First of all, notice that one can identify the two sides of

$$\lambda\{(\varphi+\overline{\varphi})\frac{\nu}{\sqrt{2}}+\varphi\overline{\varphi}\}^2 \equiv \mu^2(\varphi+\frac{\nu}{\sqrt{2}})^2+\lambda(\varphi+\frac{\nu}{\sqrt{2}})^4-\frac{\mu^4}{4\lambda},$$

provided $\nu^2 = -\mu^2/\lambda$. The mass value for the Higgs particle coming from this usual expression is $M_H = \nu\sqrt{\lambda}$. Notice also that the left hand side contains no additive constant (absence of cosmological term).

In our case the Higgs potential itself coming from $\frac{1}{g^2}Tr(\overline{\mathcal{F}}\mathcal{F})$ reads

$$V(\Phi) = \frac{2}{g^2}Tr((\Phi+\Phi^\dagger)\nu/\sqrt{2}+\Phi\Phi^\dagger)((\Phi+\Phi^\dagger)\nu/\sqrt{2}+\Phi\Phi^\dagger)^\dagger.$$

If we now express Φ in terms of the component Higgs fields, and in terms of the matrices of Yukawa couplings then remove the factor $1/g^2$ in front by rescaling the fields, we see that $V(\Phi)$ contains a term equal to $\nu^2\varphi_0\overline{\varphi}_0 Tr(M_U M_U^\dagger + M_D M_D^\dagger)$, but the term $Tr\nabla\Phi\nabla\Phi^\dagger$ leads to a kinetic term for ϕ_0 equal to $2\nabla\varphi_0\nabla\varphi_0(TrM_U M_U^\dagger + TrM_D M_D^\dagger)$, so that the mass of the Higgs field does not depend on the mass of fermions and stays undetermined (remember that ν is a free parameter). Other authors [17], using a different formalism, find quite stringent constraints relating M_H to the fermionic masses.

The full bosonic interaction contains also a term $\mathcal{F}^{LR\dagger}\mathcal{F}^{RL}$; using the previous expression for $\mathcal{F}^{LR}$ implies that the field $L-R$ becomes massive, as it should. Indeed it corresponds to the Z and W bosons. One may adopt the point of view that the present formalism dictates a particular value for the Weinberg angle; this value turns out to depend upon the fermionic content of the theory. Indeed, the gauge fields L and R consist of three copies of

$$\begin{aligned} L &= \tfrac{i}{\sqrt{2}}\vec{\tau}\vec{W} && +\tfrac{i}{\sqrt{x}}\begin{pmatrix} y & \\ & y \end{pmatrix} B \\ R &= 0 && +\tfrac{i}{\sqrt{x}}\begin{pmatrix} y+1 & \\ & y-1 \end{pmatrix} B \end{aligned}$$

Here $y = 1/3$ for quarks, since their weak hypercharge is equal to $(y = \frac{1}{3}, y = \frac{1}{3}\,; y+1 = \frac{4}{3}, y-1 = -\frac{2}{3})$, and $y = 0$ for leptons, since their weak hypercharge is equal to $(y = -1, y = -1; y+1 = 0, y-1 = -2)$. We are introducing here right neutrinos that are isospin singlets and for which $y = 0$.

For colourless quarks alone, the normalization

$$\left(\frac{1}{\sqrt{x}}\right)^2\left[\frac{1}{3}^2 + \frac{1}{3}^2 + \frac{4}{3}^2 + \left(-\frac{2}{3}\right)^2\right] = 1$$

would lead to $x = 22/9$ and $tg\theta = 3/\sqrt{11}$.

For leptons alone, the normalization

$$\left(\frac{1}{\sqrt{x}}\right)^2\left[(-1)^2 + (-1)^2 + 0^2 + (-2)^2\right] = 1$$

would lead to $x = 6$ and $tg\theta = 1/\sqrt{3}$.

More generally, if one uses an arbitrary representation and normalize fields L and R to 1 as above, one finds

$$tg^2\theta = 4TrI_3^2/TrY^2,$$

which, in the case of three families of quarks (with color) and leptons, gives $tg^2\theta = 3/5$ (or $sin^2\theta = 3/8$) as it is in the unified $SU(5)$ theory. This would be therefore the "predicted" value for the Weinberg angle. However, in the usual approach, and even without $SU(5)$ unification, one would obtain exactly the same value by postulating that the gauge group is not an arbitrary group isomorphic with $SU(3) \times SU(2) \times U(1)$, but a group *metrically* isomorphic with the $SU(3) \times SU(2) \times U(1)/(\mathbb{Z}_2 \times \mathbb{Z}_3)$ subgroup of $SU(5)$. In absence of a principle based on the ideas of group symmmetries (or a generalization of such a principle), one could then ask on which grounds one should postulate such a property. The same argument (or objection) holds here. Indeed gauge invariance alone allows for the introduction of arbitrary constants in front of the individual components of the gauge group. The conclusion is therefore that, although the value $tg^2\theta = 3/5$ appears quite "naturally" in this formalism, it should not be taken as an unescapable consequence of the construction.

A last possible "constraint" concerns the mass of the W (or Z) particle. Indeed, from the expression of $\mathcal{F}$ we obtain a term $\frac{1}{4}(\frac{\nu}{\sqrt{2}})^2 g^2 Tr(L-R)^2$ that gives a mass to the W and the Z. The trace itself reads

$$L - R = (\tfrac{i}{\sqrt{2}}(\tau_1 W_1 + \tau_2 W_2) + \tfrac{i}{\sqrt{2}} W_3 \begin{pmatrix} 1 & 0 \\ 0 & -1 \end{pmatrix} + \tfrac{i}{\sqrt{x}} B \begin{pmatrix} y & 0 \\ 0 & y \end{pmatrix})$$
$$-\tfrac{i}{\sqrt{x}} B \begin{pmatrix} y+1 & 0 \\ 0 & y-1 \end{pmatrix}$$
$$= \tfrac{i}{\sqrt{2}}(\tau_1 W_1 + \tau_2 W_2) + \tfrac{i}{\sqrt{2}} \left(W_3 - \sqrt{\tfrac{2}{x}}\, B \right) \begin{pmatrix} 1 & 0 \\ 0 & -1 \end{pmatrix},$$

so

$$Tr(L-R)^2 = W_1^2 + W_2^2 + \left(W_3 - \sqrt{\frac{2}{x}}\, B \right)^2 = 2W_+W_- + \frac{1}{\cos^2\theta} \cdot Z^2.$$

This gives the relation $M_W = \nu g/2$ which is well known in the standard model. In general, we have $M_W = g M_H/(2\sqrt{\lambda})$, and this becomes only a constraint (namely $M_H = 2M_W$), if we set λ to the "natural" value $\lambda = g^2$ as discussed before. One could hope that such relations could hold at a scale, where the previous value for θ is experimentally satisfied (maybe at some grand unification scale). Notice that other authors [17], using a different formalism (relying upon the choice of another differential algebra), obtain another type of relations. Of course, we cannot (and will not) pretend that other approaches should, or not, lead to the same "numerical" relations. Existence of constraints such as the above ones can anyway be criticized, since gauge invariance alone allows us to multiply terms of the bosonic Lagrangian by arbitrary constants; this possibility can be related to the choice of particular scalar products in the space of forms [6]), and there are no compelling reasons to set such constants equal to one (although it may look quite natural in this formalism).

The main conclusion of this section is that the structure of the whole bosonic Lagrangian of the standard model can be obtained from the formalism of non commutative geometry. Whether or not one should look for constraints and take them seriously is another matter. Our opinion is that, before reaching any conclusion on this line, one should wait till we have a full understanding of the fully quantized field theory in terms of non commutative geometry.

7. Higgs Fields and Superalgebras

The space where Ψ lives is naturally $\mathbb{Z}_2$-graded by L and R, i.e. Ψ can be decomposed into a left and a right part. Therefore transformations, that map Ψ fields to themselves, fall naturally into 2 kinds: those mapping L to L (and R to R) – we call them "even"– and those mapping L to R (and conversely) – we call them "odd". Mathematically speaking, the space of these transformations can be considered as an associative $\mathbb{Z}_2$-graded matrix algebra, whose corresponding Lie super-algebra is usually denoted by $GL(p|q)$, where p (resp. q) is the number of left Weyl (resp. right) fermions entering the Lagrangian. The usual Yang-Mills fields can be decomposed onto the even part whereas the Higgs fields can be decomposed onto the odd part of this algebra. This is a rather trivial remark, since any Yang-Mills theory (and not only the standard model) defined on an even dimensional space-time can be analysed along the same lines. Another way to express the same idea is to say, that any Yang-Mills theory with p left Weyl fermions and q right Weyl fermions can be formulated in terms of representation theory of *some* Lie superalgebra possessing a representation on a graded vector space of dimension $p+q$. In the case of the standard model (with right neutrinos), and because all the fermionic species are coupled to the *same* gauge and Higgs bosons, the matrix describing this interaction can be decomposed on a subset of the generators of $GL(24|24)$. Since we have only 4 gauge bosons and 4 Higgs bosons, we need only to use 8 generators (4 even and 4 odd ones); in other words we only need to use (or to recognize) the Lie superalgebra $Sl(2|1)$. The physical representations of interest (namely leptons, quarks, and possibly right neutrinos) correspond to direct sums of $Sl(2|1)$ representations of dimension $3 = 2+1, 4 = 2+2$, or 1. This fact was actually observed long ago [27, 28] and sometimes perceived as a kind of "miracle"; for us, we consider this property as almost tautological. The emergence of Lie superalgebras could lead people to think that one should try to enlarge the formalism of gauge theory to accomodate Lie superalgebras... Such attempts have been investigated in the past and shown to lead to serious problems and have, in any case, nothing to do with the standard model itself and even less with the non commutative geometry presentation of the standard model. In order to stress this point, let us consider the following analogy: one can observe that Dirac spinors form a representation of the Clifford algebra (the Dirac algebra of γ-matrices); this is well known; as a consequence it is also true that the spinors with four complex components also provide a representation for the (non simple) Lie algebra generated by taking commutators of *arbitrary* products of γ matrices;

this does not mean that the Lagrangian of quantum electrodynamics should be invariant (globally or locally) under such transformations. The fact, that an algebra (like the full algebra of γ matrices) is not directly related with an invariance of the Lagrangian, does not make it useless (the spin group and its Lie algebra can of course be expressed in terms of the γ's, but the Clifford algebra itself is much bigger). Not all algebras related to the mathematics of a physical model need to describe "invariances" or "symmetries"; the fact that they do not does not make them useless! The same thing is also true here for the superalgebra, along the representation of which one can decompose the matrices acting on the vector space spanned by the multi-component spinor fields $\Psi = (\Psi_L \Psi_R)$. This useful algebra is spanned by 8 generators. The first four are matrices that, in the interaction term of the Lagrangian describing interaction between fermions and gauge bosons, appear as coefficients of the Yang-Mills fields $W_\pm, W_3$, and B; they are denoted as usual by $I_\pm, I_3$, and Y. The last four are matrices that appear as coefficients of the Higgs fields $\varphi_+, \varphi_-, \varphi_0$, and $\overline{\varphi}_0$; they give rise (after having added the hermitian conjugate) to the Yukawa and mass interaction term. We call them $-\Omega'_+$, $-\Omega_-$, Ω'_-, and Ω_+. More precisely, consider the following (block) matrices:

$$\Omega_+ = \begin{pmatrix} 0 & 0 & g & 0 \\ 0 & 0 & 0 & 0 \\ 0 & 0 & 0 & 0 \\ 0 & b & 0 & 0 \end{pmatrix} \qquad \Omega'_- = \begin{pmatrix} 0 & 0 & 0 & 0 \\ 0 & 0 & 0 & -e \\ a & 0 & 0 & 0 \\ 0 & 0 & 0 & 0 \end{pmatrix}$$

$$\Omega_- = \begin{pmatrix} 0 & 0 & 0 & 0 \\ 0 & 0 & g & 0 \\ 0 & 0 & 0 & 0 \\ -b & 0 & 0 & 0 \end{pmatrix} \qquad \Omega'_+ = \begin{pmatrix} 0 & 0 & 0 & 0 \\ 0 & 0 & 0 & e \\ 0 & a & 0 & 0 \\ 0 & 0 & 0 & 0 \end{pmatrix},$$

where a, b, g, e are themselves square matrices, for example of size 3×3, if we consider only quarks coming in 3 families. In this case we decide to label the basis as follows: $\Psi = (u, c, t)_L\,(d, s, b)_L\,(u, c, t)_R\,(d, s, b,)_R$. Let us define $\sqrt{2}\,I_\pm = \{\Omega_\pm, \Omega'_\pm\}$, $2I_3 = \{\Omega_+, \Omega'_-\} - \{\Omega_-, \Omega'_+\}$, and $Y = \{\Omega_+, \Omega'_-\} + \{\Omega_-, \Omega'_+\}$. The electric charge is $Q \doteq I_3 + Y/2 = \Omega_+, \Omega'_-$. Then, provided matrices e, b, g, a satisfy the relation $eb + ga = 1$, one can show (it is straightforward but cumbersome) that the Ω matrices satisfy the relations

$$\begin{aligned} [I_3, I_\pm] &= \pm I_\pm & [I_+, I_-] &= I_3 \\ [Y, \Omega_\pm] &= -\Omega_\pm & [Y, \Omega'_\pm] &= \Omega'_\pm \\ [I_3, \Omega'_\pm] &= \pm\tfrac{1}{2}\Omega'_\pm & [I_\pm, \Omega_\mp] &= \tfrac{1}{\sqrt{2}}\Omega_\pm \\ [I_\pm, \Omega_\pm] = [I_\pm, \Omega'_\pm] &= 0 & \{\Omega_\pm, \Omega_\pm\} = \{\Omega_\pm, \Omega_\mp\} &= 0 \\ \{\Omega'_\pm, \Omega'_\pm\} = \{\Omega_\pm, \Omega'_\pm\} &= \sqrt{2}I_\pm & \{\Omega_\pm, \Omega'_\mp\} &= \pm I_3 + Y/2 \end{aligned}$$

$$\begin{aligned} [Y, I_\pm] = [Y, I_3] &= 0 \\ [I_3, \Omega_\pm] &= \pm\tfrac{1}{2}\Omega_\pm \\ [I_\pm, \Omega'_\mp] &= \tfrac{-1}{\sqrt{2}}\Omega'_\pm \\ \{\Omega'_\pm, \Omega'_\mp\} &= 0. \end{aligned}$$

One recognizes here the usual relations defining the Lie superalgebra of $SL(2|1)$. In the case of quarks, one furthermore imposes the following constraints for the hypercharge generator: $Y = (Y_L^U Y_L^D Y_R^U Y_R^D)$ with $Y_L^U = Y_L^D = 1/3\mathbf{1}_{3\times 3}$, $Y_R^U = 4/3\mathbf{1}_{3\times 3}$, and $Y_R^D = -2/3\mathbf{1}_{3\times 3}$. These constraints are satisfied if and only if, on top of the relation $eb + ga = \mathbf{1}$, the matrices e, b, g, a satisfy also the relations $-eb + ga = (1/3)\mathbf{1}$, $2ag = (4/3)\mathbf{1}$, and $-2be = -(2/3)\mathbf{1}$. Indeed one finds $Y_L^U = Y_L^D = -eb + ga$, $Y_R^U = 2ag$, and $Y_R^D = -2be$. This implies in particular $ga = ag$ and $eb = be$. One can check, that matrices $I_\pm$ and I_3 are then automatically what they should be.

One may notice that the above expressions for Ω matrices, describing the gauge and Yukawa couplings of the quark family, define a Lie superalgebra representation, which is equivalent to the sum of (three) irreducible representations (each irreducible itself splits into the direct sum of a doublet and two singlets under the branching to the Lie algebra of $SU(2) \times U(1)$).

Define now $\phi = \overline{\varphi}_0 \Omega_+ + \varphi_0 \Omega'_- - \varphi_+ \Omega'_+ - \varphi_- \Omega_-$, and write $\mathcal{C} = \overline{\Psi}\phi\Psi$. This expression can not be real, indeed $\mathcal{C} = \mathcal{C}^\dagger$ would imply $g = a^\dagger$, $b = -e^\dagger$, but the other constraints would lead to a contradiction ($ee^\dagger = -(1/3)\mathbf{1}$). To obtain a real expression, one has to add $\mathcal{C}$ and $\mathcal{C}^\dagger$. Writing $\mathcal{L} = \mathcal{C} + \mathcal{C}^\dagger$ gives

$$\begin{aligned}\mathcal{L} &= (\overline{U}_L(g + a^\dagger)\overline{\varphi}_0 U_R + \overline{U}_R(g^\dagger + a)\varphi_0 U_L) \\ &+(\overline{D}_L(-e + b^\dagger)\varphi_0 D_R + \overline{D}_R(-e^\dagger + b)\overline{\varphi}_0 D_L) \\ &-(\overline{U}_L(e - b^\dagger)\varphi_+ D_R - \overline{D}_R(e^\dagger - b)\varphi_- U_L) \\ &-(\overline{U}_R(g^\dagger + a)\varphi_+ D_L - \overline{D}_L(g + a^\dagger)\varphi_- U_R),\end{aligned}$$

and we recognize the expression of $\mathcal{L}$ given in section 5, with the identification $\mathcal{M}_U = g + a^\dagger$ and $\mathcal{M}_D = -e + b^\dagger$. Warning: The matrix ϕ defined previously in terms of the Ω matrices is not equal to the matrix Φ defined in section 5; in order to compare the two expressions, one has to first add the conjugated expressions $\mathcal{C}$ and $\mathcal{C}^\dagger$. Taking into account the constraints on blocks a, g, e, and b, one obtains the relations: $\mathcal{M}_U = g + (2/3)(g^{-1})^\dagger$ and $\mathcal{M}_D = -e + (1/3)(e^{-1})^\dagger$. These relations do not imply any "new" constraints on mass matrices $\mathcal{M}_U$ and $\mathcal{M}_D$, since g and e are themselves arbitrary. The main interest of those formulae is to provide a new parametrisation for mass matrices or matrices of Yukawa couplings. This could in turn suggest a new phenomenological ansatz for them and may even give us more insight into the structure of fermionic mass matrices. Such an ansatz was analysed in [6] in the case of two families and leads to a phenomenological expression of the Cabibbo angle in terms of quark masses.

Remark: The quantity $\mathcal{C}$ may be thought as the contribution to the Lagrangian of a particular representation of $Sl(2|1)$. One can think of $\mathcal{C}^\dagger$ as the contribution of the antiquark representation to the Lagrangian. However this identification is a little bit tricky and may lead to possible mistakes of interpretation; indeed, $\mathcal{C}$ is not hermitian, but $\mathcal{C}^\dagger$ is not the charge conjugate representation (in any case weak interactions usually violate charge conjugation and one should not build a Lagrangian that would be C-even

!). Given $\Omega_\pm$ and $\Omega'_\pm$ as before, one can define the following "hatted" Ω matrices: $\hat{\Omega}'_- = \Omega_+^\dagger, \hat{\Omega}'_+ = \Omega_-^\dagger, \hat{\Omega}_+ = \Omega_-'^\dagger$, and $\hat{\Omega}_- = \Omega_+'^\dagger$. It is then straightforward to check that these hatted Ω matrices generate (thanks to the same commutation relations) matrices $\hat{Y}$, $\hat{I}_\pm$, and $\hat{I}_3$, with, for example, $\hat{Y} = Diag(a^\dagger g^\dagger - b^\dagger e^\dagger, a^\dagger g^\dagger - b^\dagger e^\dagger, 2g^\dagger a^\dagger, -2e^\dagger b^\dagger$. We obtain in this way a new representation (the relation $b^\dagger e^\dagger + e^\dagger b^\dagger = \mathbf{1}$ being automatically satisfied since $eb + be = \mathbf{1}$). With $\mathcal{C}$ as before, we can rewrite $\mathcal{C}^\dagger$ as $\mathcal{C}^\dagger = \overline{\Psi}\phi^\dagger\Psi$, and $\phi^\dagger$ as

$$\begin{aligned}\phi^\dagger &= \varphi_0\Omega_+^\dagger + \overline{\varphi}_0\Omega_-'^\dagger - \varphi_-\Omega_+'^\dagger - \varphi_+\Omega_-^\dagger \\ &= \varphi_0\hat{\Omega}'_- + \overline{\varphi}_0\hat{\Omega}_+ - \varphi_-\hat{\Omega}_- - \varphi_+\hat{\Omega}_+'^\dagger,\end{aligned}$$

so that $\phi^\dagger$ itself appears as the contribution associated with the "hatted" representation. If one wishes to use $\mathcal{C}^\dagger$ in terms of a contribution of antiparticles, for instance $\overline{d}_R s_L$ as $-\overline{s}_R^c d_L^c$, one can do it, modulo proper care, but it may be misleading.

For leptons the idea is the same as for the quarks, and, in order to straigthen even more the analogy, we add right neutrinos to the standard model (they will turn out to be iso-singlets, as they should be). We shall order the basis as follows: $\Psi = (\Psi_L\Psi_R)$ with $\Psi_L = (\nu_e,\nu_\mu,\nu_\tau)_L\,(e,\mu,\tau)_L$, $\Psi_R = (\nu_e,\nu_\mu,\nu_\tau)_R\,(e,\mu,,\tau)_R$, and define matrices Ω as previously, in terms of new 3×3 block matrices e, b, g, a. However, in the case of leptons, the constraints for the hypercharge generator are different. Indeed, $Y = (Y_L^\nu Y_L^E Y_R^\nu Y_R^E)$ with $Y_L^\nu = Y_L^E = -1$, $Y_R^\nu = 0$, and $Y_R^E = -2$. These constraints are satisfied if and only if, on top of the relation $eb + ga = \mathbf{1}$ (which ensures that commutation relations for $SL(2|1)$ hold), the matrices e, b, g, a satisfy also the relations $eb = \mathbf{1}$ and $ag = 0$. With these constraints, one can then check that matrices $I_\pm$, I_3, and Y defined as before in terms of the matrices Ω are automatically what they should be.

One may notice that the above expressions for Ω matrices describing the gauge and Yukawa couplings of the lepton family (including right neutrinos) define a Lie superalgebra representation, which is equivalent to the sum of (three) reducible indecomposable representations (each of them splits into the direct sum of a doublet, a singlet, and the trivial representation under the branching to the Lie algebra of $SU(2)\times U(1)$).

Again we define $\phi = \overline{\varphi}_0\Omega_+ + \varphi_0\Omega'_- - \varphi_+\Omega'_+ - \varphi_-\Omega_-$, and write $\mathcal{C} = \overline{\Psi}\phi\Psi$. This expression cannot be real, and, in order to obtain a real expression, one has to add as before $\mathcal{C}$ and $\mathcal{C}^\dagger$. Writing $\mathcal{L} = \mathcal{C} + \mathcal{C}^\dagger$ gives

$$\begin{aligned}\mathcal{L} &= (\overline{\nu}_L(g + a^\dagger)\overline{\varphi}_0\nu_R + \overline{\nu}_R(g^\dagger + a)\varphi_0\nu_L) \\ &+(\overline{E}_L(-e + b^\dagger)\varphi_0 E_R + \overline{E}_R(-e^\dagger + b)\overline{\varphi}_0 E_L) \\ &-(\overline{\nu}_L(e - b^\dagger)\varphi_+ E_R - \overline{E}_R(e^\dagger - b)\varphi_-\nu_L) \\ &-(\overline{\nu}_R(g^\dagger + a)\varphi_+ E_L - \overline{E}_L(g + a^\dagger)\varphi_-\nu_R).\end{aligned}$$

We recognize the expression of $\mathcal{L}$ given in section 5, with the identification $\mathcal{M}_\nu = g + a^\dagger$ and $\mathcal{M}_E = -e + b^\dagger$, but the matrices a, g, e and b are not totally arbitrary, since they should here satisfy the constraints $eb = \mathbf{1}$ and $ag = 0$. These relations do not imply any constraints on mass matrices $\mathcal{M}_\nu$ and $\mathcal{M}_E$, but provide a new parametrisation for them. This parametrization in terms of a, g, e, g may in turn suggest a new phenomenological ansatz (for instance one can see what happens if these matrices a, g, e, g have particularly simple forms). Such an ansatz should then be considered as educated guess, but not as "prediction".

Before ending this section, we would like to notice that there exists still another interesting family of parametrizations for matrices a, g, e, and b. The reader can indeed check that, if we chose arbitrary (3×3) matrices N_L, N_{UR}, N_{DR} and choose a, g, e, and b in such a way that $2ga = (4/3)\mathbf{1} + N_L$, $2ag = (4/3)\mathbf{1} + N_{UR}$, $2eb = (2/3)\mathbf{1} - N_L$, and $2be = (2/3)\mathbf{1} - N_{DR}$, then all commutation relations for Ω matrices are still satisfied. The generators I_3, I_+, and I_- obtained from them are also equal to what they should be. However, the obtained hypercharge generator Y is not diagonal (and not necessarily hermitian), but equal to $(1/3)\mathbf{1} + N_L, (1/3)\mathbf{1} + N_L, (4/3)\mathbf{1} + N_{UR}, -(2/3)\mathbf{1} + N_{DR}$. In other words, this describes a family of quark-like objects which are not eigenstates of hypercharge (hence of charge). The Lie superalgebra specialist may relate this possibility to the existence of reducible indecomposable representations of $SL(2|1)$ whith non diagonal Cartan subalgebra [14] (take N_L, N_{UR}, N_{DR} nilpotent matrices). Relation between family mixing and existence of such representations was suggested in [6], but was leading to difficulties (emergence of flavour changing neutral currents in the quark sector), that could only be cured by a rather *ad hoc* treatment of the definition of charge conjugacy. Here we just notice that, after having defined ϕ and $\mathcal{C}$ as before and added the (usual) complex conjugate, one obtains a real expression, and one can choose to diagonalize simultaneously I_3 and Y. The rotated quark-like objects become now hypercharge (and charge) eigenstates, but the values of their charges are not standard and deviate from their usual values by corrections encoded in matrices N_L, N_{UR}, N_{DR}. This last family of parametrization leads therefore to something that deviates from the standard model, and we shall not elaborate more on this topic.

The $\mathbb{Z}_2$-graded algebra discussed in this section is not usually mentioned in textbooks explaining the construction of the standard model. However, if one decides to rewrite the Lagrangian in terms of multicomponent spinor fields $\Psi = (\Psi_L \Psi_R)$ gathering all left and right fermionic species in this way, this algebra (or better representations of it) appears naturally. It plays a role very similar to the (Clifford) Dirac algebra itself. We suggest to call it the "Yukawa algebra". Again, one should not consider this algebra as a "symmetry" of the model and it is probably better to avoid the word "symmetry" in this context in order to avoid possible misunderstandings.

8. Comments

- The roads towards noncommutative geometry and the standard model: All approaches use three ingredients: an associative algebra A describing "space", a module over this algebra describing "matter" and, finally, a differential algebra where generalized differentials live. In the very first papers on the subject, A. Connes uses the algebra ΩA of universal forms. He then introduced [2] one of its quotients, called it $\Omega_D A$, and used it in the sequel; this approach was followed in particular by [16, 17, 18, 23]. In [9, 13] a differential algebra $\Omega_{Der} A$ was built from A-valued derivations of the algebra A and subsequently used in various papers [10, 11, 12]. The formalism of A. Connes is very general and should be able to handle many problems of differential calculus in non commutative geometry. The formalism proposed in [4] is not that general, but if the purpose is to consider Yang-Mills fields and Higgs fields as different components of the same mathematical object (an algebraic connection), and to recover the Lagrangian of the standard model in a very simple way, this formalism is sufficient and does not require any mathematics beyond an elementary knowledge of $n \times n$ matrices. The remark at the origin of [4, 6, 7] is that it is not necessary to use a $\mathbb{Z}$-graded differential algebra to perform the analysis. The situation is similar to what happens with ordinary differential forms: one studies usually the theory of connections (Yang-Mills fields) and covariant derivatives in terms of (usual) differential forms like $A_\mu dx^\mu$; however, in order to write a Lagrangian describing the interactions with spinor fields, it is enough to represent these forms (in particular Yang-Mills potential and curvature) in terms of γ matrices like $A_\mu \gamma^\mu$ (which build a $\mathbb{Z}_2$-graded algebra). The same thing is true here: rather than using a $\mathbb{Z}$-graded algebra of generalized differential forms, we only use its representation on the $\mathbb{Z}_2$-graded vector space spanned by left and right spinor fields. This is why we only use $\mathbb{Z}_2$-graded algebras in [4] (not $\mathbb{Z}$-graded ones). We find this approach more familiar to physicists, and rich enough to illustrate the idea of considering Higgs fields as gauge bosons associated with the gauging of discrete directions (namely the jump between left and right chiralities).
- The general Connes' formalism involves three steps: One starts with a piece of data containing the (generalized) Dirac operator $\mathcal{D}$ – actually the Dirac-Yukawa operator– acting on a Hilbert space $\mathcal{H}$, together with an associative algebra $\mathcal{A}$ describing "space" and also acting on $\mathcal{H}$. In the case of a pure Abelian theory, $\mathcal{A}$ would just be equal to the algebra of complex functions on space-time. Actually, the formalism of [1] takes space-time as Euclidean and the Lorentz signature is recovered only at the end, thanks to a Wick rotation. In the case of a $U(1) \times U(1)$ theory with symmetry breaking, $\mathcal{A}$ is equal to the direct sum of two algebras of complex functions on space-time (one his labelled by "left" and the other by "right"); this algebra is still commutative and can be written as the space of 2×2 diagonal matrices with entries that are functions on space-time. In a more compli-

cated setting, where the gauge group is not Abelian, one has just to replace the previous $\mathcal{A}$ by its tensor product with an appropriate matrix algebra. In the second step one has to construct a differential algebra Ω_D (whose definition relies on the choice of $\mathcal{D}$) out of which one defines the generalized connections and curvatures. The third step is the construction of the Yang-Mills (or generalized Yang-Mills) Lagrangian itself and involves the so-called Dixmier trace as a substitute for integration. The triple $(\mathcal{A}, \mathcal{H}, D)$ is called a K-cycle or a spectral triple. The formalism presented in [4] starts with the same Dirac-Yukawa operator but does not require the construction of the algebra Ω_D and the use of the Dixmier trace. Its clear advantage is simplicity but it is lacking the character of generality expressed in [2, 3]. The only quantity to be computed is the expression for the generalized curvature $\mathcal{F}$ in terms of the generalized connection $\mathcal{A}$, but this can be done once and for all.

– In order to build any example of non commutative geometry, one needs an associative algebra (replacing "space"), an algebra of generalized forms, and a module (the space of matter fields). However, for a given "space" and a given kind of matter fields, the choice of the differential algebra Ω is not unique (all possible choices can ultimately seen as quotients of a so-called "universal" one). The choice used by A.Connes in [1] is not the same as the choice made by the same author in [2], and none of them coincide either with the differential algebra introduced by [9] or with ours. The choice described in [5] is a differential algebra Ξ equal to the tensor product of usual differential forms times the differential algebra built in section 4 for the space with 2 elements $\{L, R\}$. One takes $\mathcal{A}$ as 1-form and, using an appropriate d-operator, defines $\mathcal{F}$ as the 2-form $d\mathcal{A} + \mathcal{A}\mathcal{A}$. The expression previously given at the beginning of section 6 is nothing else than the representation of $\mathcal{F}$ in the fermionic space. The calculations can be done very simply at the representation level (this is recalled in section 8), and this is the reason why we skip the discussion concerning the actual choice of the $\mathbb{Z}$-graded differential algebra. Our point of view is that this freedom in the choice of the differential algebra does not matter (in the case of the physical system studied here), because all such choices lead *essentially* to the same result, namely to the expression of $\mathcal{F}$ similar to the one given in section 6, and to the usual Lagrangian of the standard model. *"Essentially"* means that all the results agree, up to factors of normalization. For instance, the expression of the Higgs potential is exactly the same, but the overall coefficient may change. For instance, in the Connes' approach (where one has to divide Ω by "junk forms" to obtain Ω_D), the kinetic term $\overline{D\varphi}D\varphi$ is proportional to $\mathrm{tr}(MM^\dagger)$, whereas the Higgs potential is proportional to $\mathrm{tr}[((MM^\dagger)_\perp)^2]$, with $MM^\dagger)_\perp = MM^\dagger - \frac{1}{n} tr\, MM^\dagger$, where M is a $n \times n$ fermionic mass matrix. This potential vanishes whenever $MM^\dagger$ is proportional to the unit matrix. In our case, this was not so (see section 6). The coefficients appearing in front of the kinetic term for scalar fields and of the Yukawa potential depend on the specific choice of the differential algebra. Spaces $\Omega_D^0(A)$, $\Omega_D^1(A)$, and $\Omega_D^2(A)$ for $A = C(M) \oplus C(M)$ have been computed in [2], and it happens that (when $MM^\dagger \neq \mathbf{1}$) they are respectively isomorphic, as vector spaces, to the Ξ^0, Ξ^1, and Ξ^2 described in [5], but the algebraic structure differs. More generally, the structure of Ω_D, when A is an arbitrary tensor product of algebras, was investigated in [19, 15, 21]. Introducing an extra arbitrary constant in front of the whole potential amounts to disregard possible mass relations. In such a case the several approaches become completely equivalent, as far as physics and the standard model of electroweak interactions are concerned.

– Calculation of the curvature: In [4, 6] the calculation of the coefficients of the generalized curvature is performed in the representation space (where only the $\mathbb{Z}_2$-grading matters). The whole $\mathbb{Z}$-graded differential algebra ΩA of universal forms is therefore mapped onto an algebra that can be taken as the $\mathbb{Z}_2$-graded tensor product of the usual algebra of differential forms times another $\mathbb{Z}_2$-graded differential algebra built in terms of even dimensional square matrices. Here we forget about the $\mathbb{Z}$-grading of forms and remember only their parity, or equivalently, we represent forms by using γ matrices belonging to the Clifford algebra. The product is

$$(a \otimes B) \odot (a' \otimes B') \doteq (-1)^{\partial B \partial a'} (aa') \otimes (B \wedge B'),$$

where a is a $2n \times 2n$ matrix and B is a form. The differential is

$$d(a \otimes B) \doteq da \otimes B + (-1)^{\partial a} a \otimes dB.$$

If we take $\mathcal{A}$ as in section 5 and define $\mathcal{F} \doteq d\mathcal{A} + \mathcal{A} \odot \mathcal{A}$, we obtain the curvature $\mathcal{F}$ given at the beginning of section 6. The reader should consult [4] for a simple exposition of this calculation. The definition and calculation of $\mathcal{A}$ and $\mathcal{F}$ in the Connes' formalism is given in [2, 3, 16].

– The reader remembers that, in case of spontaneous symmetry breaking, nature (or the formalism!) chooses a vacuum, and that there are usually infinitely many equivalent such vacua. Since the formalism of non commutative geometry brings about a formalism where the translation in the Higgs field is done automatically, one may wonder about what has happened with the previous freedom for choosing a vacuum. The answer is that the freedom of choice is encoded in the definition of the differential δ. This δ is not unique, and it can be seen that different choices for it amount to choose different minima for the Higgs potential (see [4]).

– In the formalism used by [2], the algebra of quaternions plays a very special role (it acts in particular on the algebra of generalized differential forms). It may help the reader to notice that quaternions are already present in the usual formalism of the standard model (also in ours!), but this fact is not necessarily used or recognized. The simplest way to see quaternions acting here is to look at the Yukawa interaction term given in section 5 or, equivalently, to put together in a square 2×2 matrix the two Higgs doublets coupling the left fermionic doublets to the fermionic right and left singlets. Discarding Yukawa coupling constants we have

$$\Phi = \begin{pmatrix} \overline{\varphi}_0 & \varphi_+ \\ -\varphi_- & \varphi_0 \end{pmatrix}.$$

One recognizes here a quaternion. Indeed, using Pauli matrices σ_i, set $\gamma_1 = i\sigma_1$, $\gamma_2 = i\sigma_2$, $\gamma_3 = -i\sigma_3$; then $\gamma_i^2 = -1$ and $\gamma_i\gamma_j = \epsilon_{ijk}\gamma_k$. Moreover

$$a_0\mathbf{1} + a_1\gamma_1 + a_2\gamma_2 + a_3\gamma_3 = \begin{pmatrix} a_0 - ia_3 & ia_1 + a_2 \\ ia_1 - a_2 & a_0 + ia_3 \end{pmatrix}.$$

This expression can obviously be identified with Φ.

– On the nature and values of constants appearing in the standard model: In any renormalizable quantum field theory, some parameters appear in the classical Lagrangian. These constants may be free or may be related by some kind of gauge invariance property, that one wants to enforce at the renormalized level. The free parameters have then to be fixed by an (arbitrary) renormalization prescription. We want now to stress the following. First of all, since these parameters are free, it is obvious that any kind of new constraint will be superimposed to the usual formalism. To be made widely acceptable, such a constraint should satisfy two criteria. The first is that it should physically work, the next is that it should come from a kind of aesthetical construction (usually encompassing or generalizing an inherited formalism).

  Our next comment concerns stability by renormalization. It has been shown [24], that some proposed constraints among masses of particles of the standard model may not be stable with respect to the renormalization group. This comment should be properly understood and maybe taken with a grain of salt. Indeed, the free parameters of the standard model are... free. Therefore one can renormalize them at will (at a given scale), and one can in particular renormalize them in such a way that any relation between them is satisfied. Of course it is absolutely true that a renormalization prescription involves the choice of a scale (this "substraction point" is usually chosen equal to some value of q^2, where q is a four-momentum), and that a numerical relation between renormalized parameters may be deformed by a change of scale, if the relation is not invariant under the renormalization group. But this fact does not mean that the relation itself is physically meaningless. For example, the relation $m_{muon} = 206\, m_{electron}$, which is valid in the on shell renormalization scheme of quantum electrodynamics, can be imposed and is actually imposed (because it is experimenally true on shell). However this last relation is not invariant under the renormalization group equations of QED! The third and last comment to be made about these problems of relations between constants of the standard model was already made in the text (section 6), but we repeat it here. Descriptions of the standard model based on non commutative geometry supplemented by the choice of specific scalar products in the space of fields seem to lead, at the classical level, to relations between the – otherwise arbitrary – constants of the model. Gauge invariance alone allows for more freedom; in absence of a full description of (spontaneously broken) quantum gauge field theories in terms of non commutative geometry, such constraints should be considered, in our opinion, as educated guesses. The reader may refer to [17, 18] for a detailed analysis of these constraints in the Connes framework. For us, the true power of the non commutative geometry desription of the standard model (and of quantum physics in general) is not tied up with the relevance of such constraints.

- Right neutrinos, simplicity and non trivial bundles (projectors):
  In its simplest "version", the standard model does not incorporate right neutrinos. From the point of view of non commutative geometry, and if one restricts oneself to the leptonic sector (take for instance the example of one family), lacking right neutrinos is a little bit of a nuisance. Indeed, in such a case, and in the language of A.Connes [1], the bundle of leptonic species is non trivial, and one needs to introduce a projector in the formalism, whose curvature itself enters the final expression. In the formalism explained in [4], the same phenomenon appears, because our approach (based on the tensorization of two by two complex matrices by arbitrary ones) leads to even dimensional square matrices. In order to accomodate an odd number of Weyl fermions (for instance e_L, e_R, and ν_L), one has to embed the odd dimensional matrix describing the connection into an even dimensional one by adding lines and columns of zeros. But then action of the d operator creates non zero entries in such places. The curvature is then not equal to $F = dA + A^2$, but to $F = p(dA + A^2)p + pdpdp$, where p projects on the odd dimensional subspace spanned by the Weyl fermions. The result is, as it should, the usual standard model without right neutrinos. However, introducing right neutrinos in the game (like in [6] and like in section 5 of the present paper), simplifies considerably the formalism, because one does not have to introduce such a projector. One cannot say that "non commutative geometry predicts that the neutrino has a mass", but it is clear that, from our perspective, the formalism is much simpler with a right neutrino than without. Let us remind the reader that such neutrinos are absolutely compatible with present experimental data, since the ν_R that one introduces for each family is not coupled to the (transverse part of the) gauge fields. Its main interest is to give a mass to the corresponding particle (hence a Dirac spinor), and to introduce mixing between fermionic families via a fermionic analogue of the Kobayashi-Maskawa matrix. Introduction of right neutrinos in the Connes formalism was recently investigated in [25].
- In [18] the authors raise the question: "Is any Connes-Lott model a Yang-Mills-Higgs model and conversely?" Their answer is clearly "No". However, one should not assimilate the approach initiated in [1] and further discussed

and detailed in the last chapter of [3] with non commutative geometry in general, or even with non commutatively inspired models. Other avenues are possible. Our belief is that it should be possible to distort sufficiently enough the formalism proposed initially by [1] (or [4]) to accomodate many cases of classical gauge field theory with symmetry breaking... For instance it was shown in [22] how to introduce "symmetries" into the Higgs fields of a non commutative model of the kind [1], in order to be able to accomodate scalar fields belonging to representations a priori forbidden in the initial framework. Such a distortion of the formalism allows of course the construction of more general models, but at the expense of aesthetics.

- The formalism introduced in [4] could induce one to think that this approach is a particular case of the formalism of superconnections introduced and discussed in [26]. This is actually not so. The algebraic connections defined by [1] are clearly not superconnections, since they are degree one forms in a $\mathbb{Z}$-graded differential algebra. Of course this algebra is also $\mathbb{Z}_2$-graded, since forms are even or odd, but generalized Yang-Mills potentials are defined here as one-forms and not only as odd forms. The same is true in our case, see the discussion in [5]. Notice that, when the differential algebra is represented on the space of spinors, the $\mathbb{Z}$-grading is lost and only the $\mathbb{Z}_2$-grading is left. Contrarily to what happens in the standard model, where the generalized gauge field incorporates only Higgs and usual spin one gauge fields, an algebraic superconnection would also incorporate other kinds of fields. Of course, one can consider a connection as some kind of "truncated" superconnection, but such a terminology becomes then pointless and can bring confusion.
- The ubiquitous number 24 (joke): With three families of quarks, three colors, three families of leptons including right neutrinos, we have 24 elementary right handed Weyl fermions and 24 elementary left handed Weyl fermions. The number of arbitrary parameters of the model is also equal to 24 (the gauge coupling constants g, g', the 6 quarks masses, the 6 leptonic masses, the $8 = 2 \times 4$ parameters of the leptonic and quark mixing matrices, the mass of the W, and the mass of the Higgs). We shall stop here these numerological remarks and suggest the reader interested in the beautiful properties of the number 24 to dive into almost any book of arithmetics, modular function theory, or higher dimensional cristallography.
- In the present formalism, the square of the generalized curvature gives us the whole bosonic classical Lagrangian of the standard model. One may wonder about the possibility of maintaining this unification at all orders of perturbative Q.F.T. In other words, is it possible to write the effective Lagrangian at one loop (at two loops etc) of the theory in terms of the generalized curvature $\mathcal{F}$? Such a possibility is by no means ruled out, but has not been proven yet. If true, such a property would pave the way to the next (ultimate?) goal of non commutative geometry (as far as physics of electroweak interactions are concerned): a non perturbative formulation of

a fully quantized gauge field theory. The completion of such an ambitious program would really be a truly non commutative achievement.
Final comment. Higgs fields and Yang-Mills fields can now be thought of as two particular components of the same object $\mathcal{A}$, and the square of the corresponding generalized curvature gives us the bosonic Lagrangian of the standard model. This unification is a little bit like the unification of electric and magnetic fields taken as independent components of the Faraday tensor F. By itself, such a unification is not a new theory in the sense that it was "already there". After all one can very well work with electromagnetism without using a manifestly covariant formalism! Also it does not bring about necessarily any new numerical prediction. However a unification of that kind is important at the conceptual level. Moreover it usually leads to generalizations or to new ideas that could be hardly thought of in a less unified framework. Unification of Higgs and Yang-Mills fields is for us the most important success of non commutative differential geometry in physics. This success is certainly going to stay.

Acknowledgments

We would like to thank our friends and colleagues at C.P.T. (in particular G. Esposito Farese) and at the University of Mainz (in particular R. Haussling and F. Scheck) for many discussions on those topics. One of us (R.C.) wants to thank the Erwin Schrödinger Institute of Mathematical Physics at Vienna for its hospitality.

References

1. A. Connes, J. Lott, Nucl. Phys. B **18 B** (1990) 29-47.
2. A. Connes, Les Houches Lectures (1992).
3. A. Connes, *Noncommutative Geometry*, Academic Press (1994)
4. R. Coquereaux, G. Esposito-Farese, G. Vaillant, *Higgs Fields as Yang-Mills Fields and Discrete symmetries.* Preprint Marseille (1990) CPT-90-2407. Nucl Phys B353, 689 (1991).
5. R. Coquereaux, R. Haussling, F. Scheck *Algebraic connections on parallel universes.* CPT-93/PE 2947. Int. Jour. of Mod. Phys A, Vol 10, No 1 (1995) 89-98.
6. R. Coquereaux, G. Esposito-Farese, F. Scheck, *Non Commutative Geometry and Graded Algebras in Electroweak Interactions* IHES-preprint, CPT-91/P.2464, Int. J. Mod. Phys. A, Vol 7, No 26 (1992) 6555-6593.
7. R. Coquereaux, *Higgs fields and superconnections* in *Differential Geometric Methods in Theoretical Physics*, C. Bartocci et al (eds), Proceedings of the Rapallo Conference (1990), 1991. Lecture Notes in Physics 375. Springer-Verlag.
8. R. Coquereaux, *Non Commutative Geometry: A physicist's brief survey*, Proc. of XXIIX Karpacz Winter School of Theoretical Physics: "*Fields and Geometry*" . J. of Geometry and Physics, 11 (1993) 307-324.
9. M. Dubois–Violette, *Dérivations et calcul différentiel non–commutatif.* Preprint Orsay 1987, C.R. Acad. Sci. Paris 307, I, 403–408 (1988).

10. M. Dubois–Violette, R. Kerner, J. Madore, *Non–commutative differential geometry of matrix algebras* (Orsay Preprint 1988). SLAC-PPF 88-45, J. Math. Phys. 31, 316–322 (1990).
11. M. Dubois–Violette, R. Kerner, J. Madore, *Non–commutative differential geometry and new models of gauge theory* (Orsay Preprint 1988). SLAC-PPF 88-49, J. Math. Phys., 323–329 (1990).
12. M. Dubois–Violette, R. Kerner, J. Madore, *Classical bosons in a non–commutative geometry* (1988). Class. Quantum Grav. 6, 1709–1724 (1989).
13. M. Dubois–Violette, *Non–commutative differential geometry, quantum mechanics and gauge theory*, in *Differential Geometric Methods in Theoretical Physics*, C. Bartocci et al, (eds), Proceedings of the Rapallo Conference (1990), 1991 Springer-Verlag.
14. M. Marcu, J. Math. Phys. 21 (1980) 1277.
15. W. Kalau, N.A. Papadopoulos, J. Plass, J.-M. Warzecha, MZ-TH/93-28.
16. D. Kastler *Lectures on Alain Connes' non commutative geometry and applications to fundamental interactions*, Proceedings of the First Caribbean Spring School of Mathematics and Theoretical Physics, World Scientific, to appear (1995).
17. D. Kastler, T. Schucker, *A detailed account of Alain Connes' version of the SM*, hep-th/9501077.
18. T. Schucker, B. Iochum , *Yang-Mills-Higgs versus Connes-Lott*, hep-th/9501142; to appear in Comm.Math.Phys.
19. D. Kastler D. Testard,*Quantum forms of tensor products*, Comm.Math.Phys. 155 (1993) 135.
20. R. Haussling, N.A. Papadopoulos, F. Scheck, Phys. Lett. B303 (1993) 265.
21. N.A. Papadopoulos, J. Plass and F. Scheck. *Models of electroweak interactions in non-commutative geometry: A comparison* (1993). MZ-TH/93-26.
22. A.H. Chamseddine, G. Felder and J. Fröhlich, *Grand Unification in non commutative geometry.* Nucl. Phys B395 (1993) 672-698.
23. J.C. Varilly and J.M. Gracia-Bondia, *Connes' non commutative geometry and the standard model*, J. Geom. and Physics 12 (1993) 223.
24. E. Alvarez, J.M. Gracia-Bondia and C.P. Martin, *A renormalization group analysis of the NCG constraints*, hep-th/9401099.
25. J.M.Gracia-Bondia, *Connes' interpretation of the SM and massive neutrinos* hep-th/9502120.
26. V. Matthai, D. Quillen, Topology 25 (1985) 85.
27. D.Fairlie, Phys. Lett. B82 (1979) 97.
28. Y.Ne'eman, Phys. Lett. B81 (1979) 190.

Large New Applications of Bethe Ansatz

L.D. Faddeev

St.Petersburg Branch of Steklov Mathematical Institute, Fontanka 27, St.Petersburg 191011, Russia
Research Institute for Theoretical Physics, University of Helsinki, Siltavuorenpenger 20C, SF 00014 Helsinki, Finland.

1. Introduction

Bethe Ansatz (BA) is a method of exact solution of quantum many body problems, which appeared in connection with models of quantum field theory or solid state theory in 1-dimensional space. For integrable models in $1+1$ dimensional space-time, magnetic chains are typical examples. Correspondingly, via the transfer matrix method, BA can be applied also to the 2-dimensional models of the classical statistical physics. Thus it is the main technical tool in the so-called 1-dimensional mathematical physics [1], described in details in several monographs [2], [3]. In spite of its long history (the original paper by H.Bethe [4] was published in 1931) it still shows new surprises and finds new unexpected applications. In these lectures I shall review the three of such examples:

1. Regge domain of QCD
2. Liouville model of Conformal Field Theory
3. Azbel-Hofstadter model of lattice electron in constant normal magnetic field

All these examples are important topics in modern physics, namely in the standard theory of strong interactions, string theory, and solid state physics. In the exposition I shall follow closely my recent publications in coauthorship with G.Korchemsky [5], O.Tirkkonen [6], and R.Kashaev [7], and refer to them for more details.

The topics above will be treated consecutively in Lectures 2, 3 and 4. Lecture 1 has an auxiliary character, I combine in it the necessary tools for Algebraic Bethe Ansatz (ABA), which is a modern mathematical presentation of the formalism presented by me in numerous lecture notes [8].

2. Generalization of Bethe Ansatz

We deal with the algebra of observables generated by dynamical variables $X_{n,\alpha}$, where n is an integer-valued index, usually (but not necessarily) treated as a discrete space variable, running through one-dimensional chain, and α

labels the variables in a given chain site. The chain is of finite length and closed, so that $n = 1, \ldots, N$, and the periodicity $X_{n+N,\alpha} \equiv X_{n,\alpha}$holds. Variables $X_{n,\alpha}$for different n commute,

$$[X_{n,\alpha}, X_{m,\beta}] = 0, \quad n \neq m$$

(ultralocality). Thus the total Hilbert space H has the form

$$H = \prod_{n=1}^{N} \otimes h_n,$$

where h_n is a local space, where variables $X_{n,\alpha}$ are represented.

Generic example is that of X_n being canonical variables in real or complex form. For one degree of freedom per site these are given by operators φ_n, π_n, or ψ_n, ψ^*_m with the Heisenberg commutation relations (CR)

$$[\varphi_n, \pi_n] = i\delta_{nm}\hbar$$

or

$$[\psi_n, \psi^*_m] = \delta_{nm}\hbar,$$

where $\hbar$ is a Planck constant.

The space h_n can be taken as $L_2(R)$ or l_2, or as a space of holomorphic quantization with realization $\psi^* = z$, $\psi = \frac{d}{dz}$ etc..

Another example is that of spin variables S^a_n, $a = 1, 2, 3$, with the CR

$$[S^a_n, S^b_m] = -i\hbar\epsilon^{abc} S^c_n \delta_{mn}, \tag{2.1}$$

or their q-deformation q^{S_3}, S_+ and S_-, with the generalized CR

$$q^{S_3} S_\pm = q^{\pm 1} S_\pm q^{S_3}, \quad [S_+, S_-] = \frac{q^{S_3} - q^{-S_3}}{q - q^{-1}} \tag{2.2}$$

on a given site, or analogous generators for the higher rank Lie algebras. The corresponding local Hilbert space can be a finite dimensional representation, labeled by total spin S.

One more example is given by H.Weyl type variables u_n, v_n with the exchange relations

$$u_n v_n = q v_n u_n,$$

$$[u_n, v_m] = [u_n, u_m] = [v_n, v_m] = 0, \qquad n \neq m. \tag{2.3}$$

The universality of canonical variables consists in the fact, that spin and Weyl variables can be written through them. For example

$$S^+ = \psi^*(2S - \psi^*\psi), \qquad S^- = \psi, \qquad S^3 = \psi^*\psi - S \tag{2.4}$$

realize the spin CR (2.1), and

$$\begin{aligned} S_n^{\pm} &= \frac{1}{2m\sin\gamma} e^{\pm i\frac{\pi_n}{2}} h(\pm\varphi) e^{\pm i\frac{\pi_n}{2}}, \\ q^{S_n^3} &= e^{i\gamma\varphi_n}, \\ h(\gamma) &= 1 + m^2 e^{2i\gamma\varphi} \end{aligned} \tag{2.5}$$

realize the q-deformed CR (2.2) with

$$q = e^{i\hbar\gamma}.$$

Furthermore the exponentials

$$u_n = e^{i\gamma\pi_n}, \qquad v_n = e^{i\varphi_n} \tag{2.6}$$

realize the Weyl relations (2.3) with the same q . Of course the concrete formulas (2.4), (2.5), (2.6) allowe the reduction of the canonical Hilbert space to the finite dimensional one in case, when $2S$ is integer and/or q is a root of unity.

The generating object of the Algebraic Bethe Ansatz (ABA) is a Lax operator $L_{n,a}(\lambda)$, which acts in the tensor product of local site space h_n and some auxiliary space V, so that label a stands for auxiliary. To illustrate this object I begin with examples.

Example 1. XXX case, $V = C^2$.

I take spin variables as generators, and Pauli matrices $I, \sigma_a, a = 1,2,3$, as operators in C^2. Then $L_{n,a}$ is an invariant operator in $h_n \otimes C^2$:

$$L_{n,a} = \lambda I_n \otimes I_a + i(\vec{S_n} \cdot \vec{\sigma_a}),$$

or if one take the explicit realization of Pauli matrices, $L_{n,a}$ can be written as a 2×2 matrix with operator-valued elements

$$L_n^{XXX}(\lambda) = \begin{pmatrix} \lambda + iS_n^3 & iS_n^- \\ iS_n^+ & \lambda - iS_n^3 \end{pmatrix}.$$

Example 2. XXZ case, $V = C^2$.

We take q-deformed spin variables as X_n, and put

$$L_n^{XXZ}(\lambda) = \begin{pmatrix} \sinh(\lambda + i\gamma S_n^3) & i\sin\gamma S_n^- \\ i\sin\gamma S_n^+ & \sinh(\lambda - i\gamma S_n^3) \end{pmatrix}.$$

Here we deal with the q-deformation of example 1, and it is clear that

$$\frac{L_n^{XXZ}(\gamma\lambda)}{\sin\gamma} \longrightarrow L_n^{XXX}(\lambda),$$

when $\gamma \to 0$.

Example 3.

We take Weyl pair u_n, v_n as dynamical variables, $V = C^2$, and put

$$L_n(\lambda) = \begin{pmatrix} u_n & e^{\lambda} v_n \\ e^{\lambda} v_n^{-1} & u_n^{-1} \end{pmatrix}.$$

This is rather basic example, the role of which becomes prominent in more recent time. In particular, as was shown by several groups (Korepin-Tarasov, Bazhanov-Stroganov, Volkov-Faddeev), the Lax operator of example 2 can be obtained in the form

$$L_n^{XXZ}(\lambda) = L_{n+\frac{1}{2}}(\lambda + a)L_{n-\frac{1}{2}}(\lambda - a)$$

after the reduction of the Hilbert space in the RHS.

The main property of the Lax operator is the Fundamental Commutation Relation (FCR)

$$R_{a_1,a_2}(\lambda - \mu)L_{n,a_1}(\lambda)L_{n,a_2}(\mu) = L_{n,a_2}(\mu)L_{n,a_1}(\lambda)R_{a_1,a_2}(\lambda - \mu), \qquad (2.7)$$

written in the tensor product $h_n \otimes V \otimes V$. Here $R_{a_1,a_2}(\lambda)$ is an operator in $V \otimes V$, whose matrix elements play the role of structure constants in the system of FCR (2.7).

The mathematical nature of FCR is explained in Drinfeld's theory of quasitriangular Hopf algebras [9]. I shall not give any details of this theory and bound myself by simple mnemonic rule.

In Drinfeld's theory enters an infinite dimensional algebra A, whose representations $\rho(s, \lambda)$ are labeled by spin variable s and complex parameter λ. The simplest example is a loop algebra with the corresponding evaluation representations. However the relevant examples for us are the q-deformed affine algebra $\widehat{sl_q}(2)$ (and its higher rank representations) and Yangian, which also have analogs of the evaluation representations. The universal R-matrix is an element in $A \otimes A$, satisfying the Yang-Baxter relation (YBR)

$$R_{12}R_{13}R_{23} = R_{23}R_{13}R_{12}$$

in $A \otimes A \otimes A$, with obvious notations $R_{12} = R \otimes I$, etc..

Applying representations $\rho(a_1, \lambda) \otimes \rho(a_2, \mu) \otimes \rho(s, \sigma)$ to YBR, we get FCR after the identification

$$\rho(s_1, \lambda) \otimes \rho(s_2, \mu)R = R_{s_1,s_2}(\lambda - \mu),$$
$$L_{n,a}(\lambda) = R_{\frac{1}{2},s}(\lambda); \qquad R_{a_1,a_2}(\lambda - \mu) = R_{\frac{1}{2},\frac{1}{2}}(\lambda),$$

where s defines the representation of spin variables.

If we write YBR in an equivalent form

$$R_{12}R_{32}R_{31} = R_{31}R_{32}R_{12}$$

and apply the representation $\rho(s_1, \lambda) \otimes \rho(s_2, \mu) \otimes \rho(a, \sigma)$, we get another FCR

$$R_{s_1,s_2}(\lambda - \mu)L_{s_1,a}(-\mu)L_{s_2,a}(-\lambda) = L_{s_2,a}(-\lambda)L_{s_1,a}(-\mu)R_{s_1,s_2}(\lambda - \mu),$$

which can be used to find the fundamental Lax operator $L_{n,f}(\lambda) = R_{s,n}(\lambda)$, for which me site quantum space h_n and auxiliary space V coincide, if the auxiliary operator $L_{s,\frac{1}{2}}(\lambda)$ is known.

The auxiliary Lax operator is used to write the BA equations, whereas the fundamental Lax operator generates the local Hamiltonians, which are diagonalized by means of BA. In the traditional applications of BA to spin $\frac{1}{2}$ chains these differences were not seen, as auxiliary and fundamental Lax operators coincide. However for the applications below (especially in Lecture 2) this difference is very important.

Let us explain it in more detail. Consider the Lax operator as a local holonomy along one site of the chain, and construct the corresponding monodromies

$$\begin{aligned} T_a(\lambda) &= L_{N,a}(\lambda)L_{N-1,a}(\lambda)\dots L_{1,a}(\lambda),\\ T_f(\lambda) &= L_{N,f}(\lambda)L_{N-1,f}(\lambda)\dots L_{1,f}(\lambda), \end{aligned}$$

where a stands for the auxiliary and f for the fundamental.

From FCR (2.7) and

$$\begin{aligned} R_{a,f}(\lambda-\mu)L_{n,a}(\lambda)L_{n,f}(\mu) &= L_{n,f}(\mu)L_{n,a}(\lambda)R_{a,f}(\lambda-\mu),\\ R_{f_1,f_2}(\lambda-\mu)L_{n,f_1}(\lambda)L_{n,f_2}(\mu) &= L_{n,f_2}(\mu)L_{n,f_1}(\lambda)R_{f_1,f_2}(\lambda-\mu), \end{aligned}$$

which can be readily derived in the fashion illustrated above from YBR, we get the relations

$$\begin{aligned} R_{a_1,a_2}(\lambda-\mu)T_{a_1}(\lambda)T_{a_2}(\mu) &= T_{a_2}(\mu)T_{a_1}(\lambda)R_{a_1,a_2}(\lambda-\mu),\\ R_{a,f}(\lambda-\mu)T_a(\lambda)T_f(\mu) &= T_f(\mu)T_a(\lambda)R_{a,f}(\lambda-\mu),\\ R_{f_1,f_2}(\lambda-\mu)T_{f_1}(\lambda)T_{f_2}(\mu) &= T_{f_2}(\mu)T_{f_1}(\lambda)R_{f_1,f_2}(\lambda-\mu), \end{aligned}$$

from which it follows that the invariants of monodromies commute for different values of parameter λ; in particular for

$$\mathrm{tr}_a T_a(\lambda) = F_a(\lambda), \qquad \mathrm{tr}_f T_f(\lambda) = F_f(\lambda),$$

we have

$$[F_a(\lambda), F_a(\mu)] = [F_a(\lambda), F_f(\mu)] = [F_f(\lambda), F_f(\mu)] = 0.$$

ABA allows us to diagonalize $F_a(\lambda)$ (and more implicitly $F_f(\lambda)$).

Let $T_a(\lambda)$ for $V = C^2$ be written as

$$T_a(\lambda) = \begin{pmatrix} A(\lambda) & B(\lambda) \\ C(\lambda) & D(\lambda) \end{pmatrix},$$

so that

$$F_a(\lambda) = A(\lambda) + D(\lambda).$$

We suppose that there exists a reference state Ω, such that

$$\begin{aligned} C(\lambda)\Omega &= 0,\\ A(\lambda)\Omega &= a(\lambda)^N\Omega,\\ D(\lambda)\Omega &= d(\lambda)^N\Omega. \end{aligned}$$

This condition is fulfilled in the examples 1 and 2 (see Lecture 4 for more complicated case). Then

$$\Phi(\{\lambda\}) = B(\lambda_1)\dots B(\lambda_l)\Omega$$

is an eigenstate of $F_a(\lambda)$:

$$F(\lambda)\Phi(\{\lambda\}) = \Lambda(\lambda,\{\lambda\})\Phi(\{\lambda\}), \tag{2.8}$$

with the eigenvalue

$$\Lambda(\lambda,\{\lambda\}) = \prod_{m=1}^{l} c(\lambda_m - \lambda)\, a(\lambda)^N + \prod_{m=1}^{l} c(\lambda - \lambda_m)\, d(\lambda)^N, \tag{2.9}$$

where

$$c(\lambda) = \frac{\lambda + i}{\lambda}, \quad a(\lambda) = \lambda + is, \quad d(\lambda) = \lambda - is,$$

and

$$c(\lambda) = \frac{\sinh(\lambda + i\gamma)}{\sinh\lambda}, \quad a(\lambda) = \sinh(\lambda + i\gamma s), \quad d(\lambda) = \sinh(\lambda - i\gamma s),$$

for examples 1 and 2, respectively. Here s is the spin label of relevant representation (taken to be the same for all sites of the chain). The parameters λ_m, $m = 1,\dots,l$, are obtained as the solutions of the system of algebraic equations

$$\left(\frac{a(\lambda_k)}{d(\lambda_k)}\right)^N = \prod_{j=1, j\neq k}^{m} \frac{c(\lambda_k - \lambda_j)}{c(\lambda_j - \lambda_k)}, \tag{2.10}$$

which are called the BA equations.

Local Hamiltonians are obtained from $F_f(\lambda)$ in the following manner. There exists a point λ_0 such that

$$L_{n,f}(\lambda_0) = P_{n,f},$$

where $P_{n,f}$ is a permutation operator in $h_n \otimes V_f$, both factors being similar spaces. It is clear that $F_f(\lambda_0)$ is a shift operator; then the coefficients of an expansion of $\ln F_f(\lambda)$ in the vicinity of λ_0 are local operators; in particular

$$\frac{d}{ds} \ln F(\lambda) \mid_{\lambda=\lambda_0} = H = \sum H_{i,i+1},$$

where $H_{i,i+1} = \frac{d}{ds} L(\lambda) \mid_{\lambda=\lambda_0}$ is usually taken as a physical Hamiltonian. The important statement is that local Hamiltonians have additive spectrum, allowing the quasiparticle representation; for instance, for $P = \ln F_f(\lambda)$ and H, we have

$$P\Phi(\{\lambda\}) = \sum p(\lambda_m)\Phi(\{\lambda\}),$$

$$H\Phi(\{\lambda\}) = \sum \epsilon(\lambda_m)\Phi(\{\lambda\}),$$

$$p(\lambda) = \tfrac{1}{2}\ln\tfrac{a(\lambda)}{d(\lambda)}, \quad \epsilon(\lambda) = \tfrac{d}{d\lambda}p(\lambda).$$

The proof of these statements can be found in [10]. They allow us to discuss the spectrum of physical Hamiltonians, using the information associated only with the auxiliary Lax operators, which are rather elementary.

The formulas (2.9) and (2.10) have a nice interpretation in terms of the Baxter equation. Let $Q(\lambda)$ be an operator, which can be diagonalized together with $F_a(\lambda)$, and has eigenvalues

$$q(\lambda, \{\lambda\}) = \left\{ \begin{array}{ll} \prod(\lambda - \lambda_m) & \text{Ex.1} \\ \prod \sinh(\lambda - \lambda_m) & \text{Ex.2} \end{array} \right\}.$$

Then $Q(\lambda)$ satisfies the difference equation

$$F(\lambda)Q(\lambda) = a(\lambda)^N Q(\lambda - ih) + d(\lambda)^N Q(\lambda + ih), \tag{2.11}$$

with $h = 1$ and $h = \gamma$, for examples 1 and 2, respectively. Equation (2.11) is more general than the BA equation (2.10) and can be used in the situations, when the reference state Ω is not available. The interpretation of the Baxter equation in terms of the general procedure of separation of variables is discussed by Sklyanin [11].

I finish this exposition with comments on the continuous limit of our objects, defined on the chain. Introducing the lattice spacing Δ, we want to consider the limit

$$H_{cont} = \frac{1}{\Delta} H_{chain},$$

when $\Delta \to 0$. This necessary requires that $N \to \infty$, so that the total space volume $N\Delta$ is finite or infinite. The latter case is interesting for the massive field theory, and the scaling limit is accompanied by the dimensional transmutation with the appearance of mass in the form

$$m = \frac{1}{\Delta} f(\kappa, \gamma),$$

where κ is some additional parameter entering the Lax operator. The former case is interesting for the conformal field theory; it is clear that only small eigenvalues of order $\Delta \sim \frac{1}{N}$ enter the physical spectrum. Thus this case is in the realm of finite size corrections. We shall use this in Lecture 3. In Lectures 2 and 4 we shall not have any scaling limits, and N will stay finite.

3. Bethe Ansatz and High Energy Quantum Chromodynamics

In his famous study of high energy behavior of QCD amplitudes, L.Lipatov found an effective field theory described in terms of so-called reggeized gluons. The Bethe-Salpeter equation for these quasiparticles was derived, which unexpectedly happened to be connected with the theory of spin chains. I cannot describe the long and instructive history of this development and refer to the original papers by Lipatov and his collaborators [12],[13], and a recent excellent review by M.Braun [14].

The reggeized gluon is characterized by a 2-dimensional coordinate $b = (b_1, b_2)$ (impact parameter) and a color index a. It is convenient to use complex coordinates

$$z = b_1 + ib_2, \quad \overline{z} = b_1 - ib_2.$$

The N-particle wave function $\Phi(\{z,\overline{z}\},\{a\})$ satisfies the Schroedinger equation with Hamiltonian

$$H = \sum_{i,j,a} (K_{ij} + \overline{K}_{ij}) t_i^a \otimes t_j^a, \tag{3.1}$$

with operators K_{ij} acting on coordinates, and t_i^a being color matrices, i.e. generators of the $SU(N_c)$ Lie algebra. The explicit form of K_{ij},

$$K_{ij} = -2\ln(z_i - z_j) - (z_i - z_j)\ln(-\frac{d}{dz_i}\frac{d}{dz_j})(z_i - z_j)^{-1}, \tag{3.2}$$

looks rather exotic. However in what follows this formula becomes natural.

In the limit of large number of colors $N_c \to \infty$, the Hamiltonian effectively is simplified, since for interesting components of the wave function the color part in (3.1) acts as

$$\sum_a t_i^a \otimes t_j^a \sim \delta_{i,j+1},$$

and (3.1) turns into

$$H \to K + \overline{K},$$

where K acts only on coordinate z and has the form

$$K = \sum_{i=1}^{N} K_{i,i+1},$$

typical for the spin chain. However the label i is just the number of a quasi-particle and admits no interpretation as a discrete coordinate. The analogy with the spin chain becomes even more close after the observation of Lipatov [15], that K commutes with operator A :

$$A = z_{12}z_{23}\ldots z_{N1}\frac{d}{dz_1}\cdots\frac{d}{dz_N},$$

for which he found the interpretation in terms of XXX spin chain. Consider spin operators

$$s^+ = z^2\frac{d}{dz}, \quad s^- = -\frac{d}{dz}, \quad s^3 = z\frac{d}{dz} \tag{3.3}$$

(compare with Lecture 1) with spin $s = 0$, and construct the corresponding Lax operator of Example 1. The trace of monodromy $F_a(\lambda)$ is polynomial in λ of order N :

$$F_a(\lambda) = 2\lambda^N + Q_2\lambda^{N-2} + \ldots + Q_N.$$

It can be easily seen that Q_N coincides with A. This makes one suspect that the Hamiltonian K itself is associated with this exotic example of XXX chain. And indeed it was shown in [5] (and independently by Lipatov himself in [16]), that K is the fundamental Hamiltonian, associated with the $SL(2,C)$ group and its infinite dimensional representation (3.2) with spin 0.

Let us present the corresponding arguments. It was shown in [17] how to construct the fundamental Hamiltonian for any spin S. It can be expressed as a function of the invariant expression $\left(\overrightarrow{S}_n + \overrightarrow{S}_{n+1}\right)^2$. Let J be defined by

$$J(J+1) = 2\overrightarrow{S_n} \otimes \overrightarrow{S_f} + 2S(S+1); \tag{3.4}$$

then up to normalization

$$L_{n,f} = \Gamma(i\lambda - J)\Gamma(i\lambda + J + 1).$$

For $s = 0$ the formula (3.4) is simplified, and the operator J, acting on the two complex variables z_1 and z_2, is found from the equation

$$J(J+1) = -(z_1 - z_2)^2 \frac{d}{dz_1}\frac{d}{dz_2}.$$

Correspondingly the fundamental pair Hamiltonian acquires the form

$$K = \frac{d}{d\lambda} \ln L_{n,f}(\lambda) \mid_{\lambda=0} = -\psi(-J) - \psi(J+1). \tag{3.5}$$

It is possible to check that (3.2) coincides with the expression (3.5).

Thus Lipatov's Hamiltonian is nothing but the fundamental Hamiltonian of the XXX chain, for the noncompact group $SL(2,C)$ in the representation $s = 0$. This representation corresponds to the simple action of group elements

$$g = \begin{pmatrix} a & b \\ c & d \end{pmatrix}, \quad T_g\, f(z) = f\left(\frac{az+c}{bz+d}\right).$$

As a corollary we have an important statement of the complete integrability of Lipatov's Hamiltonian. Indeed, from generalities of ABA it follows that all N commuting operators Q_i, $i = 2, \ldots N$, and not only Q_N, commute with K.

A natural question is, whether one can use the ABA to diagonalize Lipatov's Hamiltonian. This question is nontrivial due to the noncompactness of the $SL(2,C)$ group, and the answer is negative. However, a family of eigenvectors can be found. The reference state for the Lax operator $L_{n,f}$ does not exist. However a similar Lax operator

$$\widetilde{L_{n,a}} = \frac{d}{dz} L_{n,a} (\frac{d}{dz})^{-1},$$

corresponding to spin $s = -1$, produces the reference state

$$\Omega = \frac{1}{z_1^2} \cdots \frac{1}{z_N^2}$$

and leads to the BA equation

$$\left(\frac{\lambda_k - i}{\lambda_k + i}\right)^N = \prod_{j \neq k}^{l} \frac{\lambda_k - \lambda_j + i}{\lambda_k - \lambda_j - i},$$

and to the family of eigenvectors

$$\Phi = z_{12} \ldots z_{N1} B(\lambda_1) \ldots B(\lambda_l) \frac{1}{z_1^2 \cdots z_N^2},$$

where $B(\lambda)$ is the upper off-diagonal element of the monodromy $T_a(\lambda)$ in case $s = -1$. For $N = 2$ this gives eigenvectors of the form

$$\Phi(z_1, z_2) = \left(\frac{z_1 - z_2}{z_1 z_2}\right)^h \tag{3.6}$$

for $h = 2, 3, \ldots$, whereas the complete system of eigenvectors corresponds to

$$h = \frac{1+m}{2} - i\nu, \quad m = 1, 2, \ldots, \quad \nu \in R. \tag{3.7}$$

Thus one must look for another procedure to work in this case. The most promising way is to use the Baxter equation, which in our case $s = -1$ looks as

$$F(\lambda)Q(\lambda) = (\lambda + i)^N Q(\lambda + i) + (\lambda - i)^N (\lambda - i),$$

and after the change

$$S(\lambda) = \lambda^N Q(\lambda)$$

is simplified to

$$S(\lambda + i) + S(\lambda - i) - 2S(\lambda) = V(\lambda)S(\lambda),$$

where

$$V(\lambda) = \frac{Q_2}{\lambda^2} + \cdots + \frac{Q_N}{\lambda^N} .$$

For $N = 2$ the complete set of solutions can be expressed through the hypergeometric function ${}_3F_2$, and the whole family (3.6),(3.7) can be reconstructed. The case $N > 2$ is not done yet and represents a good challenge.

4. The Liouville Model

The Liouville model of 1+1 dimensional field theory appears in many applications in physics and geometry. It acquired attention recently due to Polyakov's work [19], where its importance for the theory of noncritical string was shown. The field $\Phi(x,t)$ of the model is different from usual linear fields; it is $e^{2\Phi}$, which has direct meaning as a metric (more exactly a deviation from a given background metric) of a two-dimensional "space-time" - worldsheet of the string. The equation of motion

$$\Box\Phi + 4e^{2\Phi} = 0$$

does not become homogeneous linear for small Φ. All this is to be taken into account in treating the Liouville model.

The action

$$S = \frac{1}{2\gamma}\int_0^{2\pi}\left(\frac{1}{2}(\Phi_t^2 - \Phi_x^2) - 2e^{2\Phi} - 2\Phi_{xx}\right)dx$$

of the model contains the coupling constant γ, which enters the classical Poisson brackets ($\pi = \frac{1}{2}\Phi_t$)

$$\{\pi(x), \Phi(y)\} = -\gamma\delta(x-y).$$

The pure divergence term Φ_{xx}in the action density is a remnant of the background metric in the limit when it becomes flat.

In this lecture we consider the more manageable quantum mechanical case, where Φ is purely imaginary. The hope is that this will throw new light on the model and will eventually lead to the more difficult case of real Φ.

In the imaginary case the Liouville equation can be obtained from the Sine-Gordon (SG) equation

$$\Box\Phi + 4m^2\sin\Phi = 0$$

after shift and rescaling:

$$\Phi \to \Phi + i\xi, \quad m^2e^{\xi} = 1, \quad \xi \to \infty, \quad m \to 0.$$

The SG model can be treated by the BA method [19], and it opens the possibility to quantize the Liouville model using this rescaling. We shall present here the relevant formulas, referring for the details to [6]. Let us comment that change from the hyperbolic case (real Φ) to the elliptic (imaginary Φ) must be accompanied by the change of the sign of the coupling constant γ. Thus the conformal convariance, generated by the density of energy-momentum

$$S = \frac{1}{\gamma}\left[(\Pi - \frac{1}{2}\Phi')^2 - 2e^{2i\Phi} + (\Pi - \frac{1}{2}\Phi')'\right],$$

with the Virasoro-like brackets

$$\{S(x), S(y)\} = 2(S(x) + S(y))\delta'(x-y) + \frac{1}{\gamma}\delta'''(x-y),$$

corresponds to negative central charge

$$c_{cl} = -6\frac{\pi}{\gamma}.$$

The canonical quantization, performed in [20],[21] in the beginning of 80-ties, showed that the exact quantum formula has the form

$$c = 1 - 6(\frac{\pi}{\gamma} + \frac{\gamma}{\pi} - 2).$$

The usual ingredients of ABA - lattice approximation, Lax operator along the chain, etc., are well known for the SG equation, which belongs to the family of example 2 (XXZ chain), with the quantum group generators of the form (2.5) from Lecture 1. The appropriate rescaling (during which one must be careful enough not to loose the λ–dependence) gives the Liouville model Lax operator

$$L_{n,a}(\lambda) = \begin{pmatrix} e^{i\Pi_n} & -ie^{-\lambda - i\Phi_n} \\ -2i\sinh(\lambda - i\Phi_n) & h(\Phi_n)e^{-i\Pi_n} \end{pmatrix}, \quad h(\Phi) = 1 - e^{-2i\Phi + i\gamma},$$

and the BA equation takes the form

$$e^{iN\gamma}\left(\frac{\sinh(\lambda_k - i\frac{\gamma}{2})}{\sinh(\lambda_k + i\frac{\gamma}{2})}\right)^N = \prod_{j, j\neq k} \frac{\sinh(\lambda_k - \lambda_j + i\gamma)}{\sinh(\lambda_k - \lambda_j - i\gamma)}.$$

Here N is half the length of the chain, so that in continuous limit we require

$$Na = \pi,$$

where a is lattice spacing.

One can say that these equations correspond to the XXZ spin chain with spin $-\frac{1}{2}$ and twist (quasi periodic boundary conditions), reflected in the factor $e^{i\gamma N}$ in the LHS.

Our treatment in the following is based on the string hypothesis for the solution of BA equations: The nonreal solutions of BA equations in the limit $N \to \infty$ combine into complexes $\{\lambda\}_L$, usually called strings of length L, consisting of roots λ_m of the form

$$\lambda_m = \lambda + im\gamma, \quad -\frac{L-1}{2} \leq m \leq \frac{L-1}{2},$$

m being integer for odd L, and semi-integer for even L. The center of the string λ is either real (positive string), or has fixed imaginary part $Im\lambda = \frac{\pi}{2}$ (negative string).

The complete set of BA equations is reduced to those for the centers of the strings in the form

$$e^{ip(\lambda_{K,k})N} = \prod_{L,l} S_{KL}(\lambda_{K,k} - \lambda_{L,l})$$

for a given configuration of the strings, i.e., given number of strings of length L. The explicit expressions of momentum $p(\lambda_{K,k})$ and phase-shifts S_{KL} can be found in [6].

Now we state the remarkable duality between spin $\frac{1}{2}$ and spin $-\frac{1}{2}$ XXZ chains: the set of BA equations for spin $\frac{1}{2}$ and $\gamma = \frac{\pi}{\nu+1}$, and for spin $-\frac{1}{2}$ and $\gamma = \frac{\pi\nu}{\nu+1}$, coincide, if each string of length L is substituted by that of length $\nu + 1 - L$.

Let us comment, that it is well known for spin $\frac{1}{2}$ XXZ model for rational coupling γ, that the length of the string cannot exceed ν, where $q = \nu + 1$ is the denominator in

$$\gamma = \frac{\pi p}{q}.$$

The next remark is that the eigenvalues of the local Hamiltonians for spin $\frac{1}{2}$ and spin $-\frac{1}{2}$ cases coincide up to the sign. This allows us to get all relevant information for the Liouville model with $\gamma = \frac{\pi\nu}{\nu+1}$ from that for spin $\frac{1}{2}$ XXZ model with $\gamma = \frac{\pi}{\nu+1}$. The finite size corrections for the latter are well known. The most relevant paper for our goal is that by Karowsky [22], where the case with the fixed twist $e^{2ir\gamma}$ was considered, $r \in Z_+$. In our case the twist superficially depends on the length of the chain N; however, for rational γ and even N, it has Karowsky's form, where r is the remnant of $\frac{N}{2}$ mod$(\nu+1)$; thus for different length of the chain we have all values of the twist

$$r = 0, 1, \dots \nu.$$

Karowsky investigated the asymptotic value for $N \to \infty$ of the maximal eigenvalue Λ of the transfer matrix $F(\lambda)$, for spin $\frac{1}{2}$ and $\gamma = \frac{\pi}{\nu+1}$ and the nearest eigenvalues with $\frac{1}{N}$ order of correctness. The maximal eigenvalue Λ in the case of twist r has the form

$$\ln \Lambda_{\max}(\lambda) = -NF_\infty + \frac{1}{N}\frac{\pi}{6}\left(1 - \frac{6r^2}{\nu(\nu+1)}\right)\cosh\frac{\pi\lambda}{\gamma},$$

and next to it (excitations)

$$\ln \Lambda_{\rm ex}(\lambda) - \ln \Lambda_{\max}(\lambda) = -\frac{2\pi}{N}\left((\Delta + \overline{\Delta})\cosh\frac{\pi\lambda}{\gamma} + (\Delta - \overline{\Delta})\sinh\frac{\pi\lambda}{\gamma}\right),$$

where Δ and $\overline{\Delta}$ can be calculated from the asymptotic BA equation. The "primary" Δ and $\overline{\Delta}$ correspond to the "sea" of positive strings of length 1 and one string of length K, $K = 1, 2, \dots, \nu - r$, and have the form

$$\Delta = \overline{\Delta} = \frac{((\nu+1)(K-1)+r)^2 - 1}{4\nu(\nu+1)}.$$

All other Δ and $\overline{\Delta}$ have additional integer value contributions and correspond to adding of oscillators to the primary excitations.

Thus, if we consider the chains of all possible lengths, excluding $r = 0$ (it is still not clear why this omission must be done), calculate the energy via the formula

$$E = -\frac{d}{d\lambda} \ln \Lambda(\lambda) \,|_{\lambda=i\gamma},$$

take the case $r = 1$ as a vacuum sector, and use general results of Cardy [23], we get the following results:

The central charge of the Virasoro algebra, associated with our chain, is

$$c = 1 - 6\left(\frac{\pi}{\gamma} + \frac{\gamma}{\pi} - 2\right), \qquad (4.1)$$

where $\gamma = \frac{\pi\nu}{\nu+1}$. The primary energies take the form of Kac

$$\Delta = \frac{(p(\nu+1) - q\nu)^2 - 1}{4\nu(\nu+1)}, \qquad 1 \leq q \leq p \leq \nu,$$

where p and q are expressed via the remnant r and string length K as follows:

$$p = K + r - 1, \quad q = r.$$

The case $p = q$ corresponds to the ground state in the section with values of remnant r exceeding 1; other cases have one string of higher length.

Then we see, that the BA technique allows us to get the exact results, showing that the Liouville model for coupling constant corresponds to the unitary minimal models [24], [25].

What does this experiment show about the more interesting case of real Φ? The lattice approximation, Lax operators, and FCR are purely algebraic and hold for real Φ. However, the reference state is absent, and ABA procedure is not applicable. The way out is to use the more general procedure of separation of variables by Sklyanin [11]. If we believe in the universality of (4.1), then for the central charge in the strong regime interval

$$1 \leq c \leq 25$$

the coupling constant γ must become complex. It will be interesting to derive the Baxter equation of form (2.11) in this case. Another possibility, advocated by Takhtajan, is to abandon (4.1) altogether. Only the future work will shown what is really right.

5. Azbel-Hofstadter Model

The system, consisting of one electron, confined to a plane and interacting with a constant magnetic field orthogonal to this plane, is one of the most fashionable models in solid state theory. The Hamiltonian

$$H = \frac{1}{2m}\left[(p_1 - eA_1)^2 + (p_2 - eA_2)^2\right]$$

is written as a sum of two squares

$$H = \frac{1}{2}(P^2 + Q^2),$$

with commutation between P and Q:

$$[P, Q] = -i\gamma I,$$

which formally coincides with that of canonical coordinate and momentum, with γ playing the role of "Planck constant"and being proportional to the magnetic field. Of course, in the Hilbert space of the system we have two more generators of the algebra of observables, so that the spectrum H is highly degenerate. Otherwise it has the nature of the spectrum of the harmonic oscillator (Landau levels).

A natural generalization, taking into account the lattice structure of the plane, is to consider the Hamiltonian

$$H = \cos P + \cos Q,$$

expressed via the exponents of P and Q with the Weyl type of commutation relations

$$e^{iP}e^{iQ} = e^{i\gamma}e^{iQ}e^{iP}.$$

It is also a standard system in solid state theory, usually called the Azbel-Hofstadter model. The spectrum of the Hamiltonian H, also called the Harper Hamiltonian, depends on the arithmetic nature of γ. There exists vast literature with analytic and numerical results on this. In this lecture I shall show the natural connection of the Harper Hamiltonian with the objects and methods of Bethe Ansatz. The first connection of this type was established by Wiegman and Zabrodin [26]. Here I follow my joint paper with R.Kashaev [7]. I shall point out the main ideas and refer for the technical details to [7]. I shall consider a natural generalization of the Harper Hamiltonian,

$$H = \mu(\alpha S + \alpha^{-1}S^{-1}) + \nu(\beta T + \beta^{-1}T^{-1}) + \rho(\gamma U + \gamma^{-1}U^{-1}),$$

where S, T, U are three operators with the Weyl type exchange relations

$$ST = \omega TS, \quad TU = \omega UT, \quad US = \omega SU,$$

μ, ν, ρ being real numbers, and α, β, γ having values on a circle,

$$| \alpha |=| \beta |=| \gamma |= 1.$$

The exchange constant ω (essentially the exponent of the magnetic flux) will be supposed to be rational,

$$\omega = \exp 2\pi i \frac{M}{N}.$$

We shall not fix a representation for S, T, U, require only that

$$S^N = T^N = U^N = 1,$$

and confine $\alpha, \beta . \gamma$ to the fundamental domain

$$0 \leq \arg \alpha, \arg \beta, \arg \gamma \leq \frac{2\pi}{N}.$$

The product $C = STU$ is a central element of our algebra.

We shall now show, that the above Hamiltonian describes a conservation law entering the monodromy for the chain of length 3, generated by a Lax operator $L_{n.a}(x, h)$ of the form

$$L(x, h) = \begin{pmatrix} aXY & xbY \\ xcX & d \end{pmatrix},$$

where X, Y is a Weyl pair

$$XY = \omega YX,$$

for which we shall take the standard representation in C^N:

$$\langle m \mid X \mid n \rangle = \omega^m \delta_{m,n},$$
$$\langle m \mid Y \mid n \rangle = \delta_{m,n+1},$$

so that

$$X^N = Y^N = 1.$$

x is a spectral parameter, associated with the auxiliary space C^2, and $h = (a, b, c, d)$ is a set of four complex parameters attached to a given quantum space C^N. This Lax operator is a modification of that of Example 3,

$$\widetilde{L}(\lambda) = \begin{pmatrix} u & e^{\lambda} v \\ e^{-\lambda} v^{-1} & u^{-1} \end{pmatrix}$$

with the relations $uv = qvu$, and can be obtained from it after multiplication by u, identification

$$v^{-1}u = X, \quad vu = Y, \quad \omega = q^2,$$

and rescaling operators X, Y, the spectral parameter λ, and the Lax operator itself. The fundamental commutation relations do not change under those rescalings and have the form

$$R(\frac{x}{y})L_1(x,h)L_2(y,h) = L_2(y,h)L_1(x,h)R(\frac{x}{y}),$$

where $R(x)$ given by

$$R(x) = \begin{pmatrix} x\omega - \frac{1}{x} & 0 & 0 & 0 \\ 0 & \omega(x - \frac{1}{x}) & \omega - 1 & 0 \\ 0 & \omega - 1 & x - \frac{1}{x} & 0 \\ 0 & 0 & 0 & \omega x - \frac{1}{x} \end{pmatrix}$$

takes into account the passage from $\widetilde{L}$ to L.

Now we construct the monodromy

$$T(x) = L_0(x,h_0)L_1(x,h_1)L_2(x,h_2),$$

its trace

$$F(x) = \mathrm{Tr}\ T(x),$$

and observe that $F(x)$ is a quadratic polynomial in X :

$$F(x) = F_0 + x^2 F_2,$$

where F_2 can be identified with H if we put

$$S = X_0Y_1, \quad T = Y_0X_2, \quad U = X_1Y_2,$$

and express $\alpha, \beta, \gamma, \mu, \nu, \rho$ via the coefficients a_i, b_i, c_i, d_i, $i = 0,1,2$. The representation for S, T and U has dimension N^3, but this influences only the multiplicity of the spectrum and not the eigenvalues.

Can we use ABA to diagonalize $F(x)$, and in particular H? The answer is "no", because we cannot construct the reference state. However, a more general procedure based on the Baxter equation is applicable. We shall illustrate the technique of derivation of the equation by our example. The idea is to make the local Lax operator triangular by means of the gauge transformations:

$$L_i(x,h_i) \to A_iL_i(x,h_i)A_{i+1}^{-1} = \widehat{L}_i(x,h_i).$$

We choose A_i in the form

$$A_i = \begin{pmatrix} 1 & \eta_i \\ 1 & \xi_i \end{pmatrix}, \quad \xi_i - \eta_i = 1.$$

Then $\widehat{L_i}(x,h_i)$ gets the form

$$\widehat{L_i}(x,h_i) = \begin{pmatrix} F(x,h_i \mid \eta_i,\xi_{i+1}) & -F(x,h_i \mid \eta_i,\eta_{i+1}) \\ F(x,h_i \mid \xi_i,\xi_{i+1}) & -F(x,h_i \mid \xi_i,\eta_{i+1}) \end{pmatrix},$$

where $F(x,h \mid \eta,\xi)$ is a linear form of XY, X, Y, and I,

$$F(x,h \mid \xi,\eta) = (\eta a X - xb)Y + \xi(x\eta c X - d).$$

Now we choose coefficients ξ_i, such that the equations

$$F(x, h_i \mid \xi_i, \xi_{i+1})\Phi_i = 0$$

have nontrivial solutions. For coefficients of an expansion of Φ_i on the basis $|m\rangle$,

$$\Phi_i = \sum C^i_m |m\rangle ,$$

we get the recurrence equation

$$C^i_m = C^i_{m-1}\left(-\frac{1}{\xi_i}\frac{\xi_{i+1}a_i\omega^m - xb_i}{x\xi_{i+1}c_i\omega^m - d_i}\right).$$

The periodicity $m + N \equiv m$ leads to the following conditions on ξ_i:

$$\xi_i = (-1)^N \frac{a_i^N \xi_{i+1}^N - b_i^N x^N}{c_i^N (x\xi_{i+1})^N - d_i^N}.$$

These three equations for four variables (x, ξ_1, ξ_2, ξ_3) define a point p on some algebraic curve Γ. For each such point we have local vacua $\Phi_i(p)$, $i = 1, 2, 3$. The next observation is that they are related to the diagonal elements of $\widehat{L_i}(x, h_i)$ in the following manner:

$$F(x, h_i \mid \eta_i, \xi_{i+1})\Phi_i(p) = \Delta^i_-(p)\Phi_i(\tau_-(p)),$$
$$F(x, h_i \mid \xi_i, \eta_{i+1})\Phi_i(p) = \Delta^i_+(p)\Phi_i(\tau_+(p)),$$

where $\tau_\pm$ are two automorphisms of Γ,

$$\tau_\pm(x) = \omega^{\pm 1/2}x, \quad \tau_\pm(\xi_i) = \omega^{-1/2}\xi_i,$$

and $\Delta^i_\pm(p)$ some explicit functions of p, depending also on h.

It follows, that the generalized reference state

$$\Phi(p) = \Phi_1(p) \otimes \Phi_2(p) \otimes \Phi_3(p)$$

in C^{N^3} satisfies the relation

$$F(x)\Phi(p) = \Delta_-(p)\Phi(\tau_-(p)) + \Delta_+(p)\Phi(\tau_+(p)),$$

where $\Delta_\pm(p) = \prod_i \Delta^i_\pm(p)$. This is our form of the Baxter equation, where the mere shift of the spectral parameter is substituted by an automorphism of the algebraic curve. For any projection $Q(p)$ of the vector $\Phi(p)$ on a given eigenvector φ of $F(x)$ with eigenvalue $\Lambda(x)$,

$$Q(p) = \langle \varphi \mid \Phi(p)\rangle ,$$

we have a difference equation

$$\Lambda(x)Q(p) = \Delta_-(p)Q(\tau_-(p)) + \Delta_+(p)Q(\tau_+(p)),$$

which is used to find both $\Lambda(x)$ (and in particular eigenvalues of H) and $Q(p)$. The difficult and unsolved problem is to find an appropriate parametrization of a representative family of functions on Γ, and to get from the Baxter equation an equation for the corresponding parameters - some analog of the BA equations. It follows from some general considerations, that eigenvalues of H run through N intervals ("bands" or "zones"), when the quasimomenta α, β and γ change in their fundamental domain. For the ends of these zones (and some points inside), the curve Γ degenerates to a plane C and automorphisms $\tau_\pm$ reduce to shifts, so that the Baxter equation takes the form

$$\Lambda(x)Q(x) = \Delta_+(x)Q(xq^{-1}) + \Delta_-(x)Q(xq),$$

and for $N = 2P+1$ $Q(x)$ becomes a polynom in x of degree $2P$. Thus we can get the usual Bethe-Ansatz equations. Their explicit form is given in [7]. The most interesting problem, where they could be used, is the investigation of the limit $N \to \infty$ (irrational flux), in which some universal fractal behavior of the spectrum is expected. This limit is rather different from the usual thermodynamic limit, when the length of the chain is increased; indeed, the length of the chain is fixed (equal to 3), but the order of polynomial $Q(x)$ goes to infinity. Thus the investigation of the band spectrum of the Azbel-Hofstadter model for rational flux by means of BA equations is still an interesting challenge.

6. Conclusion

In these lectures I tried to show that the Bethe-Ansatz technique, traditionally associated with magnetic chains, is a universal method with many unexpected applications. This was illustrated by consideration of three examples from different domains of modern theoretical physics. The mathematical background was put aside. For all three examples the real problems were formulated, but not solved completely.

Their solutions, such as treatment of the Liouville model in the region $1 \leq c \leq 25$, getting the eigenvalues of Lipatov's Hamiltonian for any number N of the reggeized gluons, and finding the fractal behavior of the spectrum in Azbel-Hofstadter model, will be based on a new development of the BA technique, which we expect in the near future.

This work was partly supported by the ISF grant R2H300.

References

1. E.Lieb, D.Mattis, *Mathematical Physics in One Dimension*, Academic Press, New York, N-Y (1966).

2. R.Baxter, *Exactly Solvable Models in Statistical Mechanics* , Academic Press, New York, N-Y (1982).
3. V.Korepin, N.Bogolyubov, A.Izergin, *Quantum Inverse Scattering Method and Correlation Functions* , Cambridge University Press (1993).
4. H.Bethe, Z. fur Physik **71** (1931) 205.
5. L.Faddeev, G.Korchemsky, Phys.Lett **B** 342 (1995) 311.
6. L.Faddeev, O.Tirkkonen, *Connection of the Liouville Model and XXZ spin chain*, Preprint HU-TFT-15 (1995), HEP-TH (1995).
7. L.Faddeev, R.Kashaev, Comm.Math.Phys., **169** (1995) 181.
8. L.Faddeev, Int.J.Math.Phys. **10 A** (1995) 1845.
9. V.Drinfeld, *Quantum Groups* in Proc. ICM-86 at Berkeley, **1**, 798, AMS (1987).
10. V.Tarasov, L.Takhtajan, L.Faddeev, Theor.Math.Phys. **57** (1983) 163.
11. E.Sklyanin *Functional Bethe Ansatz*, in *Integrable and superintegrable systems*, 8-33 , World Scientific, (1990); *Quantum Inverse Scattering Method. Selected Topics*, in *Quantum Groups and Quantum Integrable Systems* (Nankai Lectures in Mathematical Physics, 63-97, World Scientific, (1992).
12. L.Lipatov, in *Perturbative QCD* 411-489, ed. A.Mueller, World Scientific, Singapore (1989).
13. L.Lipatov, Phys.Lett. **251** (1990) 284; Phys.Lett. **309** (1993) 394
14. M.Braun, *The interaction of reggeized gluons and Lipatov's hard pomeron*, Lectures at University of Santiago de Compostela, Spain (1994).
15. L.Lipatov, *High-energy asymptotics of multi-color QCD and exactly solvable lattice models*, Padowa preprint DFPD/93/TH/70 (1993).
16. L.Lipatov, *High-energy asymptotics of multi-color QCD and exactly solvable lattice models*, ZhETF Lett. **59**, No 9 (1994) 571(in Russian)
17. P.Kulish, N.Reshetikhin, E.Sklyanin, Lett.Math.Phys. **5**, (1981) 393.
18. A.Polyakov, Phys.Lett **103 B** (1981) 207.
19. E.Sklyanin, L.Takhtajan, L.Faddeev, Theor.Math.Phys. **40** (1980) 688.
20. J-L.Gervais, A.Neveu, Nucl.Phys. **199 B** (1982) 50; Nucl.Phys. **257 B** (1985) 59.
21. T.Curtright, C.Thorn, Phys.Rev.Lett. **48** (1989) 1309.
22. M.Karowsky, Nucl.Phys. **300 B** (1988) 473.
23. J.Cardy, Nucl.Phys. **270 B** (1986) 746.
24. A.Belavin, A.Polyakov, A.Zamolodchikov, Nucl.Phys. **241 B** (1984) 333.
25. D.Friedan, Z.Oin, S.Shenker, Phys.Rev.Lett. **52** (1984) 1575.
26. P.Wiegman, A.Zabrodin, Nucl.Phys. **422 B** (1994) 495.

Universality in Turbulence: An Exactly Solvable Model

K. Gawedzki and A. Kupiainen

I.H.E.S., C.N.R.S., F-91440 Bures-sur-Yvette, France,
Math. Department, Helsinki University, PO Box 4, 00014 Helsinki, Finland

1. Introduction

One of the basic and most successful ideas in theoretical physics has been that of universality. It states that many systems with large or infinite number of degrees of freedom in certain asymptotic regimes display similar behaviors falling into general types. Such a situation has been encountered

in statistical mechanics of 2nd order phase transitions, where the universality applies to the long distance asymptotics of correlation functions characterized e.g. by scaling exponents, independent of numerous microscopic details of the systems [1],

in field theory, where the universality implies the cutoff independence of the effective low energy description [1],

in the theory of 1st order phase transitions exhibiting universal shapes of critical droplets [2],

in many-body condensed matter, where universality limits possible behaviors of superconductors [3] or Hall fluids [4],

in dynamical systems, where it describes types of behavior of families of maps under iteration [5],

in non-linear PDE's, where it determines space and/or time asymptotics of solutions [6].

The basic idea behind universality of certain behaviors of systems with many degrees of freedom is, that such behaviors are governed by few relevant degrees of freedom with simple dynamics. Those may be often exhibited using renormalization group transformations wiping out the irrelevant details of the system [7][8].

The question addressed in the present lectures is to what extent such a picture applies to the fully developed hydrodynamic turbulence. It should be mentioned at the very beginning that the problem is still wide open, although the first theory arguing for universality in turbulence, due to Kolmogorov [9], dates from 1941. In the 1st lecture we shall sketch, how the main claims of the Kolmogorov theory relate to the cascade picture of energy transfer in a turbulent flow. In the 2nd lecture, we shall briefly present the functional-integral approach to the statistical theory of turbulence governed by the Navier-Stokes equation, trying to stress the similarities and the differences with field theory. In the 3rd and the 4th lecture, we shall discuss a simple model of turbulent advection of a scalar quantity, known under the name of passive scalar (PS). The model is a good playground for testing the idea of universality in turbulence, since one may obtain closed expressions for the stochastic initial data evolution and for the steady state correlation functions of the scalar. In the absence of external sources, the PS with deterministic initial data undergoes a non-universal diffusion, which averaged over initial data distributed in a homogeneous way becomes a universal super-diffusion. In the presence of random sources, the steady state 2-point correlator exhibits an energy cascade of the type expected in the Navier-Stokes turbulence. The exact solutions for the higher point correlators are more difficult to analyze. We examine the origin of possible violations of strict universality of these correlators. What emerges is a picture of restricted universality applying to infrared renormalized correlators, from which few universal terms multiplied by non-universal coefficients were removed. In Conclusions, we summarize our discussion and point to possible directions of the future research.

The PS model, although introduced long time ago [10], has become recently a subject of intensive study, see [11][12] and also [13] (the latter for a mathematical work on a simplified version of the model). In the final stages of the work on these notes, we have received preprints [14][15] and [16]. The analysis of the first two has large overlaps with ours, although we disagree with some conclusions of [14] and differ in some interpretations from [15].

We would like to thank the Mittag-Leffler Institute, where our work on the passive scalar model was started, for hospitality. Discussions with Uriel Frisch, Robbert Kraichnan, Itamar Procaccia, and Achim Wirth are kindly acknowledged.

2. Energy Cascade and Kolmogorov Scaling

The evolution of the local velocity field $\mathbf{v}(t,\mathbf{x})$ of the incompressible fluid at points $\mathbf{x}$ of the 3-dimensional space is described by the Navier-Stokes (NS) equation

$$\partial_t \mathbf{v} + (\mathbf{v}\cdot\nabla)\mathbf{v} - \nu\Delta\mathbf{v} \;=\; -\nabla p + \mathbf{f}, \tag{2.1}$$

supplemented with the incompressibility condition

$$\nabla\cdot\mathbf{v} \;=\; 0\,. \tag{2.2}$$

The presence of the dissipative term multiplied by the viscosity ν of the fluid (with the dimension $\frac{length^2}{time}$) distinguishes the NS equation from the Euler equation. $\mathbf{f}(t,\mathbf{x})$ denotes the external force. $p(t,\mathbf{x})$ is the pressure (divided by the constant density) and may be eliminated from both equations:

$$p(t,\mathbf{x}) \;=\; \frac{1}{4\pi}\int\frac{\nabla\cdot[\mathbf{f}(t,\mathbf{y})-(\mathbf{v}(t,\mathbf{y})\cdot\nabla)\mathbf{v}(t,\mathbf{y})]}{|\mathbf{x}-\mathbf{y}|}\,d^3\mathbf{y}\,. \tag{2.3}$$

If the flow takes place in a finite volume, boundary conditions should be specified. The size of the volume introduces an "integral scale" L into the problem, and we shall assume that the external force $\mathbf{f}$ acts only at length scales comparable with L. Alternatively, we may consider flows in infinite space excited by external force, with Fourier components concentrated in wavenumbers $\mathbf{k}$ with $k \equiv |\mathbf{k}| \leq L^{-1}$. Under the rescaling

$$\begin{aligned} \widetilde{\mathbf{v}}(t,\mathbf{x}) &= \sigma\,\mathbf{v}(\tau t, s\mathbf{x})\,, \\ \widetilde{\mathbf{f}}(t,\mathbf{x}) &= \sigma\tau\,\mathbf{f}(\tau t, s\mathbf{x})\,, \\ \widetilde{p}(t,\mathbf{x}) &= \sigma\tau s^{-1} p(\tau t, s\mathbf{x}), \end{aligned} \tag{2.4}$$

with $\sigma s/\tau = 1$, the viscosity ν is replaced in Eq. (2.1) by $\widetilde{\nu} = \tau s^{-2}\nu$. It is then convenient to introduce a dimensionless quantity, the Reynolds number

$$R = \frac{V\,L}{\nu}\,, \tag{2.5}$$

where V stands for the typical value of the velocity differences over the integral scale L. R is invariant under the rescalings (2.4).

The basic phenomenological observation is, that the hydrodynamic flows have very different behavior for small values of R (of order 1, say), where flows are regular (laminar), and large values of R (of order 10^3 and more), where the flows are very chaotic (turbulent), with complicated phenomena occurring at intermediate values of R. Here we shall be interested in the regime of fully developed turbulence with very large Reynolds numbers, ideally in the limit $R \to \infty$. It is sensible to use a statistical description of the flows in this regime. The stochasticity may be introduced into the description by considering random external force $\mathbf{f}$, or studying the evolution of random initial data, or both. In a statistical description, the basic objects to look at will be the velocity correlation functions given by the expectation values of products of components of velocities. We shall mostly look at equal time correlators

$$< \prod_{n=1}^{N} v^{i_n}(t, \mathbf{x}_n) >, \tag{2.6}$$

which, in the stationary state of turbulence sustained by steady external forces, should be time independent. Besides we may expect that, far from the fluid boundaries, the correlators (2.6) are translationally and rotationally invariant, the property that expresses the homogeneity and isotropy of the fully developed turbulence. The most commonly studied velocity correlators are the so called structure functions

$$S_N(x) = < (\mathbf{v}(\mathbf{x}) - \mathbf{v}(0))^N > . \tag{2.7}$$

One of the simplest consequences of the NS equation (2.1) may be obtained by taking its scalar product with $\mathbf{v}$ and integrating the result over the space. This results in the relation

$$\frac{d}{dt}\frac{1}{2}\int \mathbf{v}^2 = -\frac{\nu}{2}\int (\nabla \mathbf{v})^2 + \int \mathbf{v}\cdot\mathbf{f}, \tag{2.8}$$

which expresses the energy balance: the time derivative of energy on the right hand side is equal to the difference of the injection rate $\int \mathbf{f}\cdot\mathbf{v}$ and the dissipation rate $\frac{\nu}{2}\int(\nabla\mathbf{v})^2$ of energy. Taking averages in the stationary state, we obtain

$$<\mathbf{v}\cdot\mathbf{f}> = <\tfrac{\nu}{2}(\nabla\mathbf{v})^2> \equiv \epsilon\,, \tag{2.9}$$

i.e. the equality of the (intensive) mean injection and the mean dissipation rates of energy. How is this energetic balance realized over different modes of the fluid motion? The energy injection takes place at the distances of order of the integral scale by induction of big scale L eddies. According to the picture of the turbulent flow proposed in 1922 by Richardson [17], the big eddies induce smaller eddies which, in turn, induce still smaller eddies and so on, transferring energy from large to small distance scales. This process should not lead to a loss of energy, until sufficiently small distance scales, say, smaller than η, are reached. On scales smaller than η, the dissipative term $\nu\Delta\mathbf{v}$ of the NS equation becomes important. One may formulate a more quantitative version of this picture. We shall follow here the discussion of the treatise [18], which is an excellent reference to the problems of well developed turbulence. The idea is to localize the velocity field in the Fourier space. For $\widehat{\mathbf{v}}(\mathbf{k}) \equiv \int e^{-i\mathbf{k}\cdot\mathbf{x}}\,\mathbf{v}(\mathbf{x})\,d^3\mathbf{x}$, define

$$\mathbf{v}_{\leq K}(\mathbf{x}) = \int_{|k|\leq K} e^{i\mathbf{k}\cdot\mathbf{x}}\,\widehat{\mathbf{v}}(\mathbf{k})\,\frac{d^3\mathbf{k}}{(2\pi)^3}\,. \tag{2.10}$$

A more detailed version of the energy balance equation (2.8) reads

$$\frac{d}{dt}\frac{1}{2}\int\mathbf{v}_{\leq K}^2 = -\frac{\nu}{2}\int(\nabla\mathbf{v}_{\leq K})^2 + \int\mathbf{v}_{\leq K}\cdot\mathbf{f} - \int\Pi_K\,, \tag{2.11}$$

where Π_K is the density of energy flux out of the wave numbers with $|\mathbf{k}|\leq K$. An explicit expression for Π_K may be obtained from the NS equation (2.1) and may be argued to involve mainly $\widehat{\mathbf{v}}(\mathbf{k})$ with $|\mathbf{k}|\cong K$, i.e. to be aproximately local in the wavenumber space. Taking the averages in the stationary state, we obtain the relation

$$<\mathbf{v}_{\leq K}\cdot\mathbf{f}> = <\tfrac{\nu}{2}\int(\nabla\mathbf{v}_{\leq K})^2>, + <\Pi_K>, \tag{2.12}$$

which states that the mean injection rate of energy into the modes with $|\mathbf{k}|\leq K$ is equal to the mean rate of dissipation of energy $\epsilon_{\leq K} \equiv <\frac{\nu}{2}\int(\nabla\mathbf{v}_{\leq K})^2>$ in those modes plus the mean flux of energy out of modes with $|\mathbf{k}|\leq K$. According to the cascade picture of the energy transfer, the dissipation rate ϵ_K is essentially zero for $K \ll \eta^{-1}$. Since the injection takes place only around the integral scale, the change of the mean injection rate $\varphi_{\leq K} \equiv\, <\mathbf{v}_{\leq K}\cdot\mathbf{f}>$ should be negligeable for $K \gg L^{-1}$, so that $\varphi_{\leq K}$ should be equal to its $K=\infty$ value ϵ for such K. It follows from Eq. (2.12) that in the so called **inertial range** of scales $L^{-1} \ll K \ll \eta^{-1}$, the mean energy flux in the wavenumber space $\pi_K \equiv\, <\Pi_K>$ is constant. This is an important quantitative consequence of the cascade picture of developed turbulence, which, however, is not easily verifiable directly.

Kolmogorov went further, postulating that in the inertial range $\eta \ll x \ll L$, the velocity structure functions may be expressed solely as universal functions of the mean dissipation rate ϵ and the distance x, due to the locality and self-similarity of the cascade. The shape of the structure functions is then dictated by the dimensional considerations. One obtains

$$S_N(x) = C_N\, \epsilon^{N/3} x^{N/3} \tag{2.13}$$

with universal coefficients C_N (the right hand side is the only function of ϵ and x with dimension $(\frac{length}{time})^N$). Indeed, this behavior is consistent with the cascade picture. With the typical velocity of size l eddies $v_l \sim S_N(l)^{1/N} \sim \epsilon^{1/3} l^{1/3}$, their typical energy density $\frac{1}{2} v_l^2 \sim \epsilon^{2/3} l^{2/3}$ and their typical turnover time $t_l \sim l/v_l \sim \epsilon^{-1/3} l^{2/3}$, one may estimate the energy flux from scales longer than l to those shorter than l as $\sim v_l^2/t_l \sim \epsilon$ with no scale dependence.

The scaling (2.13) of the 2nd structure function translates under the Fourier transform to the relation

$$\begin{aligned} \tfrac{1}{4\pi k^2} E(k) \equiv < \mathbf{v}(\mathbf{k}) \cdot \mathbf{v}(-\mathbf{k}) > &= \textstyle\int \mathrm{e}^{-i\mathbf{k}\cdot\mathbf{x}}\, S_2(x)\, d^3\mathbf{x} \\ &\sim \epsilon^{2/3} k^{-11/3}, \end{aligned} \tag{2.14}$$

so that the Kolmogorov theory prediction for the energy spectrum $E(k)$ is

$$E(k) \sim \epsilon^{2/3}\, k^{-5/3}\,. \tag{2.15}$$

The theory allows also to locate the scale η at which the dissipative effects become important, called usually the Kolmogorov scale. Estimating the dissipation rate on scale l as $\sim \nu(\frac{v_l}{l})^2$, η should correspond to the scale at which $\nu(\frac{v_\eta}{\eta})^2 \cong \epsilon$. Using the relations

$$v_\eta \sim \epsilon^{1/3} \eta^{1/3}\,, \qquad v_L \sim \epsilon^{1/3} L^{1/3}\,, \tag{2.16}$$

we infer that

$$\eta \sim (\tfrac{\nu}{v_L L})^{3/4} L = R^{-3/4} L\,. \tag{2.17}$$

As should be expected, the Kolmogorov scale decreases with growing Reynolds numbers, i.e. with decreasing viscosity. In the limit $R \to \infty$, the inertial range should invade all short scales. It does not mean, however, that the behavior in that range may be described by the Euler equation: the presence of even very small viscosity is essential for the existence of the stationary state of the turbulence by providing a mechanism for removal from the system of the energy injected by external forces.

The verifications of the Kolmogorov theory become difficult by the fact, that both in experiments in the atmosphere, in wind tunnels, or in liquid flows, and in numerical simulations it is hard to obtain inertial ranges extending over many orders of magnitude necessary to extract the characteristic exponents of the structure functions. Nevertheless, it seems that the behavior of low structure functions well agrees with Kolmogorov's predictions, whereas for higher structure functions with $N \sim 10$ one observes values of exponents lower than predicted (by $\sim$ 10% or more). A multitude of more or less *ad hoc* mechanisms has been proposed to explain the possible departures from the Kolmogorov scaling for the higher structure functions. The underlying idea is that of intermittency limiting the turbulent activity to a subset of temporal and spatial modes. We shall not discuss the intermittent models, here referring the interested reader to [18].

3. Functional Methods in Turbulence

The Kolmogorov theory has a phenomenological character. It is strongly related to the cascade picture of turbulence, which was postulated essentially independently of the NS equations. It may be compared to the mean-field approach to critical phenomena, which leads to correct results only in limited situations, and which, in general, has to be replaced by a more sophisticated theory, taking into account the role of statistical fluctuations. It is then essential to try to build a theory of fully developed turbulence starting from the basic equations.

For any functional evolution equation of the type

$$\partial_t \Phi = -\mathcal{F}(\Phi) + F\,, \tag{3.1}$$

where F is a functional Gaussian process, a Martin-Siggia-Rose (MSR) formalism [19] permits to obtain formal functional integral expressions for the expectation values of functionals $\mathcal{A}(\Phi)$ in the stationary state. This is done as follows. We rewrite

$$\begin{aligned} < \mathcal{A}(\Phi) > \;&=\; \frac{1}{\mathcal{N}} \int \mathcal{A}(\Phi)\ \delta(\partial_t \Phi + \mathcal{F}(\Phi) - F)\ \det(\partial_t + \tfrac{\delta \mathcal{F}(\Phi)}{\delta \Phi}) \\ & \quad \mathrm{e}^{-\frac{1}{2}(F,\, C^{-1}F)}\ D\Phi\, DF, \end{aligned} \tag{3.2}$$

where C is the covariance operator of the Gaussian process F, and $\mathcal{N}$ is the normalizing factor given by a similar integral with $\mathcal{A} = 1$. The role of the Φ integral is to express Φ as a solution of the evolution equation (3.1) (with a fixed initial condition). If the functional $\mathcal{F}(\Phi)$ is local in time, then, formally,

$$\begin{aligned} \det(\partial_t + \tfrac{\delta \mathcal{F}(\Phi)}{\delta \Phi}) \;&=\; \det(\partial_t)\ \det(1 + \partial_t^{-1} \tfrac{\delta \mathcal{F}(\Phi)}{\delta \Phi}) \\ &=\; \det(\partial_t)\ \mathrm{e}^{\frac{1}{2}\,\mathrm{tr}\ \delta \mathcal{F}(\Phi)/\delta\Phi}\,, \end{aligned} \tag{3.3}$$

since $\partial_t^{-1}(t,t') = \theta(t-t')$ and $\delta\mathcal{F}(\Phi)/\delta\Phi$ is proportional to $\delta(t-t')$. Alternatively, one may use the (Berezin) functional integral over anticommuting fields to exponentiate the determinant. Rewriting also the delta function $\delta(\partial_t \Phi + \mathcal{F}(\Phi) - F)$ as an oscillatory integral, we obtain

$$\begin{aligned} < \mathcal{A}(\Phi) > = \frac{1}{\mathcal{N}} \int \;& \mathcal{A}(\Phi)\ \mathrm{e}^{\,i\,(A,\, \partial_t\Phi + \mathcal{F}(\Phi) - F) + \frac{1}{2}\mathrm{tr}\ \delta\mathcal{F}(\Phi)/\delta\Phi - \frac{1}{2}(F,\, C^{-1}F)} \\ & DA\ D\Phi\ DF\,. \end{aligned} \tag{3.4}$$

Finally, the Gaussian integral over F leads to the relation

$$< \mathcal{A}(\Phi) > = \frac{1}{\mathcal{N}} \int \mathcal{A}(\Phi)\, \mathrm{e}^{i\,(A\,,\,\partial_t\Phi+\mathcal{F}(\Phi)) + \frac{1}{2}\mathbf{tr}\,\delta\mathcal{F}(\Phi)/\delta\Phi - \frac{1}{2}(A, CA)} \; DA\, D\Phi\,. \tag{3.5}$$

The MSR formalism applies to the Langevin equation, describing the approach to equilibrium in statistical mechanics and Euclidean field theory. In this case, the functional $\mathcal{F}(\Phi)$ appearing in the evolution equation (3.1) is of the gradient type:

$$\mathcal{F}(\Phi) = \frac{\delta S(\Phi)}{\delta \Phi}, \tag{3.6}$$

and the Gaussian process F is the white noise, i.e. its covariance C is the identity operator. For example, for the scalar ϕ^4 theory,

$$S(\Phi) = \int \left(\tfrac{1}{2}\nabla\Phi)^2 + \tfrac{1}{2}m^2\Phi^2 + \lambda\Phi^4\right) dt\, d^d\mathbf{x}\,. \tag{3.7}$$

Perturbative treatment of the functional integral (3.5), based on expanding the exponential of the terms of order higher then 2 in Φ and A into a power series, gives rise to the perturbative expansion for the correlation functions. The latter reduces for the equal-time correlators to the perturbative expansion for the equilibrium expectations in the Gibbs measure $\frac{1}{\mathcal{N}}\,\mathrm{e}^{-S(\phi)}\,D\phi$, with ϕ as Φ, but without the time dependence.

One may rewrite the NS equations (2.1) and (2.2) as a single evolution equation

$$\partial_t \mathbf{v} = -P(\mathbf{v}\cdot\nabla)\mathbf{v} + \nu\,\Delta\mathbf{v} + P\mathbf{f} \tag{3.8}$$

in the space of divergence-free vector fields, where P stands for the orthogonal projection on such fields. Taking the external force $\mathbf{f}$ random Gaussian, we end up in the setup formally resembling the Langevin dynamics. The important differences between the latter and Eq. (3.8) are, however, that, first, the term $P\,(\mathbf{v}\cdot\nabla)\mathbf{v}$ is not of the gradient type, and, second, that we want to look at the noise $\mathbf{f}$ with the Fourier components concentrated in low wavenumbers, i.e. with the kernel of the covariance in the position space close to a constant, as opposed to the delta function kernel for the Langevin equation. As a result, the role of the noise in the NS equation is to inject at small wavenumbers the energy, which is then transferred to larger wavenumbers by an essentially deterministic process, leading to the non-zero energy flux in the stationary state, whereas in the Langevin dynamics the noise plays an essential role on all scales, resulting in the thermal state with no non-zero energy fluxes between scales.

It is sometimes pointed [20], that one encounters non-vanishing momentum-space fluxes also in the "equilibrium" field theory in the presence of anomalies. For the sake of illustration, let us consider the chiral anomaly in two space-time dimensions. The chiral charge density of the Dirac field is

$$J^{50}(t,x) = \bar{\psi}(t,x)\gamma^5\gamma^0\psi(t,x)\,. \tag{3.9}$$

The anomaly calculation shows, that in the external Abelian gauge field in two space-time dimensions

$$\frac{d}{dt} < \int J^{50}(t,x)\,dx >_{\leq K} = \frac{1}{2\pi} \int \epsilon^{\mu\nu} F_{\mu\nu}(t,x)\,dx\ , \tag{3.10}$$

and the right hand side is K independent. Here the subscript $\leq K$ denotes a gauge-invariant (e.g. Pauli-Villars) subtraction at momentum-scale K. In other words, the flux of the chiral charge out of the modes with momenta $\leq K$ is constant in K. It is not clear what lessons may be drawn from this analogy for understanding the inertial range of the fully developed turbulence.

Applying the MSR formalism to Eq. (3.8), we obtain for the velocity correlation functions the expression

$$< \prod_{n=1}^{N} v^{i_n}(t_n, \mathbf{x}_n) >$$
$$= \frac{1}{\mathcal{N}} \int \prod_{n=1}^{N} v^{i_n}(t_n, \mathbf{x}_n)\ e^{i(\mathbf{a},\,\partial_t \mathbf{v} + (\mathbf{v}\cdot\nabla)\mathbf{v} - \nu\Delta\mathbf{v}) - \frac{1}{2}(\mathbf{a},\,\mathcal{C}\mathbf{a})}\ D\mathbf{a}\ D\mathbf{v}\,, \tag{3.11}$$

where the $\mathbf{a}$- and $\mathbf{v}$-integrals are over divergence-free vector fields ($\mathrm{tr}\ \frac{\delta(\mathbf{v}\cdot\nabla)\mathbf{v}}{\delta\mathbf{v}}$ drops out, as it vanishes because of symmetry reasons). Separating the expression under the exponential into the quadratic part $i(\mathbf{a},\, \partial_t\mathbf{v} - \nu\Delta\mathbf{v}) - \frac{1}{2}(\mathbf{a},\, \mathcal{C}\mathbf{a})$ and the cubic one $i(\mathbf{a},\, (\mathbf{v}\cdot\nabla)\mathbf{v})$, and expanding the exponential of the latter into the power series, we may generate the perturbation expansion for the velocity correlators, just like for the correlation functions of the Langevin evolution. The terms of the expansion may be represented by Feynman diagrams, with the propagators read off the quadratic part under the exponential, and the vertices off the cubic part. A closer look into the perturbation expansion generated this way shows, that it is plagued by infrared and ultraviolet divergences (the latter in the limit of vanishing ν). Various resummation techniques, mostly inspired by approaches used in field theory, were tried with the aim to improve the convergence, without complete success. Let us mention here another approach, originally labeled quasi-Lagrangian, which seems more promising. This approach has been developed over years by the Novosibirsk and more recently the Weizmann Institute schools [21][22][23].

The idea consists of describing the turbulent flow in the frame moving with one of its points. Suppose that, for given velocity field $\mathbf{v}(t,\mathbf{x})$, $\mathbf{x}(t)$ is the solution of the ODE

$$\frac{d\mathbf{x}}{dt} = \mathbf{v}(t,\mathbf{x}) \tag{3.12}$$

with the initial condition $\mathbf{x}(t_0) = \mathbf{x}_0$. $\mathbf{x}(t)$ is the trajectory of the material point of the fluid located at $\mathbf{x}_0$ at time t_0. The velocity field, force, and pressure in the frame moving with the fluid point are

$$\mathbf{v}'(t,\mathbf{x}) = \mathbf{v}(t, \mathbf{x} + \mathbf{x}(t))\ , \tag{3.13}$$
$$\mathbf{f}'(t,\mathbf{x}) = \mathbf{f}(t, \mathbf{x} + \mathbf{x}(t))\ , \tag{3.14}$$
$$p'(t,\mathbf{x}) = p(t, \mathbf{x} + \mathbf{x}(t))\ , \tag{3.15}$$

and they satisfy the equations following from the NS equations for $\mathbf{v}$:

$$\partial_t \mathbf{v}' + ((\mathbf{v}' - \mathbf{v}_0') \cdot \nabla)\mathbf{v}' - \nu \Delta \mathbf{v}' = -\nabla p' + \mathbf{f}' , \tag{3.16}$$
$$\nabla \cdot \mathbf{v}' = 0 , \tag{3.17}$$

where $\mathbf{v}_0'(t) \equiv \mathbf{v}_0'(t,0)$. Eqs. (3.16) and (3.17) may be again written as a single evolution equation in the space of divergence-free vector fields:

$$\partial_t \mathbf{v}' = -P((\mathbf{v}' - \mathbf{v}_0') \cdot \nabla)\mathbf{v}' + \nu \Delta \mathbf{v}' + P\mathbf{f}', \tag{3.18}$$

to which the MSR formalism may be applied, if we assume that $\mathbf{f}'$ is a Gaussian random field. This is not the same as a similar assumption about $\mathbf{f}$, but as long as the Fourier components of the force are concentrated in small wavenumbers, the details of the force distribution should not matter in the inertial range. In such a situation, one may expect that the equal-time velocity correlators in the stationary state of the evolution equation (3.18) coincide in the inertial range with those corresponding to the equation (3.8). The perturbation expansion generated by the MSR technique for the $\mathbf{v}'$-correlators has the same propagators, but modified vertices which break the translation invariance and make this way the quasi-Lagrangian expansion more difficult to analyze. Nevertheless it seems, that the quasi-Lagrangian approach leads to drastic improvement in convergence of the Schwinger-Dyson resummation of the perturbative expansion [21][23], and it is possible that it provides an important clue to the understanding of the statistical properties of the turbulent flow.

4. Passive Scalar

4.1 Definition of the Model

In view of the difficulties encountered by the theory of the Navier-Stokes turbulence, it may be useful to search for simpler systems, exhibiting some of the aspects of full-fledged turbulence, but easier to control. Such a system seems to be provided by a model of passive advection in a random velocity field $\mathbf{v}(t,\mathbf{x})$ of a scalar quantity T, whose density $T(t,\mathbf{x})$ satisfies the equation

$$\partial_t T + (\mathbf{v} \cdot \nabla)T - \nu \Delta T = f , \tag{4.1}$$

where now ν denotes the molecular diffusivity of the scalar T, and $f(t,\mathbf{x})$ describes the external sources. In the ideal situation, we should consider $\mathbf{v}$ as the velocity of the NS turbulent flow, but we shall, instead, assume that $\mathbf{v}(t,\mathbf{x})$ is a centered Gaussian field with the covariance

$$< v^i(t,\mathbf{x})\, v^j(t',\mathbf{x}') > = \delta(t-t')\, D^{ij}(\mathbf{x}-\mathbf{x}') , \tag{4.2}$$

i. e. white noise in time. The spatial part of the covariance will be taken, for concreteness, as

$$D^{ij}(\mathbf{x}) = D_0 \int \mathrm{e}^{i\mathbf{k}\cdot\mathbf{x}}\, (\mathbf{k}^2 + m^2)^{-(3+\kappa)/2}\, (\delta^{ij} - k^i k^j/\mathbf{k}^2)\, \frac{d^3\mathbf{k}}{(2\pi)^3} , \tag{4.3}$$

where the transverse projector in the Fourier space ensures the incompressibility of $\mathbf{v}$. Small m^2 should be viewed as an infrared cutoff making the integral convergent for $0 < \kappa < 2$. Note that the dimension of D_0 is $\frac{length^{2-\kappa}}{time}$. For the 2nd velocity structure function we obtain

$$< (\mathbf{v}(t,\mathbf{x}) - \mathbf{v}(t,0))^2 > \sim D^{ii}(0) - D^{ii}(\mathbf{x}) \equiv \widetilde{D}^{ii}(\mathbf{x}), \tag{4.4}$$

which, for $0 < \kappa < 2$, has the $m \to 0$ limit proportional to $|\mathbf{x}|^\kappa$, i.e. growing with the distance. More exactly,

$$D^{ij}(0) = \frac{16\,\Gamma(\kappa/2)}{3\pi^{3/2}\,\Gamma(\kappa/2)}\,\delta^{ij}\,D_0\,m^{-\kappa}, \tag{4.5}$$

i.e. it diverges with $m \to 0$, but

$$\lim_{m\to 0}\,\widetilde{D}^{ij}(\mathbf{x}) = D_1\left((2+\kappa)\,\delta^{ij}\,|\mathbf{x}|^\kappa \;-\; \kappa\,x^i x^j\,|\mathbf{x}|^{\kappa-2}\right), \tag{4.6}$$

and it is a homogeneous function of $\mathbf{x}$. $D_1 \equiv \frac{\Gamma((2-\kappa)/2)}{2^{2+\kappa}\pi^{3/2}\kappa(3+\kappa)}\,D_0$. Note the singularities at 0 and 2 in the κ-dependence. The Kolmogorov scaling (2.13) corresponds to $\kappa = \frac{2}{3}$, which lies in the interval $0 < \kappa < 2$ that we shall consider. It should be mentioned, however, that the time-decorrelation, evenness and Gaussian character of the velocity distribution is a rather drastic departure from the statistical properties of the turbulent NS velocities. Nevertheless, the model, dating back to 1949 [10], possesses an interest of its own and has been used to describe a variety of phenomena like poluant or tracer transport, and is closely related to models describing forced diffusion through porous media [24].

We shall take the field $f(t,\mathbf{x})$ describing the external sources also random Gaussian, independent of $\mathbf{v}$, with mean zero and covariance

$$< f(t,\mathbf{x})\,f(t',\mathbf{x}') >= \delta(t-t')\,\mathcal{C}(\tfrac{\mathbf{x}-\mathbf{x}'}{L}) \equiv \delta(t-t')\,\mathcal{C}_L(\mathbf{x}-\mathbf{x}'), \tag{4.7}$$

where $\mathcal{C}$ is a real positive-definite function from the Schwartz space $\mathcal{S}(\mathbf{R}^3)$, and L is the integral scale. Sometimes it will be useful to consider also the advection of a complex scalar, in which case $\mathcal{C}$ may be a complex positive-definite function. We would like to study the statistical properties of the solutions of Eq. (4.1) in the regime of small ν, small m (which may viewed as the inverse of another integral scale), and large L (in the limit $L \to \infty$, the field f has only the zero wavenumber component). In particular, the universality question for the PS may be formulated as inquiring about the existence of the limit of the correlation functions

$$< \prod_{n=1}^{N} T(t_n, \mathbf{x}_n) > \tag{4.8}$$

in a stationary state of the system, when $\nu, m, L^{-1} \to 0$, and about the independence of such a limit of the shape of the source covariance $\mathcal{C}$.

4.2 Exact Solution

The big advantage of the PS model over the NS case is, that one may obtain closed expressions for the correlation functions of the scalar (the assumed time-decorrelation of the velocity field plays here the crucial role). For the 2-point function, this was first done in [25][1]. In this sense the model is exactly solvable. In order to work out its exact solution, let us rewrite the basic evolution equation (4.1) of the scalar in a general form

$$\partial_t T + \beta T + \nu\alpha T = f \,, \tag{4.9}$$

where β is a skew-symmetric random operator ($\beta = \mathbf{v}\cdot\nabla$ in our case), and α is a positive operator ($\alpha = -\Delta$). The β term in the equation is conservative, whereas the α-one is dissipative (for the energy $\frac{1}{2}\|T\|^2$). For the sake of simplicity, we shall argue in the finite-dimensional case, i. e. when $T(t)$ and $f(t)$ take values in $\mathbf{R}^d$, and $\beta(t)$ and α are, respectively, skew-symmetric and positive symmetric $d\times d$ matrices. For the covariances of independent centered Gaussian processes $\beta(t)$ and $f(t)$, we shall write

$$\begin{aligned} <\beta_{ij}(t)\,\beta_{kl}(s)> &= \delta(t-s)\,\mathcal{B}_{ij,kl}\,, \\ <f_i(t)\,f_j(s)> &= \delta(t-s)\,\mathcal{C}_{ij}\,. \end{aligned} \tag{4.10}$$

In finite dimensions, the original stochastic PDE, becomes a stochastic ODE, which is easier to handle. Let us simplify the matter further by assuming, for a moment, that $\beta(t)$ and $f(t)$ are continuous functions. Then the solution of Eq. (4.9) with the initial condition T_0 at $t = t_0$ takes the well known form

$$T(t) = R(t,t_0)\,T_0 + \int_{t_0}^{t} R(t,s)\,f(s)\,ds\,, \tag{4.11}$$

where $R(t,t_0)$ is given by the ordered exponential ($t \geq t_0$)

$$R(t,t_0) = \mathcal{T}\,\mathrm{e}^{-\int_{t_0}^{t}(\nu\alpha+\beta(\tau))\,d\tau} \tag{4.12}$$

($\mathcal{T}$ stands for the time ordering). $R(t,t_0)$ may be also expressed by the following limiting formula:

$$\begin{aligned} R(t,t_0) &= \lim_{Q\to\infty} \exp[-\int_{I_Q}(\nu\alpha+\beta(\tau))\,d\tau]\,\exp[-\int_{I_{Q-1}}(\nu\alpha+\beta(\tau))\,d\tau] \\ &\quad\cdots\exp[-\int_{I_1}(\nu\alpha+\beta(\tau))\,d\tau] \equiv \lim_{Q\to\infty} R_Q(t,t_0)\,, \end{aligned} \tag{4.13}$$

where (I_q) is a time ordered partition of the interval $[t_0,t]$ into subintervals of length $(t-t_0)/Q$.

In the random case, the white noise stochastic processes $\beta(t)$ and $f(t)$ do not have continuous realizations. They need smearing to give genuine random variables. It is easy to see that $\int_I \beta(\tau)\,d\tau \equiv \beta^I$, where I are intervals of time, are already genuine Gaussian random variables, which are mutually independent for non-intersecting or adjacent intervals:

[1] We thank U. Frisch for explaining to us the 4-point function case.

$$< \beta^I_{ij}\, \beta^J_{kl} >= |I \cap J|\; \mathcal{B}_{ij,kl}\ . \tag{4.14}$$

Let us denote by $\mathcal{B}$ the covariance operator, acting on skew-symmetric matrices by

$$(\mathcal{B}\,\beta)_{ij} = \sum_{ijkl} \mathcal{B}_{ij,kl}\, \beta_{kl}\ . \tag{4.15}$$

Necessarily, $\mathcal{B} \geq 0$. We shall also need below a contracted version B of $\mathcal{B}$ with the matrix elements

$$B_{ij} = -\tfrac{1}{2} \sum_k \mathcal{B}_{ik,kj}\ . \tag{4.16}$$

B is a symmetric operator acting in $\mathbf{R}^d$. Note that

$$< ((\beta^I)^t\, \beta^I)_{ij} > = - < (\beta^I\, \beta^I)_{ij} > = 2|I|\, B_{ij}, \tag{4.17}$$

from which it follows that $B \geq 0$. In the random case with white noise $\beta(t)$, the approximate evolution operators $R_Q(t,t_0)$ under the limit on the right hand side of Eq. (4.13) still make sense for each Q as random (matrix valued) variables, and one can show that

Proposition 1.

$$\lim_{Q\to\infty} < R_Q(t,t_0) > = \mathrm{e}^{-(t-t_0)(\nu\alpha+B)} \;\equiv\; \mathrm{e}^{-(t-t_0)\mathcal{M}_1}\ . \tag{4.18}$$

In order to indicate how the above result arises, consider the perturbation expansion for $R(t,t_0)$ in powers of β,

$$\begin{aligned} R(t,t_0) = \sum_{m=0}^{\infty} (-1)^m \int_{t_0}^{t} d\tau_m \int_{t_0}^{\tau_m} d\tau_{m-1} \cdots \int_{t_0}^{\tau_2} d\tau_1\, \mathrm{e}^{-(t-\tau_m)\,\nu\alpha} \\ \beta(\tau_m)\, \mathrm{e}^{-(\tau_m-\tau_{m-1})\,\nu\alpha}\, \beta(\tau_{m-1}) \cdots\, \beta(\tau_1)\, \mathrm{e}^{-(\tau_1-t_0)\,\nu\alpha}\ . \end{aligned} \tag{4.19}$$

In the white noise expectation value of the right hand side computed with the use of the Wick Theorem, only neighboring $\beta(\tau_m)$ may be paired, forcing the respective times τ_m to be equal. One obtains this way the perturbation expansion of $\mathrm{e}^{-(t-t_0)(\nu\alpha+B)}$ in powers of B. Proposition 1 is a rigorous version of this result, not very difficult to prove [26].

Similarly, one may calculate expectations of products of matrix elements of $R_Q(t,t_0)$. We shall use the tensor product notation $R_Q(t,t_0)^{\otimes N}$ as a bookkeeping device for all such products. Let us denote by $\mathcal{K}$ the symmetric operator on $\mathbf{R}^d \otimes \mathbf{R}^d \cong \mathrm{Mat}_{d\times d}$, built of the coefficients $\mathcal{B}_{ij,kl}$ so that

$$(\mathcal{K}\,g)_{ik} = \sum_{jl} \mathcal{B}_{ij,kl}\, g_{jl}\ . \tag{4.20}$$

Note the difference between $\mathcal{K}$ and the covariance operator $\mathcal{B}$ of the process $\beta(t)$ defined in Eq. (4.15). In what follows, an important role will be played by the N-body symmetric operators $\mathcal{M}_N$ acting in $(\mathbf{R}^d)^{\otimes N} \cong \mathbf{R}^{Nd}$, defined by

$$\mathcal{M}_N \equiv \sum_{n=1}^{N} (\mathcal{M}_1)_n - \sum_{1\leq n<n'\leq N} (\mathcal{K})_{n,n'} , \tag{4.21}$$

where $\mathcal{M}_1 = \nu\alpha + B$, $(\mathcal{M}_1)_n$ denotes $\mathcal{M}_1$ acting in the nth factor of the tensor product $(\mathbf{R}^d)^{\otimes N}$, and, similarly, $(\mathcal{K})_{n,n'}$ stands for $\mathcal{K}$ acting in the nth and n'^{th} factors of $(\mathbf{R}^d)^{\otimes N}$.
Proposition 2. For $N = 2, 3, \ldots$,

$$\lim_{Q\to\infty} < R_Q(t,t_0)^{\otimes N} > = \mathrm{e}^{-(t-t_0)\mathcal{M}_N} . \tag{4.22}$$

Again, this result is not surprising, if we use the perturbation expansion (4.19) for each $R(t,t_0)$ and note, that the white noise expectation may pair now, besides the nearest neighbor $\beta(\tau_m)$ in the perturbative terms of the same $R(t,t_0)$, also $\beta(\tau_m)$ corresponding to different $R(t,t_0)$, resulting in the perturbative expansion of $\mathrm{e}^{-(t-t_0)\mathcal{M}_N}$ in powers of operators B and $\mathcal{K}$. In fact, one may show [26] that $R_Q(t,t_0)$ converge to a limit in all L^p spaces, $p < \infty$, of the white noise β. Naturally, we shall denote the limit by $R(t,t_0)$. $R(t,t_0)$ are contracting operators, they are continuous in t, t_0 in the L^p-norms and satisfy the composition law. We shall take then $T(t)$ given by Eq. (4.11) with the limiting $R(t,t_0)$ as the solution of the stochastic equation (4.9) with initial value T_0.

In the absence of the external sources the time evolution of the equal-time correlators of the scalar T simply reduces to the relations

$$< T(t)^{\otimes N} > = \mathrm{e}^{-(t-t_0)\mathcal{M}_N} T(t_0)^{\otimes N} , \tag{4.23}$$

hence it is given by the semigroups $(\mathrm{e}^{-t\mathcal{M}_N})_{t\geq 0}$, which may be studied by looking at the Fourier transforms of the semigroups, i.e. at the resolvents $(i\omega + \mathcal{M}_N)^{-1}$. In any case, the properties of the time evolution of the statistics of initial data are determined by the spectral properties of the symmetric operators $\mathcal{M}_N$. It is easy to see that the operators $\mathcal{M}_N$ satisfy the inequality

$$\mathcal{M}_N \geq \nu \sum_{n=1}^{N} (\alpha)_n . \tag{4.24}$$

An important consequence of this inequality is the exponential time decay of the correlation functions $< T(t)^{\otimes N} >$, as long as the dissipation is present in the equation (4.9) (recall that we are looking at the finite-dimensional case). The decay rate is given by the lowest eigenvalues $\lambda_{N,0}$ of $\mathcal{M}_N$ and is bounded below by $N\nu\mu_0$, where $\mu_0 > 0$ is the lowest eigenvalue of α. If we remove the dissipative term from the equation by setting $\nu = 0$, then the long time behavior of $< T(t)^{\otimes N} >$ will have a persistent tail, since the operators $\mathcal{M}_N$ develop zero modes, when ν is taken to zero. Suppose now that the initial data $T(t_0)$ are also random and independent of $\beta(t)$ for $t \geq t_0$. Averaging Eq. (4.23) over T_0, we obtain

$$< T(t)^{\otimes N} > \; = \; \mathrm{e}^{-(t-t_0)\mathcal{M}_N} < T(t_0)^{\otimes N} > \; . \tag{4.25}$$

A natural problem is the existence of an invariant measure on the space of initial data. It follows from relations (4.24) and (4.25), that there are no invariant measures with finite moments on the space of initial data as long as $\nu > 0$. On the other hand, when $\nu = 0$, i. e. in the absence of dissipation, one may show that any measure preserved by a. a. rotations e^{β^I} is invariant, and that this property characterizes the invariant measures with finite characteristic functions [26]. In particular, the Gibbs measures $\sim \mathrm{e}^{-\mathrm{const.}\,\|T\|^2}\, d^dT$ are invariant under the flow (4.9), when $\nu = 0$. As we shall see, the situation is quite different for infinite number of degrees of freedom.

In the presence of white-noise sources and of dissipation, the 2-point equal time correlation function evolves according to the equation

$$\begin{aligned} < T(t)^{\otimes 2} >&= \mathrm{e}^{-(t-t_0)\,\mathcal{M}_2}\, T(t_0)^{\otimes 2} + \int_{t_0}^{t} ds\, \mathrm{e}^{-(t-s)\,\mathcal{M}_2}\, \mathcal{C} \\ &= \mathrm{e}^{-(t-t_0)\,\mathcal{M}_2}\, T(t_0)^{\otimes 2} + (1-\mathrm{e}^{-(t-t_0)\mathcal{M}_2})\, \mathcal{M}_2^{-1}\, \mathcal{C}, \end{aligned} \tag{4.26}$$

which is a solution of the differential equation

$$\partial_t < T(t)^{\otimes 2} >= -\mathcal{M}_2 < T(t)^{\otimes 2} > + \mathcal{C} \; . \tag{4.27}$$

When $t_0 \to -\infty$, the term with $T(t_0)$ disappears due to the positivity of $\mathcal{M}_2$ (for $\nu > 0$), and we obtain

$$< T^{\otimes 2} > \; = \; \mathcal{M}_2^{-1}\mathcal{C} \; . \tag{4.28}$$

This is the time independent 2-point function in a steady state, in which the quantities of the energy injected by the force and dissipated balance each other. Similarly, for the general equal time correlators in the presence of the external sources and of dissipation, the dependence on the initial conditions is wiped out when $t_0 \to -\infty$ due to the positivity of all $\mathcal{M}_N$. The higher stationary state equal-time correlation functions take the form:

$$\begin{aligned} < T^{\otimes 2M+1} > \; &= \; 0 \, , \\ < T^{\otimes 2M} > \; &= \sum_{\sigma \in \mathcal{S}_{2M} \,/\, \mathcal{S}_M \times \mathcal{S}_2^M} \sigma\, F_{2M} \, , \end{aligned} \tag{4.29}$$

where $F_{2M} \in (\mathbf{R}^d)^{\otimes 2M}$ is a single 2-particle channel contribution. Permutations σ interchange the factors in $(\mathbf{R}^d)^{\otimes 2M}$, $\mathcal{S}_M$ is viewed as a set of permutations of the pairs $(2m-1, 2m)$, $m = 1, \ldots, M$, and $\mathcal{S}_2^M$ as that of permutations acting within the pairs, so that the sum is over the pairings of $\{1, \ldots, 2M\}$ or 2-particle channels.

$$F_{2M} \; = \; \mathcal{M}_{2M}^{-1} \sum_{\sigma \in \mathcal{S}_M} \sigma \left((\mathcal{M}_{2M-2}^{-1} \otimes 1_2) \, \cdots \, (\mathcal{M}_2^{-1} \otimes 1_{2M-2}) \right) \mathcal{C}^{\otimes M} \; . \tag{4.30}$$

Exact expressions for non-equal times N-point correlators are also easy to obtain. They involve also the heat operators $\mathrm{e}^{-t\mathcal{M}_{N'}}$ with $N' < N$.

Let us discuss briefly some multibody combinatorics behind the solution (4.30). Let us denote by $\mathcal{P}_{2M}$ the set of pairs $(2m-1, 2m)$, $m = 1, \ldots, M$. We shall define the connected parts F_{2M}^c of the single channel correlator by the inductive formula

$$F_{2M} = \sum_{\substack{\text{partitions} \\ \Pi \text{ of } \mathcal{P}_{2M}}} \bigotimes_{\pi\in\Pi} F_\pi^c, \tag{4.31}$$

in the, hopefully, self-explanatory notation. The connected $2M$-correlator $< T^{\otimes 2M} >^c$ is given by Eq. (4.29), with F_{2M}^c replacing F_{2M} on the right hand side. Note that, by virtue of the definition (4.21) of $\mathcal{M}_N$, for each partition Π of $\mathcal{P}_{2M}$ we may write

$$\mathcal{M}_{2M} = \sum_{\pi\in\Pi} \mathcal{M}_\pi - \sum_{\{\pi,\pi'\}\subset\Pi} \mathcal{L}_{\pi,\pi'}, \tag{4.32}$$

where $\mathcal{M}_\pi$ is $\mathcal{M}_{2|\pi|}$ acting on the tensor factors of $(\mathbf{R}^d)^{\otimes 2M}$ labeled by the pairs of π, and

$$\mathcal{L}_{\pi,\pi'} = \sum_{\substack{n\in\pi \\ n'\in\pi'}} (\mathcal{K})_{n,n'} \,. \tag{4.33}$$

The basic equation for the connected single 2-particle channel correlation functions F_{2M}^c is given by
Proposition 3. For $M > 1$,

$$\mathcal{M}_{2M}\, F_{2M}^c = \sum_{\Pi=\{\pi,\pi'\}} \mathcal{L}_{\pi,\pi'}\, (F_\pi^c \otimes F_{\pi'}^c)\,. \tag{4.34}$$

Note the connected character of the right hand side. Let us prove this relation. From the definition (4.30) of the single channel function,

$$\mathcal{M}_{2M}\, F_{2M} = \sum_{\substack{\Pi=\{\pi,\pi'\} \\ |\pi|=1}} \mathcal{C}_\pi \otimes F_{\pi'}\,. \tag{4.35}$$

On the other hand, using Eqs. (4.31) and (4.32), we obtain

$$\begin{aligned}
\mathcal{M}_{2M}\, F_{2M}^c &= \mathcal{M}_{2M}\, F_{2M} - \sum_{\substack{\Pi \\ |\Pi|>1}} \mathcal{M}_{2M} \bigotimes_{\pi\in\Pi} F_\pi^c \\
&= \mathcal{M}_{2M}\, F_{2M} - \sum_{\substack{\Pi \\ |\Pi|>1}} \sum_{\pi\in\Pi} (\mathcal{M}_\pi F_\pi^c) \bigotimes \Big(\bigotimes_{\substack{\pi'\in\Pi \\ \pi'\neq\pi}} F_{\pi'}^c \Big) \\
&+ \sum_{\substack{\Pi \\ |\Pi|>1}} \sum_{\{\pi,\pi'\}\subset\Pi} (\mathcal{L}_{\pi,\pi'}(F_\pi^c \otimes F_{\pi'}^c)) \bigotimes \Big(\bigotimes_{\pi''\neq\pi,\pi'} F_{\pi''}^c \Big)\,.
\end{aligned} \tag{4.36}$$

Notice that on the right hand side, in virtue of Eqs. (4.35) and (4.28), the 1st term cancels with the contribution to the 2nd one from π consisting of a single pair. Rewriting $\mathcal{M}_\pi F_\pi^c$ for the $|\pi| > 1$ contributions to the 2nd term, using Eq. (4.34) as the inductive hypothesis, we observe that all these contribution are cancelled by the 3rd term. What is left from the 3rd term after the cancellations gives exactly the right hand side of Eq. (4.34).

Although we worked directly with the evolution equation (4.9) in order to obtain expressions (4.28) and (4.30), one may obtain them as well using the MSR formalism, which gives

$$< T^{\otimes 2M} >= \tfrac{1}{N} \int \; T^{\otimes 2M} \, e^{i(A,\,(\partial_t+\nu\alpha+\beta)T)-\frac{1}{2}(A,CA)-\frac{1}{2}(\beta,\mathcal{B}^{-1}\beta)} \, DA \; DT \; D\beta \; . \tag{4.37}$$

Performing the Gaussian A- and T-integrals, we obtain

$$< T^{\otimes 2M} >= \tfrac{1}{N} \sum_{\substack{\text{parings} \\ \text{of } \{1,\dots,2M\}}} \int \prod_{\text{pairs}} \; (\partial_t+\nu\alpha+\beta)^{-1} C (-\partial_t+\nu\alpha-\beta)^{-1} \, e^{-\frac{1}{2}(\beta,\mathcal{B}^{-1}\beta)} \, D\beta \, . \tag{4.38}$$

If we expand

$$(\partial_t+\nu\alpha+\beta)^{-1} \; = \; \sum_{m=0}^{\infty} (-1)^m \, (\partial_t+\nu\alpha)^{-1} \left(\beta\,(\partial_t+\nu\alpha)^{-1}\right)^m , \tag{4.39}$$

and use the fact that the kernel of $(\partial_t+\nu\alpha)^{-1}$ is equal to $\theta(t-t')\, e^{-(t-t')\nu\alpha}$, the β-integral in (4.38) reduces to the previous calculation, and it reproduces the expressions (4.28) for the 2-point function, or (4.29) and (4.30) for the $2M$-correlator.

4.3 Eddy Diffusion Versus Super-Diffusion

The exact solution for the PS correlation functions obtained above in the case of finitely many degrees of freedom still makes sense for the infinite-dimensional case of the evolution equation (4.1). Instead of starting with this equation, and extending the previous analysis from the stochastic ODE case to the stochastic PDE one, which would require additional work, we shall directly study the infinite-dimensional version of the exact solution (4.29), (4.30) involving operators rather (generally unbounded) than finite matrices. The analysis of the expressions for the PS correlation functions obtained this way is more complicated than in finite dimensions, but also more interesting.

The covariance $\mathcal{B}$ of the Gaussian process $\beta = \mathbf{v}\cdot\nabla$, and the operator B related to the expectation of β^2, see Eqs. (4.16) and (4.17), are given by the kernels

$$\begin{aligned} \mathcal{B}(\mathbf{x}_1,\mathbf{y}_1;\mathbf{x}_2,\mathbf{y}_2) &= \mathcal{D}^{ij}(\mathbf{x}_1-\mathbf{x}_2)\,\partial_i\delta(\mathbf{x}_1-\mathbf{y}_1)\,\partial_j\delta(\mathbf{x}_2-\mathbf{y}_2)\,, \\ B(\mathbf{x},\mathbf{y}) &= -\tfrac{1}{2}\int \mathcal{B}(\mathbf{x},\mathbf{z};\mathbf{z},\mathbf{y})\,d^3\mathbf{z} \;=\; -\tfrac{1}{2}\mathcal{D}^{ij}(0)\,\partial_i\partial_j\,\delta(\mathbf{x}-\mathbf{y})\,. \end{aligned} \tag{4.40}$$

Since $\mathcal{D}^{ij}(0) = \frac{1}{3}\,\delta^{ij}\,\mathcal{D}^{ll}(0)$, $B = -\frac{1}{6}\,\mathcal{D}^{ll}(0)\,\Delta$, and, consequently,

$$\mathcal{M}_1 \equiv \nu\alpha + B = -(\nu + \tfrac{1}{6}\mathcal{D}^{ll}(0))\,\Delta \equiv -\nu_{\text{eff}}\,\Delta\,. \tag{4.41}$$

It follows from Eq. (4.23) with $N = 1$, that, in the absence of the sources, the expectation value of the scalar diffuses,

$$\begin{aligned} < T(t,\mathbf{x}) > &= \int \mathrm{e}^{-(t-t_0)\mathcal{M}_1}(\mathbf{x},\mathbf{y})\,T(t_0,\mathbf{y})\,d^3\mathbf{y} \\ &= \int \mathrm{e}^{(t-t_0)\nu_{\text{eff}}\,\Delta}(\mathbf{x},\mathbf{y})\,T(t_0,\mathbf{y})\,d^3\mathbf{y}, \end{aligned} \tag{4.42}$$

with the effective diffusion constant ν_{eff} composed of the molecular diffusivity ν and the **eddy diffusivity** $\frac{1}{6}\mathcal{D}^{ll}(0)$. For small m, the eddy diffusivity dominates, and the diffusion is driven by the large distance scales (recall from Eq. (4.5) that $\mathcal{D}^{ll}(0) = \mathcal{O}(D_0\,m^{-\kappa})$).

Operator $\mathcal{K}$ with kernel $\mathcal{K}(\mathbf{x}_1,\mathbf{x}_2;\mathbf{y}_1,\mathbf{y}_2) = \mathcal{B}(\mathbf{x}_1,\mathbf{y}_1;\mathbf{x}_2,\mathbf{y}_2)$ acts on functions g of six variables by

$$(\mathcal{K}g)(\mathbf{x}_1,\mathbf{x}_2) = \mathcal{D}^{ij}(\mathbf{x}_1-\mathbf{x}_2)\,\partial_{x_1^i}\partial_{x_2^j}\,g(\mathbf{x}_1,\mathbf{x}_2)\,, \tag{4.43}$$

so that for the N-body operators $\mathcal{M}_N$ given by formula (4.21) we obtain

$$\begin{aligned} \mathcal{M}_N &= -\sum_{n=1}^{N}(\nu\,\Delta_{\mathbf{x}_n} + \tfrac{1}{2}\mathcal{D}^{ij}(0)\,\partial_{x_n^i}\partial_{x_n^j}) - \sum_{n<n'}\mathcal{D}^{ij}(\mathbf{x}_n-\mathbf{x}_{n'})\,\partial_{x_n^i}\partial_{x_{n'}^j} \\ &= -\sum_{n=1}^{N}\nu_{\text{eff}}\,\Delta_{\mathbf{x}_n} - \sum_{n<n'}\mathcal{D}^{ij}(\mathbf{x}_n-\mathbf{x}_{n'})\,\partial_{x_n^i}\partial_{x_{n'}^j}\,. \end{aligned} \tag{4.44}$$

In particular,

$$\begin{aligned} \mathcal{M}_2 = -\nu(\Delta_1+\Delta_2) \quad &-\tfrac{1}{2}\mathcal{D}^{ij}(0)\,\partial_{x_1^i}\partial_{x_1^j} - \tfrac{1}{2}\mathcal{D}^{ij}(0)\,\partial_{x_2^i}\partial_{x_2^j} \\ &-\mathcal{D}^{ij}(\mathbf{x}_1-\mathbf{x}_2)\,\partial_{x_1^i}\partial_{x_2^j}\,. \end{aligned} \tag{4.45}$$

Note that $\mathcal{M}_2$ commutes with (three-dimensional) translations and in the action on translation-invariant functions of $\mathbf{x}_1 - \mathbf{x}_2 \equiv \mathbf{x}$ reduces to

$$\mathcal{M}_2 = -2\nu\,\Delta - \widetilde{\mathcal{D}}^{ij}(\mathbf{x})\,\partial_i\partial_j\,. \tag{4.46}$$

Since $\widetilde{\mathcal{D}}^{ij}(\mathbf{x}) \equiv \mathcal{D}^{ij}(0) - \mathcal{D}^{ij}(\mathbf{x})$ has an $m \to 0$ limit, so does the operator $\mathcal{M}_2$ in the action on translation-invariant functions, and when $\nu \to 0$, it becomes a singular elliptic operator

$$\mathcal{M}_2^{\text{sc}} = -D_1\left((2+\kappa)\,\delta^{ij}\,|x|^\kappa - \kappa\,x^i x^j\,|x|^{\kappa-2}\right)\partial_i\partial_j\,, \tag{4.47}$$

where $D_1 \equiv \frac{\Gamma((2-\kappa)/2)}{2^{2+\kappa}\pi^{3/2}\kappa(3+\kappa)}\,D_0$, see Eq. (4.6). The dimension of D_1, as that of D_0, is equal to $\frac{length^{2-\kappa}}{time}$.

Similarly for $N > 2$, in the action on the translationally invariant sector,

$$\mathcal{M}_N = -\nu\sum_{n=1}^{N}\Delta_{\mathbf{x}_n} + \sum_{n<n'}\widetilde{\mathcal{D}}^{ij}(\mathbf{x}_n - \mathbf{x}_{n'})\,\partial_{x^i_n}\partial_{x^j_{n'}}\,, \tag{4.48}$$

and has an $m \to 0$ and $\nu \to 0$ limit, which is a singular elliptic operator $\mathcal{M}_N^{\rm sc}$. This has important consequences. Although the time evolution given by Eq. (4.23) of fixed initial conditions is dominated for small m by the eddy diffusivity diverging as $m \to 0$, the time evolution governed by Eq. (4.25) for random initial data with a translationally invariant distribution behaves well in the limit when $m \to 0$ and $\nu \to 0$. In particular, it follows from the scaling properties of operators $\mathcal{M}_N^{\rm sc}$, that, if the translationally invariant initial moments $<\prod_n T(t_0,\mathbf{x}_n)>$ are fast decaying in the difference variables (weakly correlated initial data), then, for $m=0$, the rescaled correlations

$$\lambda^{\frac{3N}{2(2-\kappa)}} < \prod_{n=1}^{N} T(\lambda t,\, \lambda^{\frac{1}{2-\kappa}}\mathbf{x}_n) > \tag{4.49}$$

have a non-trivial limit when $\lambda \to \infty$. In other words, the equal-time correlators decay **super-diffusively** in the absence of sources, like $t^{-\frac{3N}{2(2-\kappa)}}$, i.e. faster than diffusively (like $t^{-\frac{3N}{4}}$). Similar behavior in a simplified model of passive advection was discussed in [13].

Let us look closer at the operator $\mathcal{M}_2$ in the translationally invariant sector. Since the limit $m \to 0$ presents no difficulty in that case, we may directly work in the $m=0$ case, i.e. with operator $\mathcal{M}_2 = -2\nu\Delta + \mathcal{M}_2^{\rm sc}$, see Eq. (4.47). In the radial variables

$$\begin{aligned}\mathcal{M}_2 &= -\tfrac{2}{r^2}\,\partial_r\,(\nu r^2 + D_1 r^{2+\kappa})\,\partial_r \\ &\quad + \tfrac{l(l+1)}{r^2}\,(2\nu + D_1(2+\kappa)\,r^{\kappa-2})\,,\end{aligned} \tag{4.50}$$

where l is the angular momentum ($\mathcal{M}_2$ commutes with rotations). As for the resolvents $(i\omega+\mathcal{M}_2)^{-1}$, they may be explicitly computed in the translationally and rotationally invariant ($l=0$) sector when $\nu \to 0$. In particular, one obtains for the kernel

$$(i\omega + \mathcal{M}_2^{\rm sc})^{-1}(\mathbf{x},0) = \omega^{\frac{1+\kappa}{2-\kappa}}\, f(\omega^{\frac{1}{2}}|\mathbf{x}|^{\frac{2-\kappa}{2}}) \tag{4.51}$$

with an exponentially decaying function f, $f(y) \sim y^{-\frac{2(1+\kappa)}{2-\kappa}}$ for small y. f may be expressed in terms of the Hankel functions. This result has an important implication. Recall that in finite dimensions the invariant measures for the stochastic evolution of the initial data were characterized by the symmetry generated by rotations e^{β^I}. Suppose that, as for a finite number of degrees of freedom, the evolution (4.25) preserves in the limit $\nu \to 0$ the white noise Gibbs measure $\sim\ e^{-{\rm const.}\,\|T\|^2}\,DT$, with the 2-point function

$$< T(\mathbf{x})\,T(\mathbf{y}) > \sim \delta(\mathbf{x}-\mathbf{y})\,. \tag{4.52}$$

This would imply that

$$\int g(\mathbf{x})\,(i\omega+\mathcal{M}_2^{sc})^{-1}(\mathbf{x},0)\,d^3\mathbf{x} \;=\; \frac{1}{i\omega}\,g(0) \tag{4.53}$$

for any test function g, which is in contradiction with Eq. (4.51). Taking Fourier transforms in space and time of the latter, one obtains

$$\int \mathrm{e}^{-i\mathbf{k}\cdot\mathbf{x}}\;\mathrm{e}^{-t\,\mathcal{M}_2^{sc}}(\mathbf{x},0)\,d^3\mathbf{x} \;=\; \rho(t^{\frac{1}{2-\kappa}}\,|\mathbf{k}|)\,, \tag{4.54}$$

with $\rho(0)=1$ and ρ decaying at infinity, instead of a constant as would be implied by Eq. (4.53). One should compare the relation (4.54) to the relation

$$\int \mathrm{e}^{-i\mathbf{k}\cdot\mathbf{x}}\;\mathrm{e}^{-t\Delta}(\mathbf{x},0)\,d^3\mathbf{x} = \mathrm{e}^{-t|\mathbf{k}|^2}, \tag{4.55}$$

describing the diffusive decay of the energy spectrum. For example, Eq. (4.54) implies that for homogeneous, fast decaying initial data 2-point function,

$$\int <T(t,0)\,T(t,\mathbf{x})>\;\mathbf{x}^2\,d^3\mathbf{x} \;\;\sim\;\; t^{\frac{2}{2-\kappa}}, \tag{4.56}$$

instead of being proportional to t: again a super-diffusive behavior. The reason of the failure of the Gibbs measure to be invariant under the evolution (4.25) for infinitely many degrees of freedom is the energy cascade towards degrees of freedom with smaller and smaller wavenumbers, see below. This failure shows that, for the infinite number of degrees of freedom, the symmetry group generated by the exponentials e^{β^I}, which for $\beta=\mathbf{v}\cdot\nabla$ is the group of volume preserving diffeomorphisms of $\mathbf{R}^3$, is broken. It is not excluded, however, that this symmetry may still be useful in analysis of the model.

4.4 Quasi-Lagrangian Approach to Passive Scalar

PS constitutes a perfect model to test the quasi-Lagrangian ideas mentioned at the end of Section 4. The scalar density $T'(t,\mathbf{x})\equiv T(t,\mathbf{x}+\mathbf{x}(t))$, viewed from a point moving in the fluid, satisfies the equation

$$\partial_t T' + (\mathbf{v}'-\mathbf{v}'_0)T' - \nu\Delta T' = f', \tag{4.57}$$

where $\mathbf{v}'(t,\mathbf{x})\equiv\mathbf{v}(t,\mathbf{x}+\mathbf{x}(t))$, and $f'(t,\mathbf{x})\equiv f(t,\mathbf{x}+\mathbf{x}(t))$, in the notation of Section 4. It is again a stochastic equation of the type (4.9) with $\beta=(\mathbf{v}'-\mathbf{v}'_0)\cdot\nabla$. One may show that for Gaussian homogeneous $\mathbf{v}(t,\mathbf{x})$ and $f(t,\mathbf{x})$, white noise in time, the distributions of the random fields $\mathbf{v}'(t,\mathbf{x})$ and $f'(t,\mathbf{x})$ obtained by the quasi-Lagrangian transformation become the same as those of $\mathbf{v}(t,\mathbf{x})$ and $f(t,\mathbf{x})$ asymptotically for large times. For $\mathbf{v}'(t,\mathbf{x})$ and $\mathbf{f}'(t,\mathbf{x})$ distributed as $\mathbf{v}(t,\mathbf{x})$ and $f(t,\mathbf{x})$ before, we may produce the exact solution for the correlation functions of T' by replacing the operators $\mathcal{M}_N$ by new operators $\mathcal{M}'_N$, obtained by changing the velocity correlation function (4.2) to

$$\begin{aligned}&(v'^i(t,\mathbf{x})-v'^i(t,0))\,(v'^j(t',\mathbf{x}')-v'^j(t',0))>\\&\quad=\delta(t-t')\,\big(D^{ij}(\mathbf{x}-\mathbf{x}')-D^{ij}(\mathbf{x})-D^{ij}(\mathbf{x}')+D^{ij}(0)\big)\end{aligned}$$

$$= -\delta(t-t')\left(\widetilde{D}^{ij}(\mathbf{x}-\mathbf{x}') - \widetilde{D}^{ij}(\mathbf{x}) - \widetilde{D}^{ij}(\mathbf{x}')\right) . \tag{4.58}$$

Notice that only the subtracted functions $\widetilde{D}^{ij}$ with the regular $m \to 0$ limit enter the correlator (4.58). One obtains

$$\begin{aligned}\mathcal{M}'_N &= \nu \sum_{n=1}^{N} \Delta_{\mathbf{x}_n} - \sum_{n=1}^{N} \widetilde{D}^{ij}(\mathbf{x}_n)\, \partial_{x_n^i} \partial_{x_n^j} \\ &+ \sum_{n<n'} \left(\widetilde{D}^{ij}(\mathbf{x}_n - \mathbf{x}_{n'}) - \widetilde{D}^{ij}(\mathbf{x}_n) - \widetilde{D}^{ij}(\mathbf{x}_{n'})\right) \partial_{x_n^i} \partial_{x_{n'}^j} .\end{aligned} \tag{4.59}$$

The crucial observation is, that in the action on translationally invariant functions operators $\mathcal{M}'_N$ and $\mathcal{M}_N$, the latter given by Eq. (4.48), coincide (the terms with $\widetilde{D}^{ij}(\mathbf{x}_n)$ cancel). Hence, for the PS, the stationary equal time quasi-Lagrangian correlators will be equal to the original (Eulerian) ones for the homogeneous sources. This is not the case for the non-equal time correlators, which involve the heat operators $\mathrm{e}^{-t\mathcal{M}_N}$ ($\mathrm{e}^{-t\mathcal{M}'_N}$) in the non-translationally invariant sector. The latter are governed by the eddy diffusivity, i. e. by the integral scale in the Eulerian case, but show a universal behavior in the inertial range in the quasi-Lagrangian description. The quasi-Lagrangian picture is also better behaved from the point of view of the standard perturbative approach, which consists of treating the 2-particle terms in $\mathcal{M}_N$ as a perturbation of the 1-particle ones, see Eq. (4.21). In the Eulerian approach, the unperturbed terms are dominated by the diverging eddy diffusivity, whereas in the quasi-Lagrangian one the 1-particle terms have a regular limit when $m \to 0$, compare Eqs. (4.44) and (4.59). Thus the quasi-Lagrangian perturbation expansion seems to be a better tool for the analysis of the behavior of the model.

5. Steady State of the Passive Scalar

5.1 Universality of the 2-Point Function

Let us pass to the study of the stationary state 2-point function $< T(\mathbf{x}_1)\, T(\mathbf{x}_2) > \equiv F_2(|\mathbf{x}_1 - \mathbf{x}_2|)$ in the presence of the random sources with a rotationally invariant covariance $\mathcal{C}_L = \mathcal{C}(\cdot/L)$, see Eq. (4.7). F_2 satisfies the differential equation

$$(\mathcal{M}_2 F_2)(r) = -\tfrac{2}{r^2}\, \partial_r (\nu r^2 + D_1 r^{2+\kappa})\, \partial_r F_2(r) = \mathcal{C}_L(r), \tag{5.1}$$

which is implied by Eq. (4.28). Integrating once, we obtain

$$\partial_r F_2(r) = -\frac{1}{2} \frac{\int_0^r \mathcal{C}(\rho/L)\, \rho^2\, d\rho}{\nu r^2 + D_1 r^{2+\kappa}} , \tag{5.2}$$

where the constant of integration has been chosen so as to obtain $\partial_r F_2$ vanishing at zero as required, if F_2 is to be a function on $\mathbf{R}^3$ regular at zero. One more integration gives

$$F_2(r) = \frac{1}{2} \int_r^\infty \frac{\int_0^\rho \mathcal{C}(\rho'/L)\,\rho'^2\,d\rho'}{\nu\,\rho^2 + D_1\,\rho^{2+\kappa}}\,d\rho\,, \tag{5.3}$$

where the choice of integration constant assures that F_2 decays at infinity. We want to study the behavior of $F_2(r)$, when the integral scale $L \to \infty$. To this end, let us rewrite the ρ-integral from r to ∞ as a difference of the integral from 0 to ∞ and from 0 to r:

$$\begin{aligned} F_2(r) \;&=\; \tfrac{1}{2}\,L^{2-\kappa} \int_0^\infty \frac{\int_0^\rho \mathcal{C}(\rho')\,\rho'^2\,d\rho'}{L^{-2-\kappa}\,\nu\,\rho^2 + D_1\,\rho^{2+\kappa}}\,d\rho \\ &\quad - \tfrac{1}{2} \int_0^r \frac{\int_0^\rho \mathcal{C}(\rho'/L)\,\rho'^2\,d\rho'}{\nu\,\rho^2 + D_1\,\rho^{2+\kappa}}\,d\rho\,. \end{aligned} \tag{5.4}$$

The first of the integrals is a $\mathcal{C}$-dependent constant diverging when $L \to \infty$. On the other hand, the integral from 0 to r has a limit when $L \to \infty$ equal to

$$-\tfrac{1}{6}\,\mathcal{C}(0) \int_0^r \frac{\rho^3\,d\rho}{\nu\rho^2 + D_1\rho^{2+\kappa}}\,, \tag{5.5}$$

which is universal (i.e. depends only on $\mathcal{C}(0)$ which, as we shall see in the next Section, equals twice the energy dissipation rate ϵ of the scalar). Note that the non-universal term (a constant) is annihilated by $\mathcal{M}_2$. This has to be so, if Eq. (5.1) is to be satisfied. The right hand side becomes universal in the limit $L \to \infty$, so all non-universal terms in $F_2(r)$ surviving in this limit have to be annihilated by $\mathcal{M}_2$. We shall see this general mechanism limiting possible non-universal terms also for the higher point functions. The constant term of F_2 is automatically subtracted in the 2nd structure function:

$$S_2(r) = 2(F_2(0) - F_2(r)) \;=\; \tfrac{2}{3}\,\epsilon \int_0^r \frac{\rho^3\,d\rho}{\nu\rho^2 + D_1\rho^{2+\kappa}}\,, \tag{5.6}$$

where the last equality holds for infinite L. In the $\nu \to 0$ limit,

$$S_2(r) \;=\; \frac{2\,\epsilon}{3(2-\kappa)\,D_1}\,r^{2-\kappa}\,, \tag{5.7}$$

and the same universal result holds approximately in the whole inertial range $\eta \ll r \ll L$, where at the Kolmogorov scale $\eta = \mathcal{O}((\nu/D_1)^{1/\kappa})$ the term νr^2 becomes comparable to $D_1\,\rho^{2+\kappa}$. The 2nd structure function exponent $\zeta_2 = 2 - \kappa$.

It is instructive to repeat the same analysis for anisotropic sources, i.e. without assuming that the correlation $\mathcal{C}_L$ is rotationally invariant. We shall have then to solve the equation

$$\mathcal{M}_2 F_2 = \mathcal{C}_L \tag{5.8}$$

in other angular momentum sectors. This is an easy exercise [26], and one still obtains, for m and ν equal to zero, the result

$$F_2(\mathbf{x}) - \gamma L^{2-\kappa} \underset{L\to\infty}{\longrightarrow} -\frac{\epsilon}{3(2-\kappa)D_1}|\mathbf{x}|^{2-\kappa} \equiv F_2^{sc}(\mathbf{x}), \tag{5.9}$$

where γ is a $\mathcal{C}$-dependent constant. In the case of a complex PS, however, further subtractions are needed when $0<\kappa<1$. For $F_2(\mathbf{x}) \equiv < T(0)\,\overline{T(\mathbf{x})} >$, one shows then [26], that

$$F_2(\mathbf{x}) - \gamma L^{2-\kappa} - \gamma_i x^i L^{1-\kappa} \underset{L\to\infty}{\longrightarrow} -\frac{\epsilon}{3(2-\kappa)D_1}|\mathbf{x}|^{2-\kappa}, \tag{5.10}$$

where real γ and imaginary γ_i are $\mathcal{C}$-dependent constants. The γ_i-terms are present only, if $\mathcal{C}$ has angular momentum 1 components, and this is impossible in the real case where only even l's appear[2]. Again, the non-universal terms are homogeneous zero modes of $\mathcal{M}_2$, which annihilates constant and linear functions, in accordance with the general argument. Similarly as the constant, the linear terms appearing with the coefficient $L^{1-\kappa}$ in the presence of anisotropic sources for the complex scalar drop out from the 2$^{\rm nd}$ structure function $< |T(\mathbf{x})-T(0)|^2 > = 2F_2(0) - F_2(\mathbf{x}) - F_2(-\mathbf{x})$ and do not spoil the inertial range universality of the latter.

5.2 Energy Cascade

Let us look at the energy balance of the scalar. The dynamical equation (4.27) takes the form

$$\partial_t F_2(r) = -(\mathcal{M}_2 F_2)(r) + \mathcal{C}_L(r), \tag{5.11}$$

and reduces in the stationary state to Eq. (5.1). In particular, the mean energy density $e \equiv < \frac{1}{2}T^2 >$ satisfies

$$\begin{aligned}\partial_t e = \tfrac{1}{2}\partial_t F_2(0) &= -\tfrac{1}{2}(\mathcal{M}_2 F_2)(0) + \tfrac{1}{2}\mathcal{C}(0)\\ &= \nu(\Delta F_2)(0) + \tfrac{1}{2}\mathcal{C}(0)\end{aligned} \tag{5.12}$$

(only the $-2\nu\Delta$ term of $\mathcal{M}_2$ contributes at $r=0$).

$$\epsilon \equiv -\nu(\Delta F_2)(0) = -\nu\frac{1}{r^2}\partial_r r^2 \partial_r F_2(r)|_{r=0} \tag{5.13}$$

is the mean dissipation rate, and $\frac{1}{2}\mathcal{C}(0)$ is the mean injection rate of energy (both intensive). In the steady state the energy balance reduces to the equality of the rates :

$$\epsilon = \tfrac{1}{2}\mathcal{C}(0). \tag{5.14}$$

It is, of course, satisfied by our solution for the steady state 2-point function.

[2] K.G. would like to thank R. Kraichnan for pointing this out to him.

We would like to see how the energy is distributed among different wavenumbers, following the schematic discussion of Section 1.2. For the mean energy density of modes with $|\mathbf{k}| \le K$, $e_{\le K} \equiv < \frac{1}{2} T_{\le K}^2 >$, we obtain

$$e_{\le K} = \frac{1}{2} \int\limits_{|\mathbf{k}|\le K} \left(\int e^{-i\mathbf{k}\cdot\mathbf{y}} F_2(\mathbf{y})\, d^3\mathbf{y} \right) \frac{d^3\mathbf{k}}{(2\pi)^3} \equiv \int\limits_0^K e(k)\, dk, \tag{5.15}$$

with the energy spectrum

$$e(k) \sim k^{-3+\kappa} \tag{5.16}$$

in the steady state inertial range. The dynamical equation (5.11) implies that

$$\partial_t e_{\le K} = -\epsilon_{\le K} + \varphi_{\le K} - \pi_K , \tag{5.17}$$

where

$$\begin{aligned} \epsilon_{\le K} &\equiv -\nu \int\limits_{|\mathbf{k}|\le K} \left(\int e^{-i\mathbf{k}\cdot\mathbf{y}} \Delta F_2(\mathbf{y})\, d^3\mathbf{y} \right) \frac{d^3\mathbf{k}}{(2\pi)^3} \\ &= -\frac{2}{\pi} \nu \int\limits_0^\infty \frac{\sin(Kr) - Kr\cos(Kr)}{r^3}\, \partial_r r^2 \partial_r F_2(r)\, dr \end{aligned} \tag{5.18}$$

is the mean energy dissipation rate in the modes with $|\mathbf{k}| \le K$, and

$$\begin{aligned} \varphi_{\le K} &\equiv \frac{1}{2} \int\limits_{|\mathbf{k}|\le K} \left(\int e^{-i\mathbf{k}\cdot\mathbf{y}} \mathcal{C}_L(\mathbf{y})\, d^3\mathbf{y} \right) \frac{d^3\mathbf{k}}{(2\pi)^3} \\ &= \frac{1}{\pi} \int\limits_0^\infty \frac{\sin(Kr) - Kr\cos(Kr)}{r}\, \mathcal{C}(r/L)\, dr \end{aligned} \tag{5.19}$$

is the mean energy injection rate into $|\mathbf{k}| \le K$.

$$\pi_K \equiv -\frac{2D_1}{\pi} \int\limits_0^\infty \frac{\sin(Kr) - Kr\cos(Kr)}{r^3}\, \partial_r r^{2+\kappa} \partial_r F_2(r)\, dr \tag{5.20}$$

comes from the contributions of $(\mathcal{M}_2^{sc} F_2)(r)$ to (5.11). In view of the relation (5.17), it is natural to interpret π_K as the mean energy flux out of the modes with $|\mathbf{k}| \le K$, compare Eq. (2.11) in Sect. 2. In particular, in the forced steady state, we obtain the energy balance equation

$$\varphi_{\le K} = \epsilon_{\le K} + \pi_K , \tag{5.21}$$

compare Eq. (2.12).

With the Fourier transform of $\mathcal{C}(\cdot/L)$ concentrated around the zero wavenumbers, the energy injection is essentially limited to small wavenumbers, and the mean injection rate $\varphi_{\le K}$ is approximately constant for $K \gg L^{-1}$ and equal to its $K = \infty$ value $\frac{1}{2}\mathcal{C}(0)$. On the other hand, the mean dissipation rate $\epsilon_{\le K}$ may be easily showed for $\kappa > 1$ to satisfy

$$\epsilon_{\le K} \le \mathcal{O}(\frac{\nu\epsilon}{D_1} K^\kappa) . \tag{5.22}$$

Indeed, it follows from Eq. (5.2) that $|\partial_r r^2 \partial_r F_2(r)| \leq \text{const.}\ D_1^{-1} r^{2-\kappa}$. Estimating

$$|\sin(Kr) - Kr\cos(Kr)| \leq \begin{cases} \text{const.}\ (Kr)^3 & \text{for} \quad r \leq K^{-1} \\ \text{const.}\ Kr & \text{for} \quad r \geq K^{-1} \end{cases}, \tag{5.23}$$

we obtain

$$\epsilon_{\leq K} \leq \text{const.}\ \frac{\nu\epsilon}{D_1} \left(K^3 \int_0^{1/K} r^{2-\kappa}\, dr + K \int_{1/K}^{\infty} r^{-\kappa}\, dr \right), \tag{5.24}$$

which implies the desired bound (5.22) for $\kappa > 1$. For $\kappa \leq 1$, a better estimate for $|\partial_r r^2 \partial_r F_2(r)|$ for large r, using the decay of $\mathcal{C}$, implies a somewhat worse bound

$$\epsilon_{\leq K} \leq \mathcal{O}(\frac{\nu\epsilon}{D_1} K^\kappa) + \mathcal{O}(\frac{\nu\epsilon\, L^{1-\kappa}}{D_1} K). \tag{5.25}$$

As a consequence, for $\kappa > 1$, the mean dissipation rate $\epsilon_{\leq K}$ stays very small relative to ϵ as long as $K \ll \eta^{-1}$, where $\eta = \mathcal{O}((\nu/D_1)^{1/\kappa})$ is the Kolmogorov scale. For $\kappa \leq 1$ this holds at least for $K \ll \nu^{-1} D_1 L^{\kappa-1}$. The energy balance equation (5.21) implies then that, for $\kappa > 1$, $L^{-1} \ll K \ll \eta^{-1}$, the mean energy flux π_K is almost equal to the mean injection rate $\varphi_{\leq K}$ and hence almost constant and equal to ϵ in the inertial range. For $\kappa \geq 1$ the same is true at least for $L^{-1} \ll K \ll \nu^{-1} D_1 L^{\kappa-1}$. This confirms an intuitive picture of the non-dissipative energy cascade through the decreasing distance scales from the integral scale L, where the energy is injected to the Kolmogorov scale η at least for $\kappa > 1$. At the end the energy is dissipated on distances shorter than η. What happens for $\kappa \leq 1$ and K between $\mathcal{O}(\nu^{-1} D_1 L^{\kappa-1})$ and $\mathcal{O}(\eta^{-1})$ remains to be understood.

5.3 Gaussian Case

It will be instructive to look at the case when $\kappa = 0$, with

$$\widetilde{D}^{ij} = 2\, D_1\, \delta^{ij}. \tag{5.26}$$

(Note that having finite D_1 requires infinitesimal D_0 when $\kappa \to 0$, in order to renormalize the ultraviolet divergence in the momentum-space integral (4.3); D_0 will never show up below.) Although quite trivial, the case $\kappa = 0$ appears to be instructive and will serve below as the departure point for an expansion in powers of κ. We immediately obtain for the $\kappa = 0$ operators:

$$\mathcal{M}_2^{\text{sc}} = 2\, D_1\, \nabla_{\mathbf{x}_1} \cdot \nabla_{\mathbf{x}_2} = -2\, D_1\, \Delta_{\mathbf{X}}, \tag{5.27}$$

$$\begin{aligned} \mathcal{M}_4^{\text{sc}} &= 2\, D_1 \sum_{1 \leq n < n' \leq 4} \nabla_{\mathbf{x}_n} \cdot \nabla_{\mathbf{x}_{n'}} \\ &= -2\, D_1\, (\Delta_{\mathbf{X}} + \Delta_{\mathbf{Y}} + \Delta_{\mathbf{Z}} - \nabla_{\mathbf{X}} \cdot \nabla_{\mathbf{Y}} - \nabla_{\mathbf{Y}} \cdot \nabla_{\mathbf{Z}}) \end{aligned} \tag{5.28}$$

in the difference variables $\mathbf{X} \equiv \mathbf{x}_1 - \mathbf{x}_2$, $\mathbf{Y} \equiv \mathbf{x}_2 - \mathbf{x}_3$, $\mathbf{Z} = \mathbf{x}_3 - \mathbf{x}_4$. The 2-point function

$$F_2(\mathbf{x}) = ({\mathcal{M}_2^{\mathrm{sc}}}^{-1} C_L)(\mathbf{x}) = \tfrac{1}{2D_1} \int \mathrm{e}^{i\mathbf{k}\cdot\mathbf{x}}\, |\mathbf{k}|^{-2}\, \widehat{C}_L(\mathbf{k})\, \tfrac{d^3\mathbf{k}}{(2\pi)^3}\,, \tag{5.29}$$

where

$$\widehat{C}_L(\mathbf{k}) = \int \mathrm{e}^{-i\mathbf{k}\cdot\mathbf{x}}\, C(\mathbf{x}/L)\, d^3\mathbf{x} = L^3\, \widehat{C}(L\mathbf{k})\,. \tag{5.30}$$

Hence

$$F_2(\mathbf{x}) = \tfrac{1}{2D_1}\, L^2 \int \mathrm{e}^{i\mathbf{k}\cdot\mathbf{x}/L}\, |\mathbf{k}|^{-2}\, \widehat{C}(\mathbf{k})\, \tfrac{d^3\mathbf{k}}{(2\pi)^3}\,. \tag{5.31}$$

The large L behavior may be obtained by expanding $\mathrm{e}^{i\mathbf{k}\cdot\mathbf{x}/L}$ to the second order:

$$\begin{aligned} F_2(\mathbf{x}) &= \tfrac{1}{2D_1}\, L^2 \int |\mathbf{k}|^{-2}\, \widehat{C}(\mathbf{k})\, \tfrac{d^3\mathbf{k}}{(2\pi)^3} + \tfrac{1}{2D_1}\, L x^j \int i k_j |\mathbf{k}|^{-2}\, \widehat{C}(\mathbf{k})\, \tfrac{d^3\mathbf{k}}{(2\pi)^3} \\ &- \tfrac{1}{4D_1}\, x^i x^j \int k_i k_j\, |\mathbf{k}|^{-2}\, \widehat{C}(k)\, \tfrac{d^3\mathbf{k}}{(2\pi)^3} + \mathcal{O}(L^{-1}) \\ &= \tfrac{1}{2D_1}\, ((-\Delta)^{-1} C)(0) + \tfrac{1}{2D_1}\, (\partial_i(-\Delta)^{-1} C)(0)\, x^i \\ &+ \tfrac{1}{4D_1}\, ((\partial_i \partial_j (-\Delta)^{-1} \\ &- \tfrac{1}{3}\delta_{ij})\, C)(0)\, x^i x^j + \tfrac{1}{12D_1}\, C(0)\, |x|^2 + \mathcal{O}(L^{-1})\,. \end{aligned} \tag{5.32}$$

Thus, up to a constant, and linear terms (the latter absent for real C) with C-dependent, L-diverging coefficients and a quadratic zero mode of $\mathcal{M}_2^{\mathrm{sc}}$ with a C-dependent coefficient, we obtain in the limit $L \to \infty$ the universal scaling solution $F_2^{\mathrm{sc}}(\mathbf{x}) = \frac{\epsilon}{6D_1}\, |\mathbf{x}|^2$, in accordance with the general result (5.9). Note that the separation of the scaling solution from the zero mode of the same homogeneity is ambiguous.

The 4-point function in the stationary state is given by Eqs. (4.29) and (4.30) for $M = 2$, i. e. it may be written as a sum of contributions of three 2-particle channels:

$$\begin{aligned} <\prod_{n=1}^{4} T(\mathbf{x}_n)> &= F_4(\mathbf{x}_1, \mathbf{x}_2, \mathbf{x}_3, \mathbf{x}_4) + F_4(\mathbf{x}_1, \mathbf{x}_3, \mathbf{x}_2, \mathbf{x}_4) \\ &\quad + F_4(\mathbf{x}_1, \mathbf{x}_4, \mathbf{x}_2, \mathbf{x}_3), \end{aligned} \tag{5.33}$$

with the single channel function

$$F_4 = \mathcal{M}_4^{-1}\, (\mathcal{M}_2^{-1} \otimes 1 + 1 \otimes \mathcal{M}_2^{-1})\, C_L \otimes C_L\,. \tag{5.34}$$

Working in the difference variables, we obtain for $\kappa = 0$, using the scaling operators (5.27), (5.28):

$$
\begin{aligned}
F_4(\mathbf{x}_1,\mathbf{x}_2,\mathbf{x}_3,\mathbf{x}_4) &= \tfrac{1}{4D_1^2}\int e^{i(\mathbf{k}_1\cdot\mathbf{x}_1+\ldots+\mathbf{k}_4\cdot\mathbf{x}_4)}\,(\sum_{n<n'}\mathbf{k}_n\cdot\mathbf{k}_{n'})^{-1}\\
&\quad ((\mathbf{k}_1\cdot\mathbf{k}_2)^{-1}+(\mathbf{k}_3\cdot\mathbf{k}_4))^{-1}\,\widehat{\mathcal{C}}_L(\mathbf{k}_1)\,\widehat{\mathcal{C}}_L(\mathbf{k}_3)\\
&\quad (2\pi)^3\,\delta(\mathbf{k}_1+\mathbf{k}_2)\,(2\pi)^3\,\delta(\mathbf{k}_3+\mathbf{k}_4)\prod_{n=1}^{4}\tfrac{d^3\mathbf{k}_n}{(2\pi)^3}\\
&= \tfrac{1}{4D_1^2}\int e^{i(\mathbf{k}_1\cdot(\mathbf{x}_1-\mathbf{x}_2)+\mathbf{k}_3\cdot(\mathbf{x}_3-\mathbf{x}_4)}\\
&\quad (\mathbf{k}_1^2+\mathbf{k}_3^2)^{-1}\,((\mathbf{k}_1^2)^{-1}+(\mathbf{k}_3^2)^{-1})\\
&\quad \widehat{\mathcal{C}}_L(\mathbf{k}_1)\,\widehat{\mathcal{C}}_L(\mathbf{k}_3)\,\tfrac{d^3\mathbf{k}_1}{(2\pi)^3}\,\tfrac{d^3\mathbf{k}_3}{(2\pi)^3}\\
&= F_2(\mathbf{x}_1-\mathbf{x}_2)\,F_2(\mathbf{x}_3-\mathbf{x}_4)\,, \qquad (5.35)
\end{aligned}
$$

see Eq. (5.29). It follows from Eq. (5.33), that for $\kappa=0$ the equal time 4-point function is expressed by the 2-point function by the standard Gaussian formula. This remains true for the higher point functions. Let us consider the 6-point function

$$
<\prod_{n=1}^{6}T(\mathbf{x}_n)> \;=\; \sum_{\sigma\,\in\,\mathcal{S}_6\,/\,\mathcal{S}_3\times\mathcal{S}_2^3} F_6(\mathbf{x}_{\sigma(1)},\ldots,\mathbf{x}_{\sigma(6)})\,, \qquad (5.36)
$$

where F_6 is the contribution of a single 2-particle channel,

$$
F_6 = \mathcal{M}_6^{-1}\sum_{\sigma\in\mathcal{S}_3}\sigma\left((\mathcal{M}_4^{-1}\otimes 1_2)\,(\mathcal{M}_2^{-1}\otimes 1_4)\right)\mathcal{C}_L\otimes\mathcal{C}_L\,. \qquad (5.37)
$$

Explicitly, for $\kappa=0$,

$$
\begin{aligned}
F_6(\mathbf{x}_1,\ldots,\mathbf{x}_6) &= \tfrac{1}{8D_1^3}\int e^{i(\mathbf{k}_1\cdot(\mathbf{x}_1-\mathbf{x}_2)+\mathbf{k}_3\cdot(\mathbf{x}_3-\mathbf{x}_4)+\mathbf{k}_5\cdot(\mathbf{x}_5-\mathbf{x}_6)}\\
&\quad (\mathbf{k}_1^2+\mathbf{k}_3^2+\mathbf{k}_5^2)^{-1}\\
&\quad ((\mathbf{k}_1^2+\mathbf{k}_3^2)^{-1}+(\mathbf{k}_1^2+\mathbf{k}_5^2)^{-1}+(\mathbf{k}_3^2+\mathbf{k}_5^2)^{-1})\\
&+ ((\mathbf{k}_1^2)^{-1}+(\mathbf{k}_3^2)^{-1}+(\mathbf{k}_5^2)^{-1})\\
&\quad \widehat{\mathcal{C}}_L(\mathbf{k}_1)\,\widehat{\mathcal{C}}_L(\mathbf{k}_3)\,\widehat{\mathcal{C}}_L(\mathbf{k}_5)\,\tfrac{d^3\mathbf{k}_1}{(2\pi)^3}\,\tfrac{d^3\mathbf{k}_3}{(2\pi)^3}\,\tfrac{d^3\mathbf{k}_5}{(2\pi)^3}\\
&= F_2(\mathbf{x}_1-\mathbf{x}_2)\,F_2(\mathbf{x}_3-\mathbf{x}_4)\,F_2(\mathbf{x}_5-\mathbf{x}_6)\,. \qquad (5.38)
\end{aligned}
$$

It should be clear how to extend this argument to a general $2M$-point function.

If we want to separate in F_{2M} the non-universal and the universal terms, we have to extract from equation (5.31) more detailed knowledge about the 2-point function than that given by the expansion (5.32). Since F_2 starts with an $\mathcal{O}(L^2)$ term, we have to include into the expansion of, for example, F_4 all terms of F_2 down to the order $\mathcal{O}(L^{-2})$. Since

$$
\begin{aligned}
F_2(\mathbf{x}) &= \tfrac{1}{2D_1}\sum_{n=0}^{4}\tfrac{1}{n!}\,L^{2-n}\,(\partial_{i_1}\ldots\partial_{i_n}(-\Delta)^{-1}\mathcal{C})(0)\,x^{i_1}\cdots x^{i_n}\\
&\quad+\mathcal{O}(L^{-3})\,, \qquad (5.39)
\end{aligned}
$$

$$\begin{aligned} F_4(\mathbf{X},\mathbf{Y},\mathbf{Z}) &= \frac{1}{4D_1^2}\sum_{\substack{m,n\geq 0\\ n+m\leq 4}} \frac{1}{m!}\,\frac{1}{n!}\,L^{4-m-n}\,(\partial_{i_1}\dots\partial_{i_m}\,\Delta^{-1}\mathcal{C})(0)\\ &\quad (\partial_{j_1}\dots\partial_{j_n}\,\Delta^{-1}\mathcal{C})(0)\,X^{i_1}\cdots X^{i_m}\,Z^{j_1}\cdots Z^{j_n}\\ &+\ \mathcal{O}(L^{-1})\,. \end{aligned} \tag{5.40}$$

For rotationally invariant $\mathcal{C}$, this reduces to

$$\begin{aligned} &F_4(\mathbf{X},\mathbf{Y},\mathbf{Z})\\ &= \frac{1}{4D_1^2}\Big(L^4\,(\Delta^{-1}\mathcal{C})(0)^2+\tfrac{1}{6}\,L^2\,\mathcal{C}(0)\,(\Delta^{-1}\mathcal{C})(0)\,(|\mathbf{X}|^2+|\mathbf{Z}|^2)\\ &-\tfrac{1}{144}\,(\Delta\mathcal{C})(0)\,(\Delta^{-1}\mathcal{C})(0)\,(|\mathbf{X}|^4+|\mathbf{Z}|^4)+\tfrac{1}{36}\,\mathcal{C}(0)^2\,|\mathbf{X}|^2|\mathbf{Z}|^2\Big)\\ &+\mathcal{O}(L^{-2})\,. \end{aligned} \tag{5.41}$$

Note the presence of the universal Gaussian contribution

$$F_2^{\rm sc}(\mathbf{X})\,F_2^{\rm sc}(\mathbf{Z}) = \frac{\epsilon^2}{36\,D_1^2}\,|\mathbf{X}|^2\,|\mathbf{Z}|^2$$

on the right hand side. The rest of the terms are non-universal zero modes of the operator

$$\mathcal{M}_2^{\rm sc}\otimes\mathcal{M}_2^{\rm sc}\,(\mathcal{M}_2^{\rm sc}\otimes 1+1\otimes\mathcal{M}_2^{\rm sc})^{-1}\,\mathcal{M}_4^{\rm sc}, \tag{5.42}$$

equal to $\mathcal{M}_2^{\rm sc}\otimes\mathcal{M}_2^{\rm sc} = 4D_1^2\,\Delta_{\mathbf{X}}\Delta_{\mathbf{Z}}$ in the action on $\mathbf{Y}$-independent functions. That this must be the case, may be easily seen by rewriting the relation (5.34) as the equation

$$(\mathcal{M}_2^{\rm sc}\otimes\mathcal{M}_2^{\rm sc})\,(\mathcal{M}_2^{\rm sc}\otimes 1+1\otimes\mathcal{M}_2^{\rm sc})^{-1}\,\mathcal{M}_4^{\rm sc}\,F_4 \;=\; \mathcal{C}_L\otimes\mathcal{C}_L\,. \tag{5.43}$$

Note that the separation of the universal Gaussian contribution $F_2^{\rm sc}(\mathbf{X})\,F_2^{\rm sc}(\mathbf{Z})$ from the terms with $m=n=2$ is ambiguous, since there are non-universal terms of the same homogeneity degree. Nevertheless, the calculation for $\kappa=0$ confirms the general structure of the correlations with the non-universal terms given by homogeneous zero modes of differential operators.

5.4 4-Point and 6-Point function for non-zero κ

Let us consider the 4-point function given by Eqs. (5.33) and (5.34) for general κ between 0 and 2. Recalling that $F_2=\mathcal{M}_2^{-1}\mathcal{C}_L$, it may be convenient to view Eq. (5.34) as the differential equation for F_4:

$$\mathcal{M}_4\,F_4 \;=\; F_2\otimes\mathcal{C}_L+\mathcal{C}_L\otimes F_2\,. \tag{5.44}$$

The connected part $F_4^c=F_4-F_2\otimes F_2$ of F_4 satisfies the equation

$$\mathcal{M}_4\,F_4^c \;=\; \mathcal{L}\,(F_2\otimes F_2)\,, \tag{5.45}$$

where $\mathcal{L}\equiv(\mathcal{K})_{1,3}+(\mathcal{K})_{1,4}+(\mathcal{K})_{2,3}+(\mathcal{K})_{2,4}$, see Eq. (4.34).

First, let us study the L-dependence of $F_2 \otimes F_2$. Since, as follows from Eq. (5.4) with $\nu = 0$,

$$F_2(\mathbf{x}) \;=\; \gamma\, L^{2-\kappa} - \tfrac{\epsilon}{3(2-\kappa)\,D_1}\, |\mathbf{x}|^{2-\kappa} + \mathcal{O}(L^{-2})\,, \tag{5.46}$$

we infer that the product of 2-point functions

$$\begin{aligned} F_2(\mathbf{X})\, F_2(\mathbf{Z}) \;&=\; \gamma^2\, L^{2(2-\kappa)} + \gamma\, \tfrac{\epsilon}{3(2-\kappa)\,D_1}\, L^{2-\kappa}\,(|\mathbf{X}|^{2-\kappa} + |\mathbf{Z}|^{2-\kappa}) \\ &+\; \tfrac{\epsilon^2}{9(2-\kappa)^2\,D_1^2}\, |\mathbf{X}|^{2-\kappa}\,|\mathbf{Z}|^{2-\kappa} + \mathcal{O}(L^{-\kappa})\,. \end{aligned} \tag{5.47}$$

Compare this expression with the similar expression (5.41) for $\kappa = 0$. The term $\sim$ $(|\mathbf{X}|^4 + |\mathbf{Z}|^4)$ present there comes from the $\mathcal{O}(L^{-\kappa})$ terms of the $\kappa > 0$ case.

Let us return to the connected contribution to F_4 as given by Eq. (5.45). In the action on translationally invariant functions,

$$\begin{aligned} \mathcal{M}_4 \;=\; &-\; \widetilde{D}^{ij}(\mathbf{X})\,\partial_{X^i}\,\partial_{X^j} - \widetilde{D}^{ij}(\mathbf{Y})\,\partial_{Y^i}\,\partial_{Y^j} - \widetilde{D}^{ij}(\mathbf{Z})\,\partial_{Z^i}\,\partial_{Z^j} \\ &-\; \Big(\widetilde{D}^{ij}(\mathbf{X}+\mathbf{Y}) - \widetilde{D}^{ij}(\mathbf{X}) - \widetilde{D}^{ij}(\mathbf{Y})\Big)\,\partial_{X^i}\,\partial_{Y^j} \\ &-\; \Big(\widetilde{D}^{ij}(\mathbf{Y}+\mathbf{Z}) - \widetilde{D}^{ij}(\mathbf{Y}) - \widetilde{D}^{ij}(\mathbf{Z})\Big)\,\partial_{Y^i}\,\partial_{Z^j} \\ &-\; \Big(\widetilde{D}^{ij}(\mathbf{X}+\mathbf{Y}+\mathbf{Z}) - \widetilde{D}^{ij}(\mathbf{X}+\mathbf{Y}) - \widetilde{D}^{ij}(\mathbf{Y}+\mathbf{Z}) \\ &+\; \widetilde{D}^{ij}(\mathbf{Y})\Big)\,\partial_{X^i}\,\partial_{Z^j} \end{aligned} \tag{5.48}$$

for $\nu = 0$. Note that

$$\begin{aligned} \mathcal{L}\,(F_2 \otimes F_2)(\mathbf{X},\mathbf{Y},\mathbf{Z}) \;=\; &\Big(\widetilde{D}^{ij}(\mathbf{X}+\mathbf{Y}+\mathbf{Z}) \\ &-\; \widetilde{D}^{ij}(\mathbf{X}+\mathbf{Y}) - \widetilde{D}^{ij}(\mathbf{Y}+\mathbf{Z}) + \widetilde{D}^{ij}(\mathbf{Y})\Big) \\ &(\partial_i F_2)(\mathbf{X})(\partial_j F_2)(\mathbf{Z}) \end{aligned} \tag{5.49}$$

has a well defined limit when $L \to \infty$, since it involves only the derivatives of F_2. In particular for $m = 0$ and $\nu = 0$, the $L \to \infty$ limit of $\mathcal{L}\,(F_2 \otimes F_2)$ is equal to

$$\tfrac{\epsilon^2}{9\,(2-\kappa)^2\,D_1^2}\, \mathcal{L}^{\rm sc}\, |\mathbf{X}|^{2-\kappa}|\mathbf{Z}|^{2-\kappa}, \tag{5.50}$$

and is a homogeneous (rotationally-invariant) function of $\mathbf{X}, \mathbf{Y}, \mathbf{Z}$ of degree $2 - \kappa$, whose explicit form may be easily found. It is not difficult to write down a solution $F_4^{c\,{\rm sc}}$ of Eq. (5.45) for the limiting case. Indeed, for

$$\begin{aligned} F_4^{c\,{\rm sc}} \;&=\; \tfrac{\epsilon^2}{6\,(2-\kappa)^2(5-\kappa)\,D_1^2}\, (|\mathbf{X}|^{2(2-\kappa)} + |\mathbf{Z}|^{2(2-\kappa)}) \\ &-\; \tfrac{\epsilon^2}{9\,(2-\kappa)^2\,D_1^2}\, |\mathbf{X}|^{2-\kappa}|\mathbf{Z}|^{2-\kappa} \end{aligned} \tag{5.51}$$

one obtains, using of the decomposition $\mathcal{M}_4^{\rm sc} = \mathcal{M}_2^{\rm sc} \otimes 1 + 1 \otimes \mathcal{M}_2^{\rm sc} - \mathcal{L}^{\rm sc}$, the required relation

$$(\mathcal{M}_4^{sc}\, F_4^{sc})(\mathbf{X},\mathbf{Y},\mathbf{Z}) \;=\; \frac{\epsilon^2}{9\,(2-\kappa)^2\, D_1^2}\; \mathcal{L}^{sc}\; |\mathbf{X}|^{2-\kappa} |\mathbf{Z}|^{2-\kappa}\,, \tag{5.52}$$

since $\mathcal{M}_2^{sc} \otimes 1 + 1 \otimes \mathcal{M}_2^{sc}$ does not contribute to the left hand side, and $\mathcal{L}^{sc}$ annihilates functions depending only on $\mathbf{X}$ or only on $\mathbf{Z}$. By the same argument as for the two-point function, the solution for finite but large L should differ from the universal scaling form by zero modes of $\mathcal{M}_4^{sc}$, so that

$$F_4^c \;-\; \sum_{0\le\zeta_{4,n}\le 2(2-\kappa)} L^{2(2-\kappa)-\zeta_{4,n}} \sum_m \gamma_{nm}\, F_{4,nm}^c \;\underset{L\to\infty}{\longrightarrow}\; F_4^{c\,sc}, \tag{5.53}$$

where $F_{4,nm}^c$ are homogeneous zero modes of $\mathcal{M}_4^{sc}$ of degree $\zeta_{4,n}$, and the non-universal coefficients γ_{nm} depend on the source covariance $\mathcal{C}$. In fact, a more complicated behavior with logarithmic terms,

$$\begin{aligned} F_4^c \;-\; &\sum_{0\le\zeta_{4,n}\le 2(2-\kappa)} L^{2(2-\kappa)-\zeta_{4,n}} \sum_{m,k} \gamma_{nmk}\, (\ln{(L f_{nm})})^k\, F_{4,nm}^c \\ &\underset{L\to\infty}{\longrightarrow}\; F_4^{c\,sc}, \end{aligned} \tag{5.54}$$

is possible, if $\mathcal{M}_4$ has zero modes $(\ln f_{nm})^l\, F_{nm}$, $l \le k$, for f_{nm} homogeneous functions of degree -1.

The main problem, which remains in the analysis of the universality of the 4-point functions, is to find homogeneous zero modes of $\mathcal{M}_4^{sc}$ with degree between zero and $2(2-\kappa)$, and to decide which ones appear in the relation (5.53). The study of such zero modes seems to lie beyond the scope of the techniques [27] developed for the singular elliptic operators of the type we encounter here. Note that in the full connected 4-point function only the zero modes with permutation symmetry survive. If the sources are not only homogeneous but also isotropic, only $SO(3)$-symmetric zero modes may appear. In fact, one should expect that such zero modes do enter into the expansions (5.53) or (5.54) for generic $\mathcal{C}$, whenever not forbidden by symmetries (which is a somewhat tautological statement). Existence of the zero-mode contributions to the 4-point function is important from the practical point of view. If we study a combination of 4-point functions (e.g. the 4th structure function) and fix the range of variation of distances between the points, then the non-universal terms with the smallest $\zeta_{4,n}^{\min} < 2(2-\kappa)$ dominate the behavior for large L, and we would see the anomalous exponent $\zeta_{4,n}^{\min}$ rather than $2(2-\kappa)$ as governing the distance behavior. The possible appearance of zero modes of $\mathcal{M}_4^{sc}$ in the asymptotic behavior of the 4-point function would lead to the picture of **restricted universality**, holding strictly only for the 4-point correlator, which is **infrared-renormalized** by subtracting homogeneous terms with non-universal coefficients typically divergent, when the integral scale L grows to infinity. In the presence of non-universal terms with $\zeta_{4,n} = 2(2-\kappa)$ which are L-independent, the separation between the universal and non-universal terms in the 4-point function would become ambiguous, and the Eq. (5.51) would involve an arbitrary choice.

Notice that adding the connected and disconnected contributions to F_4, one obtains (in the absence of logarithmic terms)

$$F_4(\mathbf{X},\mathbf{Y},\mathbf{Z}) \;=\; G(\mathbf{X}) + G(\mathbf{Z})$$

$$+ \sum_{\zeta_{4,n}} L^{2(2-\kappa)-\zeta_{4,n}} \sum_m \gamma_{nm}\, F^c_{4,nm}(\mathbf{X},\mathbf{Y},\mathbf{Z}) + o(L), \quad (5.55)$$

for

$$G(\mathbf{x}) = \tfrac{1}{2}\gamma^2\, L^{2(2-\kappa)} + \gamma\, \frac{\epsilon}{3(2-\kappa)D_1}\, L^{2-\kappa}\, |\mathbf{x}|^{2-\kappa} + \frac{\epsilon^2}{12(2-\kappa)^2(5-\kappa)D_1^2}\, |\mathbf{x}|^{2(2-\kappa)}.$$

Only $F^c_{4,mn}$ terms may contribute to the structure function $S_4(r)$, as all the terms dependent on only one variable difference do not survive in S_4. Hence the presence of the non-trivial contributions from the (generally) non-universal terms $F^c_{4,nm}$ is a necessary condition for the very non-triviality of S_4 for large L.

The 6 point function may be treated similarly. The connected part F^c_6 of the single 2-particle channel contribution F_6 is given by

$$\begin{aligned} F_6(\mathbf{x}_1,\ldots,\mathbf{x}_6) &= F^c_6(\mathbf{x}_1,\ldots,\mathbf{x}_6) + F^c_4(\mathbf{x}_1,\ldots,\mathbf{x}_4)\, F_2(\mathbf{x}_5,\mathbf{x}_6) \\ &+ F^c_4(\mathbf{x}_1,\mathbf{x}_2,\mathbf{x}_5\mathbf{x}_6)\, F_2(\mathbf{x}_3,\mathbf{x}_4) + F^c_4(\mathbf{x}_3,\ldots,\mathbf{x}_6)\, F_2(\mathbf{x}_1,\mathbf{x}_2) \\ &+ F_2(\mathbf{x}_1,\mathbf{x}_2)\, F_2(\mathbf{x}_3,\mathbf{x}_4)\, F_2(\mathbf{x}_5,\mathbf{x}_6)\,, \end{aligned} \quad (5.56)$$

see Eq. (4.31). It satisfies Eq. (4.34), which reads

$$\begin{aligned} \mathcal{M}_6\, F^c_6 &= ((\mathcal{L}_{1,3} + \mathcal{L}_{2,3})\,(F^c_4(1,2)\, F_2(3)) \\ &+ (\mathcal{L}_{1,2} + \mathcal{L}_{2,3})((F^c_4(1,3) F_2(2)) \\ &+ (\mathcal{L}_{1,2} + \mathcal{L}_{1,3})\,(F_2(1)\, F^c_4(2,3))\,, \end{aligned} \quad (5.57)$$

where the numbers in the paranthesis label the 3 consecutive pairs of points. In the case that the non-universal non-constant zero modes do not appear in the connected 4-point function, the right hand side of Eq. (5.57) has a finite $L \to \infty$ limit, so that the relation of the type (5.53),

$$F^c_6 \;-\; \sum_{0\leq\zeta_{6,n}\leq 3(2-\kappa)} L^{3(2-\kappa)-\zeta_{6,n}} \sum_m \gamma_{nm}\, F^c_{6,nm} \;\underset{L\to\infty}{\longrightarrow}\; F^{c\,\mathrm{sc}}_6\,, \quad (5.58)$$

with $F^c_{6,mn}$ being homogeneous zero modes of the degree $\zeta_{6,n}$ of $\mathcal{M}^{\mathrm{sc}}_6$, should hold for $m = 0 = \nu$ (with eventual additional logarithmic terms, as in (5.54)). As we shall see in the next Section, such zero modes with rotational and permutation symmetries exist at least for small κ. Hence, the non-constant non-universal terms should appear, if not already in the connected 4-point function, then in the connected 6-point function. Whether they are visible in the corresponding structure functions is a more difficult question, to which we do not have definite answer yet.

Let us end this section by stressing a difference between the infrared renormalization required by the correlation functions of the PS, and the renormalization of infrared divergences occurring in massless models of statistical mechanics or field theory. There the divergences are an artifact of the perturbative expansion and signal, that the control parameters of the system are off their critical values. The divergences may be removed by finite shifts of the control parameters, which depend non-smoothly on the coupling constants (hence the infinities in the perturbative treatment). On the contrary, in the

PS model of turbulent advection, the genuine divergences appear, when the size of the system tends to infinity, and the correlators become finite only after the subtraction of the diverging terms. A similar picture appeared in Polyakov's 2-dimensional conformal turbulence [20], where it was argued that extracting the conformal-invariant part of the correlation functions requires subtraction of non-universal polynomial terms in the position space[3]. We expect the infrared renormalization of the type described above to be the general feature of universality in turbulent systems and the most important lesson to learn from the study of the PS, even if, at present, we are unable to decide, whether non-universal terms are present in the structure functions, the most commonly studied correlators.

5.5 κ-Expansion

We may study the homogeneous zero modes of the operator $\mathcal{M}_4^{\rm sc}$ in perturbation expansion in powers of κ. We shall work here only to the first order in κ. Eq. (4.6) implies that (for $m=0$)

$$\begin{aligned}\widetilde{D}^{ij}(\mathbf{x}) &= 2D_1\,\delta^{ij} + \kappa D_1\left(2\delta^{ij}\ln|\mathbf{x}| - x^i x^j\,|\mathbf{x}|^{-2} + \delta^{ij}\right) + \mathcal{O}(\kappa^2)\\ &\equiv 2D_1\,\delta^{ij} + 2\kappa D_1\,R^{ij}(\mathbf{x}) + \mathcal{O}(\kappa^2)\,. \end{aligned}\tag{5.59}$$

As a result, to the first order in κ,

$$\begin{aligned}\mathcal{M}_4^{\rm sc} &= \mathcal{M}_{4,0}^{\rm sc}\\ &-\ 2\kappa D_1\left(R^{ij}(\mathbf{X})\,\partial_{X^i}\partial_{X^j} + R^{ij}(\mathbf{Y})\,\partial_{Y^i}\partial_{Y^j} + R^{ij}(\mathbf{Z})\,\partial_{Z^i}\partial_{Z^j}\right)\\ &-\ 2\kappa D_1\left(R^{ij}(\mathbf{X}+\mathbf{Y}) - R^{ij}(\mathbf{X}) - R^{ij}(\mathbf{Y})\right)\partial_{X^i}\partial_{Y^j}\\ &-\ 2\kappa D_1\left(R^{ij}(\mathbf{Y}+\mathbf{Z}) - R^{ij}(\mathbf{Y}) - R^{ij}(\mathbf{Z})\right)\partial_{Y^i}\partial_{Z^j}\\ &-\ 2\kappa D_1\left(R^{ij}(\mathbf{X}+\mathbf{Y}+\mathbf{Z}) - R^{ij}(\mathbf{X}+\mathbf{Y}) - R^{ij}(\mathbf{Y}+\mathbf{Z}) + R^{ij}(\mathbf{Y})\right)\\ &\qquad \partial_{X^i}\partial_{Z^j}\\ &\equiv \mathcal{M}_{4,0}^{\rm sc} + 2\kappa D_1\,V_4\,, \end{aligned}\tag{5.60}$$

where the subscript "0" refers to $\kappa=0$. $\mathcal{M}_4^{\rm sc}$ commutes with 3-dimensional translations and rotations, and with the permutations of four points. We shall search for its zero modes respecting these symmetries. Note that

$$\mathcal{M}_{4,0}^{\rm sc} = -2D_1\left(\Delta_{\widetilde{\mathbf{X}}} + \Delta_{\widetilde{\mathbf{Y}}} + \Delta_{\widetilde{\mathbf{Z}}}\right), \tag{5.61}$$

where

$$\widetilde{\mathbf{X}} = \mathbf{X}, \quad \widetilde{\mathbf{Y}} = \sqrt{2}(\mathbf{Y} + \tfrac{1}{2}\mathbf{X} + \tfrac{1}{2}\mathbf{Z}), \quad \widetilde{\mathbf{Z}} = \mathbf{Z}\,. \tag{5.62}$$

Denoting $R \equiv (\widetilde{\mathbf{X}}^2 + \widetilde{\mathbf{Y}}^2 + \widetilde{\mathbf{Z}}^2)^{1/2}$, we obtain

$$\mathcal{M}_{4,0}^{\rm sc} = -\tfrac{2D_1}{R^8}\,\partial_R R^8\,\partial_R - \tfrac{2D_1}{R^2}\,\Phi, \tag{5.63}$$

where Φ is the Laplacian on the sphere S^8 in the space of $(\widetilde{\mathbf{X}}, \widetilde{\mathbf{Y}}, \widetilde{\mathbf{Z}})$. The symmetric zero modes of the lowest homogeneity of $\mathcal{M}_{4,0}^{\rm sc}$ occur in degree

[3] K.G. thanks A. Polyakov for discussion of this point.

zero (constants) and in degree 4. The latter, when expressed in terms of point-differences, have the form

$$a\sum_{\{n,n'\}}(\mathbf{x}_n-\mathbf{x}_{n'})^4 + b\sum_{\{\{n,m\},\{n,m'\}\}}(\mathbf{x}_n-\mathbf{x}_m)^2(\mathbf{x}_n-\mathbf{x}_{m'})^2 + c\sum_{\{\{n,n'\},\{m,m'\}\}}(\mathbf{x}_n-\mathbf{x}_{n'})^2(\mathbf{x}_m-\mathbf{x}_{m'})^2, \qquad (5.64)$$

where the pairs $\{n,n'\}$ and $\{m,m'\}$ are assumed different, as well as the pairs $\{n,m\}$ and $\{n,m'\}$, and where

$$20a + 28b + 6c = 0\,. \qquad (5.65)$$

The constant survives as the eigenvalue of $\mathcal{M}_4^{\rm sc}$ for $\kappa \neq 0$. The change of the two independent zero modes $R^4 f_i$, $i = 1,2$, of the form (5.64) may be traced by the standard degenerate perturbation theory. This is done as follows. Looking for a homogeneous zero mode $R^{4+\kappa\lambda}(a_1 f_1 + a_2 f_2 + \kappa f_3)$, with a homogeneous degree zero function f_3 orthogonal to $f_{1,2}$ in $L^2(S^8)$, we obtain in the linear order in κ

$$\mathcal{M}_{4,0}^{\rm sc}\left(\lambda R^4 \ln R(a_1 f_1 + a_2 f_2) + R^4 f_3\right) + 2D_1 V_4 R^4(a_1 f_1 + a_2 f_2) = 0, \qquad (5.66)$$

or, using the form (5.63) of $\mathcal{M}_{4,0}^{\rm sc}$,

$$-15\lambda\,(a_1 f_1 + a_2 f_2) - 44 f_3 - \Phi f_3 + \tfrac{1}{R^2} V_4 R^4 (a_1 f_1 + a_2 f_2) = 0\,. \qquad (5.67)$$

Upon taking the $L^2(S^8)$ scalar products with $f_{1,2}$, f_3 drops out, resulting in the relation

$$-15\lambda \sum_{j=1,2}(f_i, f_j)\, a_j + \sum_{j=1,2}(f_i, \tfrac{1}{R^2} V_4 R^4 f_j)\, a_j = 0. \qquad (5.68)$$

Hence λ has to solve the equation

$$\det\left((f_i, \tfrac{1}{15 R^2} V_4 R^4 f_j) - \lambda\,(f_i, f_j)\right) = 0\,. \qquad (5.69)$$

It may be decided numerically, whether this equation has solutions $\lambda < -2$, so that there exist homogeneous zero mode of $\mathcal{M}_4^{\rm sc}$ with degree $\leq 4-2\kappa$ for small κ, which, multiplied by $\gamma L^{-(2+\lambda)\kappa}$, should appear as non-universal terms in the truncated 4-point function. Given λ and a non-zero solution (a_1, a_2) of Eq. (5.68), the $\mathcal{O}(\kappa)$ contribution f_3 to the zero mode should be computed from Eq. (5.67).

If Eq. (5.69) has one double root, but there is only one solution (a_1, a_2) (up to normalization) of Eq. (5.68), then one should look for an additional zero mode of $\mathcal{M}_4^{\rm sc}$ of the form

$$\kappa\,(\ln R)\, R^{4+\kappa\lambda}(a_1 f_1 + a_2 f_2 + \kappa f_3) + R^{4+\kappa\lambda}(b_1 f_1 + b_2 f_2 + \kappa f_4)\,. \qquad (5.70)$$

This leads in the first order to the relation

$$\begin{aligned} &- 15((a_1 + \lambda b_1)f_1 + (a_2 + \lambda b_2)f_2) - 44\, f_4 - \Phi\, f_4 \\ &+ \tfrac{1}{R^2} V_4 R^4 (b_1 f_1 + b_2 f_2) = 0, \end{aligned} \tag{5.71}$$

which implies that

$$-15 \sum_{j=1,2} (f_i, f_j)(a_j + \lambda b_j) + \sum_{j=1,2} (f_i, \tfrac{1}{R^2} V_4 R^4 f_j)\, b_j = 0\,. \tag{5.72}$$

The latter equation has a non-zero solution for (b_1, b_2). In this case, the logarithmic terms $\sim (\ln L)\, L^{-(2+\lambda)\kappa}$ should appear in the asymptotic expansion of F_4^c for $\kappa > 0$, if $\lambda \leq -2$.

The zero modes of the scaling operator $\mathcal{M}_6^{\rm sc}$ may be studied similarly by writing

$$\mathcal{M}_6^{\rm sc} = \mathcal{M}_{6,0}^{\rm sc} + 2\,\kappa\, D_1\, V_6\,, \tag{5.73}$$

and using the perturbation expansion. By a linear change of variables, $\mathcal{M}_{6,0}^{\rm sc}$ is equivalent to the Laplacian in $\mathbf{R}^{15}$. As for $\mathcal{M}_{4,0}^{\rm sc}$, the homogeneous zero modes of $\mathcal{M}_{6,0}^{\rm sc}$ with the lowest degree, symmetric under rotations and permutations, are constants and the $4^{\rm th}$-order polynomials (5.64) satisfying this time the relation

$$20a + 56b + 36c = 0\,. \tag{5.74}$$

For $\kappa > 0$ the constants are still annihilated by $\mathcal{M}_{6,0}^{\rm sc}$, whereas the two $4^{\rm th}$-order zero modes may change slightly the homogeneity degree (one of them may also pick a logarithm). As for small κ the deformed degree will still be smaller than $3(2-\kappa)$, the deformed zero modes, accompanied by a positive power of L and a non-universal coefficient, should appear in the asymptotic expansion of the connected 6-point function discussed above.

6. Conclusions

At the beginning of these lectures, we have sketched the main idea of Kolmogorov's picture of universality in the fully developed turbulence, related to the energy cascade. We briefly described, how the turbulence problem may be formulated in the language formally very close to that of equilibrium statistical mechanics or Euclidean field theory, stressing, however, the important differences between the two classes of problems. The bulk of the lectures was devoted to the discussion of a simple model of turbulent phenomenon, the passive scalar (PS), which we treated as a test ground for the universality ideas. The exact solution for the model in terms of singular elliptic many-body operators showed non-universal diffusive decay of correlations for deterministic initial data, but a universal super-diffusive decay for random homogeneous initial data. In the presence of random sources injecting the energy at small wavenumbers, the steady state exhibited an inertial range with the energy cascade and universal scaling in the connected equal-time two-point function of the scalar. We have argued, that the higher-point equal-time connected functions are expected to show only restricted universality, with the terms dominating for large size systems characterized by universal anomalous exponents and non-universal amplitudes. We have shown, that such terms should appear in the 6-point function, if not in the 4-point one, at least for small velocity exponent κ. Further work along well traced direction is required to decide on this question as well as on the question what terms dominate the respective structure functions. One of the open problems is to relate the terms with anomalous exponents to the physical idea of intermittency. One should also understand what role, if any, is played in the model by the volume preserving diffeomorphisms. Upon infrared renormalization, which subtracts the non-universal terms, the equal-time correlators of the scalar should become finite and universal (with normal scaling) in the limit when molecular diffusivity $\nu \to 0$ and the integral scale $L \to \infty$. In particular, no ν-dependent subtractions are needed in the steady-state equal-time correlators of the scalar, and their short-distance behavior is easy to analyze [14][15]. This behavior determines, what ultraviolet (short distance) renormalization is required by the composite operators involving the gradients of T. The basic role of ν is to assure the convergence to the steady state by providing the mechanis, which removes on small scales the energy injected by the sources at large distances.

We expect the main features of the behavior of the PS described above to be present also in other systems exhibiting a fully developed turbulence, including the Navier-Stokes flow. Still much remains to be done even for the PS. One important step could be a complete perturbative analysis with the above picture of restricted universality established order by order. The κ-expansion discussed above is a possibility, but already the analysis of the lowest terms proved to be quite complicated [26]. Other possibilities are the quasi-Lagrangian perturbation expansion, or the expansion proposed in [14]. Ultimately, if no way to compute exactly the non-universal terms in the correlators is found, a multiscale analysis employing perturbative arguments in conjunction with a renormalization group type analysis should provide a tool to fully control the behavior of correlation functions. Developing such methods for the PS may be a first step towards a complete theory of fully developed turbulence. More than half-century after Kolmogorov's work, despite further progress, a fully developed understanding of the fully developed turbulence remains a major challenge.

References

1. *Phase transitions and critical phenomena.* Vol. 6, eds. C. Doomb, M. S. Green. Academic Press, London 1976
2. R. L. Dobrushin, R. Kotecký, S. Shlosman: *The Wulff construction: a global shape from local interactions.* AMS, Translations of Mathematical Monographs 104, Providence, Rhode Island 1992
3. J. Feldman, E. Trubowitz: *The Flow of an Electron-Phonon System to the Superconducting State.* Helv. Phys. Acta **64** (1991), 213-357

4. J. Fröhlich: Lectures, Schladming, 1995.
5. M. J. Feigenbaum: *Qualitative Universality for a Class of Nonlinear Transformations.* J. Stat. Phys. **19** (1978), 25-52
6. J. Bricmont, A. Kupiainen: *Renormalizing Partial Differential Equations.* Commun. Pure. Appl. Math. **47** (1994), 893-922
7. K. G. Wilson, J. Kogut: *The Renormalization Group and the ϵ-Expansion.* Phys. Rep. **12** (1974), 75-200
8. K. G. Wilson: *The Renormalization Group and Critical Phenomena.* Rev. Mod. Phys. **55** (1983), 583-600
9. A. N. Kolmogorov: *The Local Structure of Turbulence in Incompressible Viscous Fluid for Very Large Reynolds' Numbers.* C. R. Acad. Sci. URSS **30** (1941), 301–305
10. A. M. Obukhov: *Structure of the Temperature Field in a Turbulent Flow.* Izv. Akad. Nauk SSSR, Geogr. Geofiz. **13** (1949), 58-69
11. R. H. Kraichnan: *Anomalous Scaling of a Randomly Advected Passive Scalar.* Phys. Rev. Lett. **72** (1994), 1016-1019
12. V. S. L'vov, I. Procaccia, A. Fairhall: *Anomalous Scaling in Fluid Dynamics: the Case of Passive Scalar.* Phys. Rev. **E 50** (1994), 4684-4704
13. A. Majda: *Explicit Inertial Range Renormalization Theory in a Model for Turbulent Diffusion.* J. Stat. Phys. **73** (1993), 515-542
14. M. Chertkov, G. Falkovich, I. Kolokolov, V. Lebedev: *Normal and Anomalous Scaling of the Forth-Order Correlation Function of a Randomly Advected Passive Scalar* Weizmann Institute preprint, chao-dyn/95030001
15. A. L. Fairhall, O. Gat, V. L'vov, I. Procaccia: *Anomalous Scaling in a Model of Passive Scalar Advection: Exact Results.* Weizmann Institute preprint 1995
16. R. H. Kraichnan, V. Yakhot, S. Chen: *Scaling Relations for a Randomly Advected Passive Scalar Field*, preprint 1995, submitted to Phys. Rev. Lett.
17. L. F. Richardson: *Weather prediction by numerical process.* Cambridge University Press, Cambridge 1922
18. U. Frisch: *Turbulence: the legacy of A. N. Kolmogorov.* Cambridge University Press. Cambridge 1995, to appear
19. P. C. Martin, E. D. Siggia, H. A. Rose: *Statistical Dynamics of Classical Systems.* Phys. Rev. **A 8** (1973), 423-437
20. A. M. Polyakov: *The Theory of Turbulence in Two Dimensions.* Princeton University preprint, hep-th/9212145
21. V. I. Belinicher, V. S. L'vov: *A Scale Invariant Theory of Fully Developed Hydrodynamic Turbulence.* Sov. Phys. JETP **66** (1987), 303-313
22. V. S. L'vov: *Scale Invariant Theory of Fully Developed Hydrodynamic Turbulence - Hamiltonian Approach.* Phys. Rep. **207** (1991), 1-47
23. V. S. L'vov, I. Procaccia: *Exact Resummation of the Theory of Hydrodynamic Turbulence. I. The Ball of Locality and Normal Scaling.* Weizmann Institute preprint (1995), submitted to Phys. Rev. **E**
24. J. Glimm, D. H. Sharp: *A Random Field Model for Anomalous Diffusion in Heterogeneous Porous Media.* J. Stat. Phys. **62** (1991), 415-424
25. R. H. Kraichnan: *Small-Scale Structure of a Scalar Field Convected by Turbulence.* Phys. Fluids **11** (1968), 945-963
26. K. Gawędzki, A. Kupiainen: in preparation
27. C. E. Gutiŕerrez, G. S. Nelson: *Bounds for the Fundamental Solution of Degenerate Parabolic Equations.* Commun. Partial Diff. Eq. **13** (1988), 635-649

Introduction to Integrable Models of Statistical Physics

H. Grosse

Institut für Theoretische Physik, Universität Wien

1. Introduction

This year's Schladming school was devoted to various aspects of low–dimensional models of statistical physics and quantum field theory. It reflected the fact that various subjects became interrelated recently:

Classical integrable models, quantum spin models, vertex models, Yang–Baxter relation, braids and knots, quantum groups, Virasoro algebra representations, conformal field theory, Kac–Moody algebra representations, quantum Hall effect, anyons, fractional statistics, Chern–Simons model, topological field theory, strings, non–commutative geometry, Bose–Fermi correspondence, quantum wires, Luttinger liquid, ...

We shall concentrate on quantum spin models (chapter 2 and 4), which are quantum integrable in the sense of the Yang–Baxter relation (chapter 5).

In addition we sketch the solution of the Luttinger model (part 3), which describes electron transport along one–dimensional "quantum wires". Finally, in part 6, we indicate a few of the many algebraic relationships which follow from integrability. They lead to the subject of quantum geometry.

The subject started when Lenz asked his student Ising in 1925 to study the thermodynamics of a one–dimensional spin system. The teacher became disappointed, since no nontrivial phase transition was found. Heisenberg wrote down, in 1928, a more general quantum spin model, which was supposed to describe ferromagnetism. Already in 1938 Bethe found an efficient method to obtain eigenfunctions of this Hamiltonian. The big step forward was done by Onsager in 1944, when he calculated the free energy of the two–dimensional Ising model. Already in 1950 Tomonaga formulated and solved a field theoretic model with four fermion interaction. The statistical physics model with nonlocal interaction is solvable too, as it has been shown by Luttinger in 1963. The general algebraic structure behind integrable models has been worked out first in the classical case. Gardner, Green, Kruskal, and Miura invented the inverse scattering transform method in 1967. The quantum systems were treated by Yang (1967) and Baxter (1972). A number of groups worked out the second quantization of the inverse scattering method, and all recent developments resulted.

Our aim is to study models describing critical phenomena [1]. We consider as simple example the Ising model:

Assign to cubic lattice points $j = (j_1, \ldots, j_d) \to s_j \in \{1, -1\}$ a classical spin variable s_j. The set of $\{s_j\}$ is called a spin field configuration. Define the energy of such configurations by

$$H_N(\{s_j\}) = -\sum_{\langle ij \rangle}^{N} s_i s_j - h \sum_{j}^{N} s_j, \tag{1.1}$$

where $\langle ij \rangle$ denotes nearest neighbours lattice points, and h a magnetic field. At $T = 0$ we have two ground state configurations. For $d \geq 2$ and T small, two phases show up. The discrete up–down symmetry is spontaneously broken. The mean free energy becomes non–analytic in the thermodynamic limit. The free energy is defined through the partition function

$$Z_N(h, T) = \sum_{\text{conf.}} e^{-\beta H_N(\{s_j\})} = e^{-\beta F_N}, \tag{1.2}$$

and expectation values of observables are given by

$$\langle A \rangle_N = \frac{1}{Z_N} \sum_{\text{conf.}} e^{-\beta H_N} A. \tag{1.3}$$

Well–known examples of (1.3) are magnetization M_N and susceptibility χ_T^N:

$$M_N = \left\langle \sum_{i=1}^{N} s_i \right\rangle_N = \frac{1}{\beta} \frac{\partial}{\partial h} \ln Z_N, \qquad \chi_T^N = \frac{\partial M_N}{\partial T}, \tag{1.4}$$

as well as the inner energy U_N and the specific heat C_h^N:

$$U_N = \langle H_N \rangle_N, \qquad C_h^N = \frac{\partial U_N}{\partial T}. \tag{1.5}$$

The connection to thermodynamics is given by $F_N = U_N - T \cdot S_N$, where S_N denotes the entropy.

The thermodynamic limit of these quantities,

$$c_h = \lim_{N \to \infty} \frac{C_h^N}{N}, \qquad m = \lim_{N \to \infty} \frac{M_N}{N}, \qquad \text{etc.}, \tag{1.6}$$

shows a typical power law behaviour in the reduced temperature $\tau = (T - T_c)/T_c$ if we approach the critical point T_c:

$$c_h \simeq |\tau|^{-\alpha}, \qquad m_S \simeq |\tau|^{\beta}, \qquad \chi_T \simeq |\tau|^{-\gamma}, \tag{1.7}$$

where we introduced the spontaneous magnetization $m_S = \lim_{h \searrow 0} m$, and defined critical exponents α, β, γ. m_S is called an order parameter, since it allows to distinguish between the low temperature ordered phases, where it is not vanishing, and the high temperature phase, where it is zero.

A more general class of models is given by

$$Z_N = \int \prod_{j=1}^{N} (d\mathbf{s}_j \; e^{-V(\mathbf{s}_j)}) e^{-\beta H(\{\mathbf{s}_j\})}; \tag{1.8}$$

they have either a discrete or a continuous symmetry.

Two more exponents (ν, η) are defined through cluster properties:

$$\langle s_i s_j \rangle^c \equiv \langle s_i s_j \rangle - \langle s_i \rangle \langle s_j \rangle \stackrel{|i-j| \to \infty}{\longrightarrow} \begin{cases} e^{-|i-j|/\xi(T)}, & T \neq T_c \\ \dfrac{\text{const}}{|\mathrm{i}-\mathrm{j}|^{\mathrm{d}-2+\eta}}, & T = T_c \end{cases}, \tag{1.9}$$

where $\xi(T)$ denotes the correlation length, which diverges at $T = T_c$:

$$\xi(T) \stackrel{|\tau| \searrow 0}{\simeq} |\tau|^{-\nu}.$$

We quote examples of critical exponents:

	Mean Field	$d = 2$ Ising	$d = 3$ Ising
α	0	0	0.11
2β	1	1/4	0.65
γ	1	7/4	1.24
η	0	1/4	0.03

There exist inequalities and relations among them. At $T = T_c$ we get long range order.

Such models describe gas–liquid transition, ferromagnetism, superconductivity, suprafluidity, structural phase transitions, and many more.

Quantum spin 1/2 models can be defined by assigning σ–matrices to each lattice point $j \to (\sigma_j^x, \sigma_j^y, \sigma_j^z)$, with $\sigma_j^\alpha = \mathbf{1} \otimes \ldots \otimes \sigma^\alpha \otimes \mathbf{1} \otimes \ldots \otimes \mathbf{1}$ in $\bigotimes_{j=1}^{N} (\mathbf{C}^2)$, and the σ^α are at the j–th place. The Hamiltonian for the XYZ model is given by

$$H = -\sum_{\langle ij \rangle} (J_x \sigma_i^x \sigma_j^x + J_y \sigma_i^y \sigma_j^y + J_z \sigma_i^z \sigma_j^z) + \mathbf{h} \sum_{j=1}^{N} \sigma_j. \tag{1.10}$$

The partition function and expectation values are determined through a trace:

$$Z_N = \text{Tr}\; e^{-\beta H}, \qquad \langle A \rangle_N = \frac{1}{Z_N} \text{Tr}\; e^{-\beta H} A. \tag{1.11}$$

Some general results are:

In $d = 1$ one cannot break spontaneously a discrete symmetry, as long as short range interactions are considered; this follows from an energy–entropy argument.

The Ising model in $d \geq 2$ has a phase transition at finite temperature; this follows from Peierls contour estimates.

The Mermin–Wagner–Hohenberg theorem states: A continuous symmetry cannot be broken spontaneously in $d = 1$ and $d = 2$; there exist thick Bloch walls, and $m_S = 0$.

Due to a Goldstone mode, there exist nontrivial phase transitions in $d \geq 3$.

Quantum spin models may have long range order at $T = 0$ or at nonzero temperature. Three possibilities may occur: there is no gap; there exists a gap, and the isolated ground state is non–degenerate or degenerate. We know examples to all three cases.

2. Quantum Spin Models

The deep connection between models of statistical physics and quantum field theory is also reflected in very simple relationships. There is, for example, the (Klein)–Jordan–Wigner transformation, which allows to map the spin algebra for N–spins to the algebra of the canonical anticommutation relations. We compare

$$[S_n^\alpha, S_m^\beta] = 0, \qquad n \neq m; \qquad [S_n^\alpha, S_n^\beta] = i\varepsilon^{\alpha\beta\gamma} S_n^\gamma, \tag{2.1}$$

which is a partly commutative algebra, with $\{c_n, c_m\} = 0$, and $\{c_n, c_n^\dagger\} = 1, \ldots$, the CAR algebra.

For $N = 1$ it is easy to identify this mapping, since

$$S^+|\downarrow\rangle = |\uparrow\rangle, \qquad S^+|\uparrow\rangle = 0, \qquad \text{and} \qquad c^+|0\rangle = |1\rangle, \qquad c^+|1\rangle = 0, \ldots .$$

Both algebras can be realized in $\mathbf{C}^2$.

As for the general case we need a **disorder**– or **kink**–operator,

$$K_n = \exp\left(i\pi \sum_{j=1}^{n-1} S_j^+ S_j^- \right), \tag{2.2}$$

which rotates spins at lattice places $j = 1, 2, \ldots, n-1$, around the z-axis; more precisely, let $\sigma^1|\pm\rangle = \pm|\pm\rangle$, and denote the product state by $|+, +, \ldots, +\rangle$ etc.. Then

$$K_n|+, \ldots, +\rangle = |-, -, -, \ldots, -, \overbrace{+,}^{n\text{–th place}} \ldots, +\rangle. \tag{2.3}$$

Define next

$$c_n = K_n S_n^-, \qquad c_n^+ = K_n^+ S_n^+.$$

It is tedious, but simple, to prove that the two algebras are mapped into each other. This transformation can be used to map spin models into fermion models. For example, the XX–model, with Hamiltonian

$$H^{xx} = \sum_{n=1}^{N} (S_n^x S_{n+1}^x + S_n^y S_{n+1}^y) \tag{2.4}$$

and suitable boundary conditions, is mapped into the free fermion Hamiltonian

$$H^{xx} = \frac{1}{2}\sum_{n}^{N}(c_n^\dagger c_{n+1} + h.c.) = \int_{-\pi}^{\pi} dk\ \varepsilon_k c_k^\dagger c_k. \tag{2.5}$$

Through Fourier transformation the Hamiltonian is diagonalized, and the dispersion relation $\varepsilon_k = \cos k$ results. Half–filling corresponds to filling all negative energy levels (Dirac sea). Occupation number distribution is given by the step function: one for energies below the Fermi energy, and zero otherwise. Small interactions give small deviations.

The system is gapless and critical; no massive excitations exist. Correlation functions have power law fall off and no long range order, in the sense that

$$\lim_{|m-n|\to\infty} \langle s_n^+ s_m^- \rangle \to 0$$

occurs.

For a study of the continuum model one identifies the even–site operators as the first components $c_{2n} \propto \psi_{1,n}$, and the odd–site operators as the second components $c_{2n+1} \propto \psi_{2,n}$ of a spinor field. The Hamiltonian

$$H_0 = \int dx : \psi^\dagger \alpha \frac{1}{i}\frac{\partial}{\partial x}\psi :$$

results.

The same method can be applied to the XY–model [2]:

$$H_\gamma = \sum_{n=1}^{N}\{(1+\gamma)S_n^x S_{n+1}^x + (1-\gamma)S_n^y S_{n+1}^y\}, \tag{2.6}$$

which interpolates between the isotropic ($\gamma = 0$) XX–model and the ($\gamma = 1$) Ising model. For $\gamma \to 0$, the x–component of the spin is completely ordered, while the y–component is completely disordered.

H_γ is transformed into

$$H_\gamma = \frac{1}{2}\sum_{n}^{N}\{c_n^\dagger c_{n+1} + \gamma c_n^\dagger c_{n+1}^\dagger + h.c.\} \qquad (+\text{ boundary conditions}). \tag{2.7}$$

To diagonalize H_γ, a Bogoliubov transformation of the type

$$\begin{aligned} D_k &= A_{km}c_m + B_{km}c_m^\dagger, \\ D_k^\dagger &= B_{km}c_m + A_{km}c_m^\dagger, \end{aligned} \tag{2.8}$$

may be applied. H_γ becomes

$$H_\gamma = \sum_k E_k^\gamma D_k^\dagger D_k, \qquad E_k^\gamma = \sqrt{1-(1-\gamma^2)\sin^2 k}. \tag{2.9}$$

$\gamma = 0$ gives the previous result. $\gamma = 1$ yields $E_k^1 = 1$ for $|k| \leq k_F$, and $E_k^1 = 0$ for $|k| > k_F$. For all $\gamma \neq 0$ a gap exists.

In the continuum limit the Hamiltonian of the massive Dirac operator results.

We discuss next the, more complicated, XXX antiferromagnetic Heisenberg chain.

$$H = \sum_{n=1}^{N} \mathbf{S}_n \cdot \mathbf{S}_{n+1} \tag{2.10}$$

with periodic boundary condition. The total spin operator $\mathbf{S} = \sum_{n=1}^{N} \mathbf{S}_n$, as well as the shift operator (on the ring) Π, are conserved. The ferromagnetic ground state $|\Omega\rangle = |\uparrow, \uparrow, \ldots, \uparrow\rangle$ is easy to guess. A basis for all states is obtained from $|\Omega\rangle$ by acting with S_n^- operators: $S_n^-|\Omega\rangle, \ldots, S_{n_1}^- \ldots S_{n_M}^-|\Omega\rangle$. It is tempting to diagonalize $\mathbf{S}$ and Π. Spin wave or magnon eigenstates of H are obtained of the form

$$|k\rangle = \sum_{n=1}^{N} e^{ikn} S_n^-|\Omega\rangle, \qquad k = \frac{2\pi}{N}p, \qquad p \in \mathbf{Z}. \tag{2.11}$$

An ansatz for two–magnon scattering [3] is

$$|k_1 k_2\rangle = \sum_{n_1 < n_2} \psi_{n_1 n_2}^{k_1 k_2} S_{n_1}^- S_{n_2}^- |\Omega\rangle \tag{2.12}$$

with

$$\psi_{n_1 n_2}^{k_1 k_2} = \{e^{i\Theta_{k_1 k_2}/2} e^{i(k_1 n_1 + k_2 n_2)} + k_1 \leftrightarrow k_2\}.$$

Nontrivial scattering of magnons occurs. $|k_1 k_2\rangle$ is an eigensolution, if the phase shift solves the transcendental equation

$$2 \cot \frac{\Theta_{k_1 k_2}}{2} = \cot \frac{k_1}{2} - \cot \frac{k_2}{2}. \tag{2.13}$$

Repulsion occurs, and $\psi_{n_1 n_2}^{k_1 k_2} = 0$.

For M magnons

$$|k_1, \ldots, k_M\rangle = \sum_{n_1 < \ldots < n_M} \psi_{\underline{n}}^{\underline{k}} S_{n_1}^- \ldots S_{n_M}^- |\Omega\rangle, \tag{2.14}$$

the ansatz

$$\psi_{\underline{n}}^{\underline{k}} = \sum_{\text{perm } \pi} \exp[i \sum_{\alpha} k_{\pi_\alpha} n_\alpha + \frac{i}{2} \sum_{\alpha < \beta} \Theta_{k_{\pi_\alpha} k_{\pi_\beta}}] \tag{2.15}$$

leads to eigensolutions. The M–magnon S–matrix is factorized into a product of two–"particle" S–matrices. $\psi_{\underline{n}}^{\underline{k}}$ yields a solution, if $\Theta_{k_1 k_2}$ solves again the before–mentioned transcendental equation. But, in addition, the so–called Bethe equations have to be fulfilled:

$$k_\alpha N = \sum_{\beta \neq \alpha} \Theta_{k_\alpha k_\beta} + 2\pi I_\alpha, \qquad I_\alpha \in \mathbf{Z}. \tag{2.16}$$

Exponentiating the last relation yields

$$e^{ik_\alpha N} = \prod_{\beta \neq \alpha} S_{k_\alpha k_\beta} \tag{2.17}$$

which has an interesting interpretation, connected to periodicity. One mode with quasimomentum k_α scatters on all others while returning back.

3. Luttinger Liquid

In order to solve (the simplest) model with a quartic interaction, we should first briefly discuss the bosonization formulae [4]. The simplest situation concerns massless chiral fermions on the circle S^1. We study a representation of the CAR on the Fock space $\mathcal{F}(h)$, with $h = L^2(S^1)$: Let $\{\psi_k, \psi_q^\dagger\} = \delta_{k,q}$, with $k, q \in \{\frac{2\pi}{L} n | n \in \mathbf{Z}\}$ $(2\pi = l)$. Second quantization maps operators

$$h \xrightarrow{A} h \qquad \text{to} \qquad Q(A) = \sum_{k,q} : \psi_k^\dagger A_{kq} \psi_q : . \tag{3.1}$$

Normal ordering is understood with respect to the vacuum Ω, and

$$\psi_k \Omega = 0, \quad k > 0, \qquad \text{but} \quad \psi_{-k}^\dagger \Omega = 0, \quad k \leq 0. \tag{3.2}$$

The Hamiltonian

$$H = \int_{S^1} dx : \psi^\dagger \frac{1}{i} \frac{\partial}{\partial x} \psi := \sum_k : \psi_k^\dagger \psi_k : \tag{3.3}$$

and the current components

$$J_k = \int dx\, e^{-ikx} : \psi^\dagger \psi := \sum_q : \psi_{q-k}^\dagger \psi_q : \tag{3.4}$$

are standard observables; $J_0 = Q$ denotes the charge operator. From the definition of Ω we deduce that $J_k \Omega = 0$, $\forall\, k \geq 0$. Normal ordering implies the

Theorem:.

$$[J_k, J_q] = k \cdot \delta_{k+q} \, ! \tag{3.5}$$

The r.h.s. represents the FT of the $\delta' - U(1)$ Schwinger term. (Similar cocycles occur for Kac–Moody and the Virasoro algebra.) (3.5) allows to map from fermions to bosons:

$$J_{-q}/\sqrt{q} = a_q^\dagger, \qquad J_q/\sqrt{q} = a_q, \quad q > 0 \Longrightarrow [a_q, a_k^\dagger] = \delta_{q,k}. \tag{3.6}$$

The full boson $\leftrightarrow$ fermion correspondence will be discussed next. We first construct a dense set of vectors of $\mathcal{F}(h)$. There exists a unitary operator R, such that $R\psi_k R^\dagger = \psi_{k+1}$, or $R\psi(x)R^\dagger = e^{ix}\psi(x)$. R induces a large gauge transformation. From the definition, the transformation properties of J_k and H result: $RJ_kR^\dagger = J_k - \delta_{k,0}$, $RHR^\dagger = H - Q$. We define now subsets

$$\mathrm{d}^{(n)} = \text{linear span } \{J_{-k_1} \ldots J_{-k_m} R^n \Omega\}; \tag{3.7}$$

a technical proof shows that $\overline{\bigoplus_n \mathrm{d}^{(n)}} = \mathcal{F}(h)$, and therefore with $a\Omega = 0$ and $[R, a] = [J_k, a] = 0 \Rightarrow a = 0$. We note that

$$RHR^\dagger - H = -Q, \qquad [J_k, H] = kJ_k. \tag{3.8}$$

In addition, with $T = \sum_q J_{-q}J_q$ we obtain almost similar relations:

$$RTR^\dagger - T = 0, \qquad [J_k, T] = kJ_k. \tag{3.9}$$

Since, with $t = \frac{1}{2}(Q^2 + Q) \Rightarrow RtR^\dagger - t = Q$, $T\Omega = t\Omega = 0$ and $[J_k, t] = 0$, we obtain the (old) Kronig identity, which is nowadays part of the Sugawara relation [4]:

$$H = \frac{1}{2}(Q^2 + Q) + \sum_{k>0} k a_k^\dagger a_k. \tag{3.10}$$

The full reconstruction of the spinor field $\psi_\varepsilon = e^{iQx} R V_+ V_-$ is given in terms of vertex operators

$$\begin{aligned} V_+ &= \exp[\sum_{q>0} \frac{J_{-q}}{q} e^{iqx} e^{-\varepsilon x}], \\ V_- &= \exp[\sum_{q>0} \frac{J_q}{q} e^{-iqx} e^{-\varepsilon x}]. \end{aligned} \tag{3.11}$$

The check, that the CAR is fulfilled, is tedious.

The Luttinger model is now given by

$$H = H_0 + V \qquad \text{with} \qquad H_0 = \int_{S^1} dx : \psi^\dagger \alpha \frac{1}{i} \frac{\partial}{\partial x} \psi :, \tag{3.12}$$

where ψ denotes a two–component spinor. Since we have a linear dispersion law, filling the negative energy states means, that Ω is defined by

$$\psi_{1,k}\Omega = \psi^\dagger_{2,k}\Omega = 0, \quad k>0 \quad \text{and} \quad \psi^\dagger_{1,k}\Omega = \psi_{2,k}\Omega = 0, \quad k \le 0. \tag{3.13}$$

We use Kronig's relation for both chiralities and note that one zero energy state is filled:

$$H_0 = \frac{1}{2}(Q_1^2 + Q_1) + \sum_{k>0} k a^\dagger_k a_k + \frac{1}{2}(Q_2^2 - Q_2) + \sum_{k>0} k b^\dagger_k b_k. \tag{3.14}$$

The density–density interaction is defined by

$$V = \int_{S^1} dx \int_{S^1} dy : \psi_1^\dagger\psi_1(x)V(x-y)\psi_2^\dagger\psi_2(y) := \sum_k \widetilde{V}_k(a^\dagger_k b^\dagger_k + a_k b_k). \tag{3.15}$$

There exists again a Bogoliubov transformation, which allows to diagonalize the total Hamiltonian.

$$\begin{aligned} a_k &= \cosh\lambda_k A_k - \sinh\lambda_k B^\dagger_k, \\ b^\dagger_k &= -\sinh\lambda_k A_k + \cosh\lambda_k B^\dagger_k. \end{aligned} \tag{3.16}$$

It is a unitary transformation, generated from $U_k = \exp(\lambda_k(a_k b_k - a^\dagger_k b^\dagger_k))$, and mapping a_k, b_k to $A_k = U_k a_k U^\dagger_k$, $B_k = U_k b_k U^\dagger_k$; diagonalization of H is obtained if $\widetilde{V}_k = \tanh 2\lambda_k$, and

$$U(H_0+V)U^\dagger = \sum_{k>0}\{\varepsilon_k(A^\dagger_k A_k + B^\dagger_k B_k) - \varepsilon^0_k\}, \tag{3.17}$$

$$\varepsilon_k = k\sqrt{1-\widetilde{V}_k^2}, \qquad \varepsilon^0_k = k(1-\sqrt{1-\widetilde{V}_k^2}).$$

The Bogoliubov transformation mixes 1– and 2–modes, and

$$U\psi_1 U^\dagger = e^{iQx} R_1 V_1^+ V_1^- V_2^+ V_2^- \tag{3.18}$$

results. With the help of the algebra of vertex operators,

$$V_1^-(u)V_1^+(v) = \exp\left[\sum_{q>0}\frac{1}{q}\left(\frac{v}{u}\right)^q\right] V_1^+(v)V_1^-(u), \tag{3.19}$$

it is easy to obtain the correlation functions of the model. The two point function

$$\langle\Omega|\psi_1^\dagger(x)\psi_1(0)|\Omega\rangle = \langle\Omega_0|\psi_1^\dagger(x)\psi_1(0)|\Omega_0\rangle \exp\left[\sum_{q>0}\frac{\sinh\lambda_q}{q}(e^{ixq}+e^{-ixq}-2)\right] \tag{3.20}$$

is proportional to the free Greens function. The additional factor implies that

$$G(x)/G_0(x) \overset{|x|\to\infty}{\simeq} 1/|x|^\alpha$$

with $\alpha = \lambda^2/4\pi^2 v^2(0)$, and the fermi surface becomes modified.

4. From Integrable Models to Quantum Spin Models

We are very much familiar with classical (finite dimensional) integrable models. Consider a system with $2N$–dimensional phase space and local coordinates (q_i, p_i). Let $\omega = \sum_i dq_i \wedge dp_i$ be the canonical two–form, and $\{f, g\} = \omega(X_f, X_g)$ be the Poisson bracket of functions f and g. Equs. of motion are then given by $\dot{q}_i = \{q_i, H\}$, $\dot{p}_i = \{p_i, H\}$. The Liouville–Arnold theorem states:

If there exist N globally conserved quantities K_i, which are in involution, $\{K_i, K_j\} = 0$, then there exists a map from (q_i, p_i) to new coordinates (φ_i, I_i), which are cyclic action angle variables: $\varphi_i(t) = \omega_i t + \varphi_i(0)$, $I_i(t) = I_i(0)$.

For ∞ dimensional systems, a Lax pair or a zero curvature condition implies integrability [3]. Examples are the KdV or the nonlinear Schrödinger equation, the Sine–Gordon equation, and many more. For the nonlinear Schrödinger equation, for example, L is given by

$$\left[i\frac{\partial}{\partial t} + M, i\frac{\partial}{\partial x} + L\right] = 0 \qquad \text{with} \qquad L = \begin{pmatrix} \lambda & \psi^\dagger \\ \psi & -\lambda \end{pmatrix}. \tag{4.1}$$

There exists a matrix M such that the consistency implies the nonlinear equation $i\frac{\partial}{\partial t}\psi = -\psi_{xx} + (\psi^\dagger \psi)\psi$. One goes over to scattering data of the L–operator. They have a free time evolution. The inverse problem allows to go back to the original field variables at time t. We remark that the monodromy matrix $g(x) = P\exp\int_{-L}^{x} L$ obeys quadratic Poisson brackets of the type: $\{g^1, g^2\} = [r_{12}, g^1 g^2]$, where

$$g^1 = g \otimes \mathbf{1}, \qquad g^2 = \mathbf{1} \otimes g, \qquad r_{12} = \frac{1}{\lambda} P_{12}, \qquad P = \left(\begin{array}{cc|cc} 1 & & & \\ & 0 & 1 & \\ \hline & 1 & 0 & \\ & & & 1 \end{array}\right), \tag{4.2}$$

with P being the permutation matrix. The r–matrix fulfills the classical Yang–Baxter relation:

$$[r_{12}, r_{23}] + [r_{23}, r_{31}] + [r_{31}, r_{12}] = 0. \tag{4.3}$$

Returning to Quantum Spin Models, we solve the one–dimensional Ising model by going to a QSM at **one** point:

$$Z_N = \sum_{S_1,\dots,S_N} e^{\beta S_1 S_2} \dots e^{\beta S_N S_1} = \operatorname{Tr} T^N, \qquad T = \begin{pmatrix} e^{\beta} & e^{-\beta} \\ e^{-\beta} & e^{\beta} \end{pmatrix},$$

$$e^{-\beta F_N} = \lambda_1^N \left(1 + \left(\frac{\lambda_2}{\lambda_1}\right)^N\right), \qquad f = \lim_{N\to\infty} \frac{F_N}{N} = -\frac{1}{\beta} \ln \lambda_1, \tag{4.4}$$

where λ_1 denotes the larger eigenvalue. This result holds in general. (T_c is here zero.) T denotes the transfer matrix. The two–dimensional analogue connects vertex models to quantum spin models like the XXZ (or XYZ) model, to which we turn next.

There exists a class of ice–like models. Ice forms a lattice of H_2O molecules. A simplified $d = 2$ model is obtained by placing O–atoms at sites of a quadratic lattice and H–atoms, between them. Assume two possible positions for the H–atoms, being close to a site atom or far away. Surround each O–atom by exactly two H–atoms. This allows for six vertices to which we assign weights ω_i, $i = 1, \ldots, 6$. Assuming a symmetry under reflections one obtains three independent constants $\omega_1 = \omega_2 = a$, $\omega_3 = \omega_4 = b$, $\omega_5 = \omega_6 = c$. The partition function is given by

$$Z = \sum_{\text{conf } \{\omega_j\}} \prod_{n,m} \omega_j(n,m). \tag{4.5}$$

The model becomes critical, if $|a^2 + b^2 - c^2| \leq 2|ab|$. Correlation functions decay then algebraically. Adding a vertex with a source and one with a sink yields the eight vertex model. For the six vertex model we define a matrix

$$L^{\alpha_1 \alpha_2}_{\beta_1 \beta_2'} = \left(\begin{array}{cc|cc} a & 0 & 0 & 0 \\ 0 & b & c & 0 \\ \hline 0 & c & b & 0 \\ 0 & 0 & 0 & a \end{array} \right)_{01} \qquad \text{in } \mathbf{C}^2_0 \otimes \mathbf{C}^2_1, \tag{4.6}$$

describing one vertex. N vertices of a row are encoded by the monodromy matrix

$$M_N \equiv L_{0N} \cdot \ldots \cdot L_{02} \cdot L_{01} \qquad \text{in } \mathbf{C}^2_0 \otimes \mathbf{C}^2_1 \otimes \ldots \otimes \mathbf{C}^2_N. \tag{4.7}$$

With periodic boundary conditions, $\text{tr}_0\, M_N = T_N$, the transfer matrix, describing the interaction from row to row, is obtained. Finally $Z = \text{Tr}\, T_N^M$ for an $N \times M$ lattice. Tr denotes the trace in $\bigotimes_{j=1}^N \mathbf{C}^2$.

As a suitable parametrization we choose $a = \sin(\gamma - \lambda)$, $b = \sin\lambda$, $c = \sin\gamma$, where λ is called the spectral parameter.

We introduce the XXZ–Hamiltonian

$$H_{XXZ} = \frac{1}{2} \sum_{j=1}^{N} (\sigma_j^x \sigma_{j+1}^x + \sigma_j^y \sigma_{j+1}^y + \Delta(\sigma_j^z \sigma_{j+1}^z + 1)), \tag{4.8}$$

and claim that

$$H_{XXZ} = \sin\gamma \frac{d}{d\lambda} \ln T \bigg|_{\lambda=0} \qquad \text{with } \Delta = -\cos\gamma. \tag{4.9}$$

To show (4.9) we first note that $L_{0n}(0) = \sin\gamma P_{0n}$, where P_{0n} denotes the permutation operator for vector spaces $\mathbf{C}^2_0$ and $\mathbf{C}^2_n$. Next

$$(\sin\gamma)^{-N} T(0) = \text{tr}_0\, P_{0N} \ldots P_{01} = \text{tr}_0\, P_{12} P_{23} \ldots P_{N-1,N} P_{N0} = U,$$

where U denotes the shift operator. Let $(\partial L_{0n}/\partial\lambda)|_{\lambda=0} \equiv \bar{L}_{0n}$. We calculate the logarithmic derivative of the transfer matrix at $\lambda = 0$ and obtain

$$\sin\gamma \frac{d}{d\lambda} \ln T\Big|_{\lambda=0} = U^{-1}\sum_{j=1}^{N} \mathrm{tr}_0\,(P_{01}\dots P_{0j-1}\bar{L}_{0j}P_{0j+1}\dots P_{0N})$$

$$= \sum_{j=1}^{N} P_{j-1j}\bar{L}_{j-1j} \tag{4.10}$$

$$= \sum_{j=1}^{N} \left(\begin{array}{cc|cc} -\cos\gamma & 0 & 0 & 0 \\ 0 & 0 & 1 & 0 \\ \hline 0 & 1 & 0 & 0 \\ 0 & 0 & 0 & -\cos\gamma \end{array}\right)_{j-1j} . \tag{4.11}$$

The indices indicate the vector space in which the matrices act.

5. Quantum Integrability — Yang–Baxter Relation

We take γ fixed and consider $L_{0n}(\lambda)$ as a function of λ. We take the tensor product of two auxiliary spaces and consider $L_{0n}(\lambda) \otimes L_{\bar{0}n}(\mu)$ acting in $\mathbf{C}_0^2 \otimes \mathbf{C}_{\bar{0}}^2 \otimes (\bigotimes_{j=1}^{N} \mathbf{C}^2)$. It is remarkable that there exits an operator $R \in \mathrm{End}\,(\mathbf{C}_0^2 \otimes \mathbf{C}_{\bar{0}}^2)$ such that the Yang–Baxter relation [5]

$$R_{0\bar{0}}(\lambda-\mu)L_{0n}(\lambda)L_{\bar{0}n}(\mu) = L_{0n}(\mu)L_{\bar{0}n}(\lambda)R_{0\bar{0}}(\lambda-\mu) \tag{5.1}$$

holds. The solution for the quantum R–matrix is given in our case by $R = PL$, where P denotes the permutation matrix. If we introduce $L = PR$ into (5.1), push all P operators on one side, and change the notation of indices, we obtain

$$R_{23}(\lambda-\mu)R_{12}(\lambda)R_{23}(\mu) = R_{12}(\mu)R_{23}(\lambda)R_{12}(\lambda-\mu) \tag{5.2}$$

in $\mathbf{C}_1^2 \otimes \mathbf{C}_2^2 \otimes \mathbf{C}_3^2$. We remind the reader that R_{23} means really $\mathbf{1}_1 \otimes R_{23}$ etc.. (5.1) and (5.2) are identical to the S–matrix factorization conditions used by many authors.

There exist rational, trigonometric, elliptic, and hyperelliptic solutions to (5.2). If we expand R around the unit operator: $R(\lambda) \simeq \mathbf{1} + i\hbar r(\lambda)$, $r(\lambda)$ obeys the classical Yang–Baxter relation. (5.2) is closely related to the braid relation, which will be explained later.

A number of consequences follow. Since L_{0n} and $L_{\bar{0}m}$ with $n \neq m$ commute, it follows from (5.1) that

$$L(\lambda-\mu)M(\lambda)M(\mu) = M(\mu)M(\lambda)L(\lambda-\mu), \tag{5.3}$$

and taking the trace in $\mathbf{C}_0^2 \otimes \mathbf{C}_{\bar{0}}^2$ we conclude that

$$[T(\lambda), T(\mu)] = 0. \tag{5.4}$$

We have therefore obtained an infinite number of conserved quantities for H_{XXZ}, a signal for integrability. The first nontrivial conserved quantity for H_{XXX} is given, for example, by

$$F = \sum_{\ell} \boldsymbol{\sigma}_\ell \cdot (\boldsymbol{\sigma}_{\ell+1} \times \boldsymbol{\sigma}_{\ell+2}), \tag{5.5}$$

which can be used to explain the degeneracies in the spectrum.

Relations (5.3) contain much more information: Let us introduce $2^N \times 2^N$ matrices $A_\lambda, B_\lambda, C_\lambda, D_\lambda$ as the elements of

$$M(\lambda) = \begin{pmatrix} A_\lambda & B_\lambda \\ C_\lambda & D_\lambda \end{pmatrix}. \tag{5.6}$$

Remember, that we intend to diagonalize the transfer matrix, which is now expressed in terms of the new quantities as $T(\lambda) = A_\lambda + D_\lambda$. Equations (5.3) give 16 relations among these quantities. Let us quote one of them:

$$A_\lambda B_\mu = \frac{a_{\mu-\lambda}}{b_{\mu-\lambda}} B_\mu A_\lambda - \frac{c_{\mu-\lambda}}{b_{\mu-\lambda}} B_\lambda A_\mu, \ldots . \tag{5.7}$$

The strategy is now to choose (or to guess) a kind of vacuum (or highest weight vector) state. In our case it will be the ferromagnetic state. It is automatically an eigenstate of $T(\lambda)$:

$$T(\lambda) |\uparrow \ldots \uparrow\rangle = (a_\lambda^N + b_\lambda^N) |\uparrow \ldots \uparrow\rangle. \tag{5.8}$$

Next we define vectors by acting with products of $B(\lambda)$'s on the chosen reference vector: $|\lambda_1 \ldots \lambda_M\rangle = B_{\lambda_1} \ldots B_{\lambda_M} |\uparrow \ldots \uparrow\rangle$. We require that these states become eigenstates of $T(\lambda)$. This would be the case, if (5.7) contains no second term on the right hand side. The eigenvalue becomes then

$$\varepsilon_{(\lambda_1 \ldots \lambda_M)} = a_\lambda^N \prod_{i=1}^{M} \frac{a_{\lambda_i - \lambda}}{b_{\lambda_i - \lambda}} + b_\lambda^N \prod_{i=1}^{M} \frac{a_{\lambda - \lambda_i}}{b_{\lambda - \lambda_i}}. \tag{5.9}$$

Next one requires that the additional terms vanish. They cancel, if the transcendental Bethe equations

$$e^{ip_\ell N} \equiv \left(\frac{a_{\lambda_\ell}}{b_{\lambda_\ell}}\right)^N = \prod_{j \neq \ell}^{M} \frac{a_{\lambda_\ell - \lambda_j} b_{\lambda_j - \lambda_\ell}}{b_{\lambda_\ell - \lambda_j} a_{\lambda_j - \lambda_\ell}} \equiv \prod_{j \neq \ell}^{M} e^{i\phi_{\lambda_j - \lambda_\ell}}, \qquad \ell = 1, \ldots, M, \tag{5.10}$$

are fulfilled. Solutions to (5.10) can be obtained for small as well as large N. We note that $b_{\lambda_\ell - \lambda}$ vanishes at $\lambda = \lambda_\ell$, but the transfer matrix should be regular. Vanishing of the residuum at these poles explains (5.10), which rewritten is identical to Bethe's equations:

$$N p_\ell = 2\pi I_\ell + \sum_{j \neq \ell} \phi_{\lambda_j - \lambda_\ell}, \qquad I_\ell \in \mathbf{Z}. \tag{5.11}$$

Many models can be treated using the above scheme or a generalization of it (RSOS, XYZ, Hubbard, ...).

6. Algebraic Relations

A knot is a closed line embedded into $\mathbf{R}^3$ without crossings. A link is a set of knotted knots. A complete classification of links has not been achieved. In order to define braids [6], we choose n points in $\mathbf{R}^2$ and consider mappings γ_i from $[0,1] \to \mathbf{R}^2 \times [0,1] \subset \mathbf{R}^3$, $i = 1,\dots,n$, with the properties that $\dot{\gamma}_i(t) > 0$ for $t \in [0,1]$, $\gamma_i(t) \neq \gamma_j(s)$ for $i \neq j$ and all $s,t \in [0,1]$, and $\gamma_i(0) = x_i$, $\gamma_i(1) = x_{\pi(i)}$, where π denotes a permutation from the symmetric group of n elements S_n. We identify all such mappings which can be obtained through continuous deformations. The resulting object forms a braid of n strings $\in B_n$. Algebraically the group B_n consists of words generated from $\{e, b_i, b_n^{-1} | i = 1,\dots,n-1\}$. b_j exchanges the j–th and the $(j+1)$–st string, so that one string lies above the other. b_j^{-1} puts them in the opposite way. There are three types of Reidemeister moves, $e = b_i b_i^{-1} = b_i^{-1} b_i$, $b_i b_j = b_j b_i$ for $|i-j| \geq 2$, and

$$b_i b_{i+1} b_i = b_{i+1} b_i b_{i+1}. \tag{6.1}$$

B_n is defined by factorizing with respect to these relations. (6.1) is called the braiding relation and related to (5.2). If we identify upper and lower ends of a braid we obtain a link. Many braids lead to the same link. The Markov theorem identifies invariant functions on links.

If we require that b_j's have only two eigenvalues $+1$ and $-q^2 \in \mathbf{C}$, we impose the quadratic conditions $(g_i - 1)(g_i + q^2) = 0$. Together with $g_i g_j = g_j g_i$, $|i-j| \geq 2$, and $g_i g_{i+1} g_i = g_{i+1} g_i g_{i+1}$, the Hecke algebra is defined. One representation is obtained from the R–matrix through gauging and taking the limit $\lambda \to -i\infty$,

$$g_i = \left(\begin{array}{cc|cc} 1 & 0 & 0 & 0 \\ 0 & 0 & q & 0 \\ \hline 0 & q & 1-q^2 & 0 \\ 0 & 0 & 1 & 1 \end{array}\right)_{ii+1}, \qquad -e^{-i\gamma} = q, \tag{6.2}$$

acting in $\bigotimes_{j=1}^{N} \mathbf{C}^2$. If we transform to e_i, putting $g_i = -(1+q^2)e_i + 1$, we obtain $e_i^2 = e_i = e_i^\dagger$ and

$$e_1 e_2 e_1 - \frac{1}{(q+q^{-1})^2} e_1 = e_2 e_1 e_2 - \frac{1}{(q+q^{-1})^2} e_2. \tag{6.3}$$

If both sides of (6.3) are put to zero, the Temperley–Lieb–Jones algebra is obtained.

From the algebraic structure behind the Yang–Baxter relation the notion of quantum groups has been deduced. We explain first the notions of coproduct, counit and coinverse within two examples:

Let G be a group, and $\mathcal{A} = C(G)$ be the abelian algebra of functions over G. Let $(m_{k\ell})$ be matrix elements of the N–dimensional defining representation. The product within G implies for $\mathcal{A}$ an algebra homomorphism to $\mathcal{A} \otimes \mathcal{A}$, called the coproduct Δ:

$$(\Delta f)(g_1, g_2) = f(g_1 \cdot g_2), \qquad \Delta(m_{k\ell}) = \sum_{j=1}^{N} m_{kj} \otimes m_{j\ell}. \tag{6.4}$$

In the same sense the inverse mapping and the unit element imply the coinverse and counit:

$$\mathcal{A} \xrightarrow{i} \mathcal{A}, \qquad (i(f))(g) = f(g^{-1}), \qquad \varepsilon(f) = f(\mathbf{1}_G). \tag{6.5}$$

Consider $U(\mathcal{G})$, the universal enveloping algebra of a Lie algebra $\mathcal{G}$ (for example $SL(2)$), with generators H and $X^{\pm}$:

$$[H, X^{\pm}] = \pm 2X^{\pm}, \qquad [X^+, X^-] = H. \tag{6.6}$$

The coproduct is now given by

$$\Delta(g) = g \otimes \mathbf{1} + \mathbf{1} \otimes g, \qquad \Delta(\mathbf{1}) = \mathbf{1} \otimes \mathbf{1}, \tag{6.7}$$

the counit by $\varepsilon(g) = 0$ for $g \neq \mathbf{1}$, and $\varepsilon(\mathbf{1}) = 1$.

These two examples will next be generalized to the q–deformed quantum group $\mathcal{A}_q(sl(2))$ and to the q–deformed universal enveloping algebra $U_q(sl(2))$.

We generalize the Yang–Baxter relation to [5]

$$\widetilde{R}_{12} T_1 T_2 = T_2 T_1 \widetilde{R}_{12} \tag{6.8}$$

with $T = \begin{pmatrix} a & b \\ c & d \end{pmatrix}$, $T_1 = T \otimes \mathbf{1}$, $T_2 = \mathbf{1} \otimes T$, and take as $\widetilde{R}$–matrix

$$\widetilde{R} = \left(\begin{array}{cc|cc} q & 0 & 0 & 0 \\ 0 & 1 & q - q^{-1} & 0 \\ \hline 0 & 0 & 1 & 0 \\ 0 & 0 & 0 & q \end{array}\right). \tag{6.9}$$

Then $a, b, c,$ and d fulfill the algebra

$$\begin{gathered} ab = qba, \quad ac = qca, \\ ad - da = (q - q^{-1}) bc, \\ bc = cb, \quad bd = qdb, \quad cd = qdc, \end{gathered} \tag{6.10}$$

which follows from (6.8). There exists a central element $\det{}_q T = ad - qbc$, being the q–deformation of the determinant. The coproduct is given like in (6.4) by

$$\Delta \begin{pmatrix} a & b \\ c & d \end{pmatrix} = \begin{pmatrix} a & b \\ c & d \end{pmatrix} \otimes \begin{pmatrix} a & b \\ c & d \end{pmatrix}, \tag{6.11}$$

and is an algebra homomorphism.

$U_q(sl(2))$ is given by generators $X^{\pm}$ and H and relations:

$$[H, X^{\pm}] = \pm 2X^{\pm}, \qquad [X^{+}, X^{-}] = \frac{q^{H} - q^{-H}}{q - q^{-1}}. \tag{6.12}$$

The coproduct for H is still the usual one, $\Delta(H) = H \otimes \mathbf{1} + \mathbf{1} \otimes H$, but those for $X^{\pm}$,

$$\Delta(X^{\pm}) = X^{\pm} \otimes q^{H/2} + q^{-H/2} \otimes X^{\pm}, \tag{6.13}$$

introduce an "asymmetry" and make the algebra non-cocommutative. The counit is given by $\varepsilon(H) = \varepsilon(X^{\pm}) = 0$, the coinverse by

$$S(H) = -H \text{ and } S(X^{\pm}) = -q^{\pm 1} X^{\pm}.$$

We note that there exists an universal R–matrix and the notion of quasi-triangularity. There exists a duality between Hopf algebras relating the two examples above.

The physical interesting application concerns the symmetry of the XXZ–chain in case special boundary terms are introduced ($\sin q(\sigma_1^z - \sigma_N^z)$). The resulting spin Hamiltonian $\widetilde{H}^{XXZ}$ commutes with the q–deformed total spin operator,

$$[\widetilde{H}^{XXZ}, \widetilde{X}^{\pm}] = [\widetilde{H}^{XXZ}, \widetilde{H}] = 0, \tag{6.14}$$

where

$$\widetilde{H} = \sum_{j=1}^{N} S_j^z, \qquad \widetilde{X}^{\pm} = \sum_{j=1}^{N} q^{S_3^1 + \ldots S_3^{j-1}} S_j^{\pm} q^{-S_3^{j+1} - \ldots - S_3^N}, \tag{6.15}$$

and $\widetilde{H}$, $\widetilde{X}^{+}$, and $\widetilde{X}^{-}$ obey the relations (6.12). We note, that after adding these boundary terms translation invariance on the ring is lost. A different model with a nonlocal boundary term, which is quantum group invariant and has a kind of translation invariance built in, has been treated and solved recently [7].

We end this chapter with a few remarks showing the relationship of the above mentioned algebraic structures to various other subjects.

For $q = 1$ the $\widetilde{R}$ matrix (6.9) becomes the unit matrix, and $R = P\widetilde{R} = P$ the permutation matrix. $R = \Pi^{S=1} - \Pi^{S=0}$, where Π^{S} denote the projection operators to spins. A similar decomposition can be achieved for $q \neq 1$:

$$R = \Pi_q^{S=1} - q^2 \Pi_q^{S=0}.$$

The space on which $\widetilde{R}$ acts can be decomposed too; this way we obtain the Manin plane:

$$\begin{aligned} \mathbf{C}\langle x, y\rangle / (\Pi_q^{S=0}(x_i \otimes x_j)) \quad &: \quad xy = qyx; \\ \mathbf{C}\langle \xi, \eta\rangle / (\Pi_q^{S=1}(\xi_i \otimes \xi_j)) \quad &: \quad \xi^2 = \eta^2 = 0, \ \xi\eta = -\frac{1}{q}\eta\xi. \end{aligned} \tag{6.16}$$

ξ and η can be identified as differentials $\xi = dx$, $\eta = dy$, and a differential calculus on the Manin plane results.

We mentioned already that link–invariants can be obtained from solutions of the Yang–Baxter equation. Following ideas of Atiyah, Witten suggested the connection to the Chern–Simons model. Topologically invariant quantities can be obtained as expectation values of products of Wilson line parallel transport operators along closed curves C_i:

$$\langle \chi_{\rho_1}(C_1) \ldots \chi_{\rho_M}(C_M) \rangle, \qquad \chi_\rho(C) = \mathrm{Tr}\, P \exp[\int_C A^i T_i^\rho]. \tag{6.17}$$

ρ denotes the representation, and the expectation value is taken within the model with action

$$S = \frac{k}{4\pi} \int d^3x \ \mathrm{Tr} \left(A \wedge dA + \frac{2}{3} A \wedge A \wedge A \right). \tag{6.18}$$

Recently we have studied this model within the Hamiltonian approach and proposed a quantization by deforming the Poisson structure. The quadratic algebra of monodromies becomes a quantum group algebra [8].

Many of the above mentioned aspects will be considered in detail in the other contributions to this volume.

Acknowledgement
I would like to thank the Theoretical Physics Group of the University of Graz, especially Ludwig Pittner, for inviting me to lecture in Schladming and to join the Organizing Committee.

References

1. H. Grosse, *Models in Statistical Physics and Quantum Field Theory*, Springer, 1988.
2. D.C. Mattis, *The Many-Body Problem*, World Scientific, 1993.
3. H. Grosse, *An Introduction to Integrable Models and Conformal Field Theory*, Proc. of the XXIX Winter Schladming School, Springer 1991, eds. H. Mitter and W. Schweiger.
4. H. Grosse, E. Langmann and E. Raschhofer, *The Luttinger–Schwinger Model*, preprint, Vienna 1995.
5. L. Faddeev, *Quantum Completely Integrable Models in Field Theory*, Les Houches Lecture, 1982.
6. J. Fröhlich, *Statistics of Fields, the Yang–Baxter Equation, and the Theory of Understand Links*, Cargése Lecture 1987.
7. H. Grosse, S. Pallua, P. Prester and E. Raschhofer, Journ. Phys. A: Math. **27** (1994) 4761.
8. A.Y. Alekseev, H. Grosse and V. Schomerus, *Combinatorial Quantization of the Hamiltonian Chern-Simons Theory. I and II*, Commun. Math. Phys., to be published 1995.

Yang-Baxter Equation and Reflection Equations in Integrable Models

P.P. Kulish

Departamento de Física Teórica and IFIC, Centro Mixto Universidad de Valencia - CSIC, 46100 Burjassot (Valencia) Spain.
St.Peterburgs Branch of Steklov Mathematical Institute of the Russian Academy of Scienes, Fontanka, 27, St.Petersburg, 191011, Russia.

1. Introduction

Integrable and/or solvable models have been always quite important for theoretical physics. One of their attractive features is the direct possibility to go beyond perturbation theory, giving a solid background for theoretical hypothesis and constructions. Another valuable characteristic is related to the combination of different mathematical methods required to solve some specific model, thus giving rise to mutual interrelations among formally separated fields of mathematics. The start of the recent activity in the quantum integrable models (see reviews [1, 2, 3, 4, 5, 6, 7, 8] and Refs. therein) was definitely related with the development of the soliton theory, although the influence of the preceding pioneering contributions is out of discussion (see [10, 11, 12, 13] and Refs. therein). This is a large field of research, and even recent monographs [11, 12, 13, 14, 15, 21, 17, 18] do not overlap much in between.

There is a variety of quantum integrable models and quite a few interrelations among them. In these lectures we restrict ourselves to those models, the integrability of which is related more or less directly to the Yang-Baxter equation and to reflection equations. More of that, we will discuss mostly the algebraic properties of these equations, their solutions and corresponding integrable models, putting aside very elaborated analytical techniques and/or problems (of great physical significance), such as the thermodynamic limits of finite size systems, massless and/or conformal field theory limits, continuous limits of the lattice models, correlation functions and form factors, critical exponents etc.. The main notions related to the Yang-Baxter equation (YBE) will be introduced: factorizable scattering on the line, the Zamolodchikov-Faddeev algebra, the fusion procedure or the bound state scattering, the Yang-Baxter or transition matrix algebra, integrals of motion. The factorizable scattering on a half-line gives rise to the reflection equation (RE) and a boundary operator. Most of the applications of the YBE can be extended to the RE case with appropriate modifications. Due to the vertex-face model correspondence of statistical mechanics we introduce an interaction round face analogue of the Zamolodchikov-Faddeev algebra and the corresponding reflection equation for the boundary weights. The super or Z_2-graded analogue of these constructions with the generalizations to the color algebras will also be given, as well as some particular integrable systems with finite degrees of freedom. The interrelations among few forms of the Bethe Ansatz: coordinate, algebraic, analytical, and functional, will be mentioned as well.

More particular applications of the Bethe Ansatz technique to specific integrable models with direct physical meaning , (a high energy limit of the QCD, the Azbel - Hofstadter problem), and of the reflection equation formalism to the Chern-Simons theory, and quantized moduli spaces of flat connections can be found in other lectures of this School.

2. Yang-Baxter Equation and Zamolodchikov-Faddeev Algebra

Let us start the presentation, which will be full of definitions and technicalities, from a specific integrable model to have in mind an example. This is the one-dimensional Bose gas consisting of n sort particles with the Dirac delta-function two-particle potential. The field operators $\psi_a(x), \psi_b^\dagger(y)$ satisfy the canonical commutation relations, $a, b = 1, 2, \cdots, n$,

$$[\psi_a(x), \psi_b^\dagger(y)] = \delta_{ab}\delta(x-y) \quad , \quad [\psi_a(x), \psi_l(y)] = 0. \tag{2.1}$$

The field Hamiltonian is

$$H = \int dx(\partial_x\psi_a^\dagger(x)\partial_x\psi_a(x) + c\psi_a^\dagger(x)\psi_b^\dagger(x)\psi_b(x)\psi_a(x)). \tag{2.2}$$

The translational invariance (on the whole line or with the periodic boundary conditions, circle) and the internal $U(n)$ invariance are obvious. The corresponding symmetry generators are

$$P = -i\int dx\psi_a^\dagger(x)\delta_x\psi_a(x), \quad U_{ab} = \int dx\psi_c^\dagger(x)(u_{ab})_{cd}\psi_d(x), \tag{2.3}$$

where u_{ab} are n^2 generators of the unitary Lie algebra $u(n)$. One of these operators is the particle number operator

$$N = \int dx\psi_a^\dagger(x)\psi_a(x)\,. \tag{2.4}$$

The Fock space of states is the direct sum of the number operator N eigenspaces,

$$\mathcal{H}_F = \sum_{M=0}^{\infty}\mathcal{H}_M. \tag{2.5}$$

The common eigenfunctions Ψ_M of operators N, P, and H,

$$\Psi_M = \int d^M x \Psi(1, \cdots, M|\lambda_1, \cdots, \lambda_M) \prod_{i=1}^{M} \psi^\dagger_{a_i}(x_i)|0\rangle, \qquad (2.6)$$

are constructed as appropriate linear combinations of the one particle eigenstates (the plane waves). The numbers $j = 1, 2, \cdots M$ in the argument of Ψ_M (2.6) refer to both coordinate x_j and isotopic a_j indices. The coefficients in the linear combination depend on the one particle parameters (momenta) λ_j and on the isotopic indices a_j,

$$\Psi(\{j\}|\{\lambda_j\}) = \sum_{\sigma \in \mathcal{S}_M} A_\sigma(\{a_j\}, \{\lambda_n\}) exp(i \sum_{m=1}^{M} \lambda_{\sigma m} x_m), \qquad (2.7)$$

where σ are all elements of the permutation group $\mathcal{S}_M$.

Such a form of the Hamiltonian eigenfunctions is known as the **coordinate Bethe Ansatz**. The conditions of the wave function continuity and its appropriate derivative jumps on the hyperplanes $x_j = x_{j+1}$ (the sewing conditions) define the coefficients $A_\sigma(\{a_j\}|\{\lambda_k\})$. The coefficients A_σ and $A_{\sigma'}$ with $\sigma' = \sigma_j \sigma$, where σ_j is the transposition of the indices $j, j+1$, are related by the two particle S-matrix [20]:

$$A_{\sigma'} = S(\lambda_j - \lambda_{j+1}) A_\sigma \, , \qquad (2.8)$$

where for the model chosen (2.2)

$$S(\lambda) = (\lambda + ic\mathcal{P})/(\lambda - ic) \, , \qquad (2.9)$$

and $\mathcal{P}$ is the permutation operator in $C^n \otimes C^n$. The periodicity condition for the system on finite interval $(0, L)$ results in the **Bethe equations** for the set of M momenta λ_j,

$$exp(i\lambda_j L) = -\prod_k S_{jk}(\lambda_j - \lambda_k) \, , \qquad (2.10)$$

where in the ordered product $k = j+1, ..., M-1, M, 1, ..., j-1$. Hence the meaning of the RHS is the scattering matrix of the j-th particle on the other $(M-1)$ particles. This would be just a phase factor for the scalar particles $(S(\lambda) = (\lambda + ic)/(\lambda - ic))$, but for the n component case one has to diagonalize the complicated scattering matrix to arrive at a system of scalar equations. All quantities in the RHS are particular values at $\lambda = \lambda_j$ of the **transfer matrix** $(k = 1, 2, ..., M)$,

$$t(\lambda; \{\lambda_m\}) = tr_a T(\lambda; \{\lambda_m\}) \equiv tr_a \prod_k S_{ak}(\lambda - \lambda_k), \qquad (2.11)$$

of the inhomogeneous $GL(n)$-spin magnet of M sites with the transition or monodromy matrix $T(\lambda; \{\lambda_m\})$. The trace in the expression for the transfer matrix $t(\lambda; \{\lambda_m\})$ is taken over the **auxiliary space** $V_a = C^n$, while

$t(\lambda;\{\lambda_m\})$ is an operator (matrix) in the space $\prod_{k=1}^{M}(C^n)_k$. The important property of the commutativity of the transfer matrix for different values of the spectral parameter

$$[t(\lambda;\{\lambda_m\}), t(\mu;\{\lambda_m\})] = 0$$

follows easily from the fundamental commutation relation for the transition matrix $T(\lambda;\{\lambda_m\})$ (see (15)).

Hence the RHS of (2.11) $t(\lambda_j;\{\lambda_m\})$ can be diagonalized simultaneously. As result the **hierarchy of the Bethe Ansatze** appears (or the nested Bethe Ansatz), and the complete parametrization of $\Psi(\{j\}|\{\lambda_j\})$ has n sets of "quasimomenta" including λ_j (see below).

The consistency condition of this system is the **Yang-Baxter equation** (YBE) for the S-matrix $S(\lambda)$:

$$S_{jk}(\lambda_j-\lambda_k)S_{jl}(\lambda_j-\lambda_l)S_{kl}(\lambda_k-\lambda_l)$$

$$= S_{kl}(\lambda_k-\lambda_l)S_{jl}(\lambda_j-\lambda_l)S_{jk}(\lambda_j-\lambda_k). \tag{2.12}$$

One uses very often in the general setting the notation R (the **R-matrix**) for the YBE solution.

The complete scattering matrix $S(\{\lambda_k\})$ of the M particle is given by the ratio of the coefficients of the incoming wave ($\lambda_1 < ... < \lambda_M$) and the outgoing wave

$$exp(i\sum_j \lambda_j x_j), \quad exp(i\sum_j \lambda_{M-j+1}x_j)\,.$$

This ratio is factorized into the ordered product of $M(M-1)/2$ two particle S-matrices (9) (the **factorizable scattering** on the line).

It was proposed for an algebraic description of the factorizable scattering in the general case to introduce a set of (annihilation) operators $Z_a(\lambda)$ [19] satisfying the commutation relations (the **Zamolodchikov algebra**)

$$Z_a(\lambda)Z_b(\nu) = S_{ab,cd}(\lambda-\nu)Z_d(\nu)Z_c(\lambda), \tag{2.13}$$

where $S(\lambda-\nu)$ is an $n^2\times n^2$ matrix. Using the associativity property of this algebra, and changing the order of the product $Z_{a_1}(\lambda_1)Z_{a_2}(\lambda_2)Z_{a_3}(\lambda_3)$ in two possible ways, one arrives at the consistency condition (12). Extending the Zamolodchikov algebra (13) by adding n more (creation) conjugated operators $Z_a^\dagger(\nu)$, one gets the **Zamolodchikov-Faddeev algebra** (ZF-algebra)

$$Z_a(\lambda)Z_b^\dagger(\nu) = \delta_{ab}\delta(\lambda-\nu) + Z_c^\dagger(\nu)\hat{S}_{ac,bd}(\lambda-\nu)Z_d(\lambda). \tag{2.14}$$

It is useful to write down the ZF-algebra in a compact matrix form by introducing the n component column $A(\lambda) = (Z_1(\lambda), ..., Z_n(\lambda))^t$ and the n

component row $A^{\dagger}(\nu) = (Z_1^{\dagger}(\nu), ..., Z_n^{\dagger}(\nu))$. Then the defining relations of the ZF-algebra are

$$A(\lambda) \otimes A(\nu) \equiv A_1(\lambda)A_2(\nu) = S_{12}(\lambda - \nu)A_2(\nu)A_1(\lambda),$$

$$A_1^{\dagger}(\lambda)A_2^{\dagger}(\nu) = A_2^{\dagger}(\nu)A_1^{\dagger}(\lambda)S_{21}^{\dagger}(\nu - \lambda),$$

$$A_1(\lambda) \otimes A_1^{\dagger}(\nu) = I_1\delta(\lambda - \nu) + A_2^{\dagger}(\nu)\hat{S}_{12}(\nu - \lambda)A_2(\lambda),$$

where subscripts refer to the corresponding isotopic spaces $C^n \otimes C^n \equiv V_1V_2$, and $S_{21} = \mathcal{P}S_{12}\mathcal{P}$, $\hat{S}_{12} = \mathcal{P}S_{12}$. Due to the unitarity property of the S-matrix: $S_{12}(\lambda - \nu)S_{21}(\nu - \lambda) = I_{12}$, and the YBE, one can construct the Fock space representation $\mathcal{H}_F$ of the ZF-algebra, using the generalizing symmetrizing operators which include $\hat{S}$ instead of the permutation operators [37].

To solve the non-linear Schrödinger equation (NS) (the Heisenberg equation of motion for $\psi_a(x,t)$ with the Hamiltonian (2)) in the framework of the quantum inverse scattering method (QISM) [1 - 7], one has to consider the corresponding **auxiliary linear problem**

$$\frac{d}{dx}T(\lambda, x) = L(\lambda, x)T(\lambda, x)\,.$$

For the model in question L depends on the spectral parameter λ and the original dynamical local variables (2.1)

$$L(\lambda, x) = \lambda J + \sum_a (c\psi_a(x)e_{a,n+1} + \psi_a^{\dagger}(x)e_{n+1,a}),$$

where J is $(n+1)\times(n+1)$ diagonal matrix:J = diag $(I_n, -1)$, and e_{ij} are the $(n+1)\times(n+1)$ basis matrices. Hence the auxiliary linear space here is C^{n+1}. This is the **L-operator** of the QISM or the classical soliton theory [13]. Solution of the operator valued matrix first order equation with normal ordering with respect to the local fields (2.1) and the vacuum $|0\rangle$,

$$T(\lambda, x) =: exp\left(\int^x L(\lambda, y)dy\right):,$$

defines the **transition or monodromy matrix** $T(\lambda, x)$. Its entries are the new variables of the model (the quantum scattering data (QSD)).

For the NS case it is natural to represent $T(\lambda, x)$ in the block form

$$T = \begin{pmatrix} A & B \\ C & D \end{pmatrix},$$

where A is an $n \times n$ matrix, B and C are n component vectors, D is a scalar. The quadratic commutation relations of the new variables are defined by the fundamental relation [1, 2, 3]

$$R(\lambda-\nu)T_1(\lambda)T_2(\nu)=T_2(\nu)T_1(\lambda)R(\lambda-\nu), \qquad (2.15)$$

where the standard QISM notations [1-7] are used, $T_1 = T\otimes I,\ T_2 = I\otimes T$, to embed matrices in C^n into $C^n\otimes C^n$. From the structure of the L-operator one can conclude that the n elements of the row $C(\lambda)$ act on the vacuum as some creation operators. This is the starting point to construct the eigenfunctions of the transfer matrix $t(\mu) = trT(\mu) = trA + D$ algebraically:

$$\Psi_M = \prod_{j=1}^{M} C_{a_j}(\lambda_j)0\rangle,$$

in the framework of the **algebraic Bethe Ansatz** [1, 2, 3].

The mentioned above commutativity of the transfer matrices $t(\mu) = trT(\mu)$ follows from (15), taking the trace of $T_1T_2 = R_{12}^{-1}T_2T_1R_{12}$ over both spaces $C^n \otimes C^n$. The trace form of the integrals of motion generating function $t(\mu)$ for continuous as NS (2.2) or chain models as (2.11) gives rise to the periodic boundary conditions. For the M site spin Hamiltonian

$$H = \left(\sum_{n=1}^{M-1} h_{n,n+1}\right) + h_{M,1},$$

the M-th site spin interacts with the first one. The treatment of the non-periodic boundary conditions in the algebraic framework of the QISM requiers the RE [29, 32, 34, 31] (for the coordinate Bethe Ansatz see [12, 28]).

Using block decomposition for the R-matrix one can rewrite the compact form of the fundamental relations (15) in terms of the blocks A, B, C, and D. Restricting to the case of finite number of particle, and after an appropriate limit to the whole line, one gets the ZF-algebra realization in terms of the QSD [50]:

$$Z(\lambda) = (A)^{-1}(\lambda)B(\lambda), \quad Z^{\dagger}(\nu) = C(\nu)D^{-1}(\nu)\,.$$

The R-matrix of (15) satisfies the YBE

$$R_{12}(\lambda-\mu)R_{13}(\lambda-\nu)R_{23}(\mu-\nu) = R_{23}(\mu-\nu)R_{13}(\lambda-\nu)R_{12}(\lambda-\mu)\,.$$

For the NS the R-matrix has the same stucture as (2.9) (but it is a $(n+1)^2 \times (n+1)^2$ matrix).

In the general situation the R-matrix depends on the spectral parameter u and some other parameters, i.e. $R(u;\eta,...)$. Although there is no complete mathematical theory of the Yang-Baxter equation, variety of solutions are known as well as different fields of their applications. In particular, many solutions are related to the simple Lie (super) algebras. They are classified by the Lie algebra, its irreducible representations Λ_j, and the spectral parameter dependence: rational, trigonometric, and elliptic ones [3, 9, 11, 48]. The

$sl(2)$ spin 1/2 R-matrix related to the XXX-magnet is used in few lectures of these Proceedings. The recent development relates the spectral parameter dependent R-matrices to the affine Lie (super-)algebras [16]. There are also solutions of the YBE with the spectral parameter on the algebraic curves of higher genus (the Potts models). The R-matrices acting in infinite dimensional spaces can be found in papers [58].

Some particular properties of the YBE solution $R(u;\eta,...)$, which are important for different applications (but not always valid for a given solution), are: regularity

$$R(0) = \phi(\eta)\mathcal{P} \ ,$$

P-symmetry

$$\mathcal{P}R_{12}(u)\mathcal{P} \equiv R_{21}(u) = R_{12}(u) \ ,$$

T-symmetry

$$R_{12}^{T}(u) = R_{12}(u) \, ,$$

unitarity

$$R_{12}(u)R_{21}(-u) = \rho(u)I \ ,$$

crossing symmetry

$$R_{12}^{t_1}(u)R_{12}^{t_1}(-u-\eta) = \xi(u)I \, ,$$

quasiclassical property

$$R(u;\eta) = I + \eta r(u) + \mathcal{O}(\eta^2) \, ,$$

where $r(u)$ is the classical r-matrix [3, 12] and $\phi(u)$, $\rho(u)$, $\xi(u)$ are some functions related to the R-matrix normalization. Many R-matrices have only PT-symmetry: $R_{12}^{T}(u) = R_{21}(u)$. The regularity is used to extract from $t(u)$ of the lattice models the integrals of motion, which are local in terms of initial spin variables, when the R-matrix itself is the L-operator. The quasiclassical property gives rise to the direct connection of the quantum model to the corresponding classical one [3, 13].

It is easy to see, that the product of the R-matrices $R_{13}(u)R_{23}(v)$ and $R_{1'3}(w)$ are intertwined (like (2.15)) by the matrix $R_{11'}(u-w)R_{21'}(v-w)$. If the original R-matrix $R_{12}(x)$ is degenerated into a projector at $x=\eta$, then one can project the above product to the corresponding subspace for the fixed difference of the spectral parameters $u-v=\eta$. For the Yang solution (2.9) at $\lambda = \pm ic$ one has the (anti)symmetrizer P_+, P_- . This is the **fusion procedure** [3] to get new R-matrices from the known ones. It has a direct physical interpretation as construction of the bound state S-matrix [19, 21]. Using the fusion procedure, the L-operators in the higher dimensional irreducible representations can be obtained, giving rise to the ZF operators for the bound states, and to the integrable lattice models of higher spins such as the spin s XXZ-model. The connection of the R-matrices and integrable models with the simple Lie (super-)algebras is reflected in the structure of the Bethe equations: they include r sets of "quasimomenta" , where r is the Lie algebra rank, and the Cartan matrix [37, 47].

Omitting the spectral parameter dependence in the YBE, one gets still a very interesting equation, the solutions of which (the **constant R-matrices**) can be considered as structure constants of the quantum groups and the quantum algebras. Then the quadratic relations (15) (without the spectral parameter) are the defining relations on the n^2 generators T_{ab} of the corresponding quantum group. The constant R-matrices $\hat{R}_{ij}$ have also direct relation to the braid group (BG), for one of the defining relations of the BG generators $\sigma_i\, \sigma_{i+1}\, \sigma_i = \sigma_{i+1}\, \sigma_i\, \sigma_{i+1}$ coincides with the YBE for $\hat{R}_{ij} = \mathcal{P}_{\rangle|}\mathcal{R}_{\rangle|}$:

$$\hat{R}_{i-1i}\, \hat{R}_{ii+1}\, \hat{R}_{i-1i} = \hat{R}_{ii+1}\, \hat{R}_{i-1i}\, \hat{R}_{ii+1}\, .$$

3. Reflection Equations and Their Covariance

Let us consider factorizable scattering of particles with internal degrees of freedom on a half-line [22]. Then even the one-particle process in nontrivial (a reflection from the wall), and it is described by an $n \times n$ matrix $K(u)$, the **reflection matrix** or the boundary S-matrix.

For an algebraic description it is useful to add to the ZF-algebra a formal boundary operator B with relation [24]

$$Z_a(u)B = K_{ab}(u)Z_b(-u)B\, .$$

Then the two particle factorizability gives rise, additionally to the YBE for the two body S-matrix (2.12), to the **reflection equation** (RE)

$$S_{12}(u-v)K_1(u)S_{21}(u+v)K_2(v) = K_2(v)S_{12}(u+v)K_1(u)S_{21}(u-v)\, . \tag{3.1}$$

There are natural properties of the reflection matrix $K(u)$, as in the preceding Sec. of $R(u;\eta,...)$: $K(0) = I$ (regularity); $K(u)K(-u) = I$ (unitarity); T-symmetry $K^T(u) = K(u)$; the crossing symmetry is more elaborated, and it involves the S-matrix itself [24]. The constructions of the quantum group invariant spin systems [30] use the RE [31, 32, 42, 54]. Recent field theoretical applications of the RE can be found in [23-27, 36].

It is interesting to point out that, as a consequence of the RE (3.1), the quadratic combination

$$\varphi(u) = \sum_{a,b} Z_a(-u)K_{ab}Z_b(u)$$

is a "local" field : $[\varphi(u),\ \varphi(w)] = 0$ [24, 52]. This property was used to construct the metric tensor field for the quantum Liouville theory [52] with a slightly different RE, and to propose a form of the boundary operator

$$B = exp\left(\int_0^\infty \varphi(u)du\right)$$

in the dual Hamiltonian picture of the boundary conformal field theory [24].

Although for many R-matrices solutions to the reflection equations were found (cf [22, 25, 32, 33]) there is no direct relation of them to the Lie algebra theory. In particular, many of them do not depend on the quasiclassical parameter η (in the fundamental representation for the R-matrix).

Let us give two examples of the RE solutions. Due to the $GL(n)$ symmetry of the Yang solution (2.9): $[R, M \otimes M], M \in GL(n)$, the corresponding K-matrix can be transformed: $K \to K' = MKM^{-1}$ with arbitrary M, and the solution of the RE is

$$K(u) = \xi I + uC, \quad C^2 = I.$$

In the elliptic case (the eight-vertex R-matrix [11]) the solution $K(u)$ is [22, 25, 33]

$$K(u) = \begin{pmatrix} x(u) & y(u) \\ z(u) & w(u) \end{pmatrix},$$

$$\begin{aligned} x(u) &= sn(\xi + u) \\ y(n) &= \mu\, sn2u(\lambda(1 - ksn^2u) + 1 + ksn^2u)(1 - k^2sn^2\xi sn^2u)^{-1}, \\ w(u) &= sn(\xi - u), \\ z(n) &= \mu\, sn2u(\lambda - 1 - (\lambda + 1)ksn^2u)(1 - k^2sn^2\xi sn^2u)^{-1}, \end{aligned}$$

where $snu \equiv sn(u; k)$ is the Jacobi elliptic function of modulus $0 < k < 1$, ξ, λ, μ are parameters.

The RE (2.16) has an important **covariance property**: if $T(u)$ and $K(u)$ satisfy the relations (2.15), (2.16) then $K'(u) = T(u)K(u)T(-u)^{-1}$ is also a RE solution, provided the entries of $K(u)$ and $T(u)$ commute, $[K_{ab}(u), T_{cd}(u)] = 0$. The proof follows easily by the substitution of $K'(u)$ into the RE, and using few times different forms of the fundamental relation (2.15), e.g.

$$T_2^{-1}(-v)R_{12}(u + v)T_1(u) = T_1(u)R_{12}(u + v)T_2^{-1}(-v).$$

This property gives rise to the **Sklyanin monodromy matrix** [29, 27]

$$\mathcal{T}(u) = T(u)K(u)T(-u)^{-1}. \tag{3.2}$$

If the matrix $T(u)$ is constructed as an ordered product of N independent L-operators, then $\mathcal{T}(u)$ can be interpreted as the monodromy matrix of N site lattice model, with a boundary condition described by the matrix $K(u)$, or a boundary interaction described by the operator valued entries of the matrix $K(u)$. To extract the corresponding Hamiltonian and other integrals of motion, the transfer matrix is constructed, using a special trace

$$\tau(u) = trK_+(u)T(u)K(u)T(-u)^{-1} = trK_+(u)\mathcal{T}(u) . \qquad (3.3)$$

An extra K-matrix $K_+(u)$ is any solution of a "conjugated" RE [29, 32, 34, 51], defined in such way to guarantee the commutativity $[\tau(u),\ \tau(w)] = 0$. In the regular case

$$R_{an}(0) \sim \mathcal{P}_{an} , \quad K(0) \sim I , \quad h_{n,n+1} = (\partial/\partial u \hat{R}_{n,n+1})(0) ,$$

the Hamiltonian is [29]

$$H = \sum_{n=1}^{M-1} h_{n,n+1} + K_1'(0) + (tr_0 K_+(0) h_{M,0})/trK_+(0) . \qquad (3.4)$$

The stucture of the **fusion procedure for the reflection equation** is similar to the R-matrix case, but the projected combination ($R_{12}(\eta) \sim$ projector)

$$K_1(u)R_{21}(2u-\eta)K_2(u-\eta)$$

includes the additional R-matrix in between the K-matrices [32, 51].

More general RE with four different R-matrices (with or without the spectral parameter),

$$R_{12}^{(1)}\, K_1\, R_{12}^{(2)}\, K_2 \ = K_2\, R_{12}^{(3)}\, K_1\, R_{12}^{(4)} ,$$

which are related among themselves by some concistency conditions similar to (2.12) and (2.15), can be found in different papers these days (see [55, 51] and Refs therein). In particular, this kind of RE was used for the quantum group invariant spin models with topological interaction [42, 43]. Some generalizations of the RE are related to the Coxeter groups [44].

The constant RE is also of interest, as it was in the YBE case. The corresponding RE-algebras, with the K-matrix entries considered as the generators, are related to the quantum group homogeneous spaces, while the c-number solutions K can be considered as representations of a special generator of the BG$^{(1)}$ in the solid handlebody [51], with the defining relation

$$\hat{R}_{12}\, K_1\, \hat{R}_{12}\, K_1 \ = K_1\, \hat{R}_{12}\, K_1\, \hat{R}_{12} .$$

In solvable models of statistical mechanics the R-matrix defines the Boltzmann weights $R_{\alpha\beta;\gamma\delta}(u;\eta,...)$ of the vertex models, where α,β,γ and δ are spin variables on the four edges around a vertex. The models on the dual lattice are known as the interaction round face models (IRF). The corresponding Boltzmann weights $w(a,b,c,d|u)$ satisfy the **star-triangular equation (relation)** (STR) or Baxter relation [11]

$$\sum_g w(a,b,g,f|u)w(f,g,d,e|u+v)w(g,b,c,d|v)$$

$$=\sum_g w(f,a,g,e|v)w(a,b,c,g|u+v)w(g,c,d,e|u).$$

The four sites surrounding a face are ordered anticklockwise from the southwest corner, and u is a complex (spectral) parameter.

Let us introduce a face analogue of the ZF-algebra: Its generators are parametrized by two indices $Z_{ab}(u)$ and satisfy the relations (see e.g. [57])

$$Z_{ab}(u)Z_{bc}(v)=\sum_d w(a,b,c,d|u-v)Z_{ad}(v)Z_{dc}(u),$$

with the only one summation over d in the RHS. As in the previous case the scattering of a one particle $Z_{da}(u)$ on N others $Z_{b_ic_i}(v_i)$ will give a transition matrix

$$T(u,v_1,\cdots,v_N)=\prod_{j=1}^{N} w(b_j,b_{j+1},c_{j+1},c_j|u-v_j).$$

By analogy with the YBE and ZF algebra constructions, one can formulate the following problems: 1) What are the conjugated (annihilation) operators and their commutation relations with $Z_{ab}(u)$? 2) How to translate the fusion procedure to the STR case? 3) What is an IRF analogue to the RE ?

Let us start from the last question and introduce a formal operator B of the boundary, which satisfies relations with $Z_{ab}(u)$:

$$Z_{ab}(u)B=Q(a,s,b;u)Z_{as}(-u)B\,,$$

where $Q(a,s,b;n)$ is a face analogue of the reflection matrix K. To derive an STR analogue of the reflection equation, the two ways of transforming the product of three operators may be considered:

$$Z_{ab}(u)Z_{bc}(v)B\longrightarrow Z_{ak}(-u)Z_{kd}(-v)B.$$

The resulting relation for the "scattering" matrix $w(\cdots)$ and the reflection matrix $Q(\cdots)$ is

$$w(a,b,c,g|u-v)Q(g,c,f|u)w(a,g,f,k|v+u)Q(k,d,f|v)$$

$$=Q(b,m,c|v)w(a,b,m,g|u+v)Q(s,d,m|u)w(a,s,d,k|n-v)\,,$$

with summation over g, f in the LHS and over m, s in the RHS .

To relate vertex models with the IRF models the intertwining vectors come into the play [11]:

$$\sum_{\beta,\nu} R_{\beta\nu;\lambda\nu}\phi_{ll'}(\beta)z_{m'l'}(\nu) = \sum_{m} w(m,m',l,l')z_{ml}(\lambda)\phi_{mm'}(\alpha) .$$

These vectors remind a combination of the operators of the ZF-algebras of the YBE and STR types. They can be used to relate the reflection matrices of the vertex and IRF models:

$$\phi_{ab}(\beta)K_{\beta\nu}(u)\phi_{bs}(\nu) = Q(a,s,b;u) ,$$

where the spectral parameter dependence of the vectors is omitted.

4. Integrable Models with Anticommuting Variables (Fermions)

One of the first integrable models of the quantum field theory - the massless Thirring model - has attracted a lot of attention of theoretical physicists during decades:

$$H = \int dx(\bar{\psi}\gamma\partial\psi + g(\bar{\psi}\psi)^2) .$$

It contains Fermi fields $\psi_\alpha(x,t)$, $\psi^+_\beta(x,t)$, satisfying anticommutation relations

$$\{\psi_\alpha(x,t), \psi^+_\beta(y,t)\} = \delta_{\alpha\beta}\delta(x-y) .$$

The addition of the mass term $m\bar{\psi}\psi$ to the Hamiltonian gives rise to the very rich dynamical properties of the massive Thirring model. In particular, it is dual to the famous sine-Gordon model, with the change of the strong coupling regime to the weak coupling one.

The famous δ-potential Bose gas model [20] (another name is quantum nonlinear Schrödinger equation) (2.2) was also generalized to the multicomponent fermionic case in the framework of the coordinate Bethe Ansatz [45] (see below).

Taking into account the "superization" procedure of the theoretical physics in the seventies, let us follow this pattern and extend most of the preceding equations and constructions to the case of anticommuting variables. Mathematically it is related to Z_2-graded vector space, Z_2-graded algebras, or as in the theoretical physics text: super-spaces, super-algebras, super-analysis...

The Z_2-graded Zamolodchikov algebra is generated by the $(m+n)$ operators $Z_i(u)$ satisfying

$$Z_i(u)Z_j(v) = (-)^{p(i)p(j)} R_{ij;kl}\,(u,v) Z_l(v) Z_k(u)\,, \tag{4.1}$$

where $i, j = 1, 2, \cdots, m+n, p(i)$ is a parity function: $p(i) = 0$ $(i = 1, \cdots, m)$ and $p(i) = 1$ $(i = m+1, \cdots, m+n)$; hence one has m boson and n fermion operators. The coefficients in (4.1) are c-numbers, and the R-matrix is even:

$$p(R_{ij;kl}) = p(i) + p(j) + p(k) + p(l) = 0\,. \tag{4.2}$$

Due to the Z_2 valuedness of the function $p(i)$, and the R-matrix structure (4.2), one can change the sign factor in (4.1) to $(-)^{p(k)p(l)}$. The super-associativity requirement for $Z_a(u)Z_b(v)Z_c(w)$ results in the graded Yang-Baxter equation (grYBE)

$$R_{ab;a'b'}\,(u,v) R_{a'c;jc'}\,(u,w) R_{b'c';hl}\,(v,w)\;(-)^{p(b')(p(j)+p(a'))} \tag{4.3}$$

$$= R_{bc;b'c'}\,(v,w) R_{ac';a'l}\,(u,w) R_{a'b';jk}\,(u,v) (-)^{p(b')(p(a)+p(a'))}\,,$$

where the condition (4.2) was used to reduce the factors in the LHS and RHS:

$$(-)^{p(a)p(l)+p(a')p(c)+p(b')p(c')}, \quad (-)^{p(b)p(c)+p(a)p(c')+p(a')p(b')}\,.$$

Recalling these sign factors, one can write the grYBE in the same matrix form as previously, but now in the Z_2-graded tensor product of three super-spaces $C^{m|n}$:

$$R_{12}\,(u,v) R_{13}\,(u,w) R_{23}\,(v,w) = R_{23}\,(v,w) R_{13}\,(u,w) R_{12}\,(u,v)\,. \tag{4.4}$$

Considering the third space as an unspecified quantum space (or substituting instead of the $(m+n) \times (m+n)$ blocks in R_{13} and R_{23} formal entries) one arrives at the graded FRT-relation

$$R_{12}\,(u,v) T_1\,(u) T_2\,(v) = T_2\,(v) T_1\,(u) R_{12}\,(u,v) \quad . \tag{4.5}$$

The latter relation has the same sign factors in the component form as (4.3). These factors can be interpreted as result of the Z_2-graded tensor product of two even matrices F and G $(p(F_{ac}) = p(a) + p(c))$:

$$(F \otimes G)_{ab;cd} = (-)^{p(b)(p(a)+p(c))} F_{ac} G_{bd}\,. \tag{4.6}$$

Hence, one gets additional sign factors for T_1 and the sign factor-free T_2 (to consider more tensor factors, the choice of T_1 sign factor-free is more convenient):

$$(T_1)_{ab;cd} = (T \otimes I)_{ab;cd} = (-)^{p(b)(p(a)+p(c))} T_{ac}\delta_{bd} , \tag{4.7}$$

$$(T_2)_{ab;cd} = (I \otimes T)_{ab;cd} = \delta_{ac}T_{bd} .$$

An additional sign factor appears also in the matrix of the permutation (flip) operator

$$\mathcal{P}(v \otimes w) = (-)^{p(v)p(w)} w \otimes v, \quad v, w \in C^{m|n},$$

$$(\mathcal{P})_{ab;cd} = (-)^{p(a)p(b)} \delta_{ad}\delta_{bc} . \tag{4.8}$$

To arrive later on at the graded reflection equation, it is convenient to write (4.7) in a matrix form with the usual tensor product

$$(T_1)_{gr} = \Gamma(T \otimes I)\Gamma, \tag{4.9}$$

where $(\Gamma)_{ab;cd} = (-)^{p(b)p(a)}\delta_{ac}\delta_{bd}$, hence $\Gamma^2 = I$ (which is not the case of the color algebras). The Z_2-graded algebraic structures were analyzed also in [35], and many R-matrices were found [48].

As in the first lecture, the simplest integrable models correspond to the rational solutions of the grYBE. The $GL(m|n)$ symmetric R-matrix is (the graded Yang solution)

$$R(u - v) = (u - v) + \eta\mathcal{P}. \tag{4.10}$$

The corresponding models (L-operators) are the $GL(m|n)$ isotropic graded magnets and the super-matrix nonlinear Schrödinger equation [37, 51]:

$$L_{gm}(u) = u + \eta \sum_{i,j} e_{ij} \otimes s_{ji}(-)^{p(j)}, \tag{4.11}$$

$$L_{NS}(x, u) = uJ + \eta \sum (e_{ab}\psi_{ba}(x) \pm e_{ba}\psi^{\dagger}_{ab}(x)), \tag{4.12}$$

where s_{ij} are the generators of the super-algebra $gl(m|n)$, $i, j = 1, 2, \cdots, m+n$; $\psi(x), \psi^{\dagger}_{ab}(x)$ are bose or Fermi fields according to $p(a) + p(b) = 0$ or 1, and $1 \leq b \leq N$, $N + 1 \leq a \leq m + n$; J is the block diagonal matrix $(I_N - I_{(m+n-N)})$.

Among integrable models related to the Z_2-graded case one can mention the $osp(1|2)$ - non-linear Schrödinger equation [37], which is described by the rational limit of the $osp(1|2)$ R-matrix and the supersymmetric sine-Gordon model with the trigonometric $sl(2|1)$ R-matrix [3]. Although the structure of the $osp(1|2)$ R-matrix is similar to the one of the $sl(2)$ spin 1 R-matrix, the solution of the corresponding $osp(1|2)$-magnet (or NS) is obtained [37] by the **analytic Bethe Ansatz**. Due to the arguments of analyticity, crossing symmetry, and the bare vacuum, the form of the transfer matrix $t(u)$ eigenvalue is the following:

$$\Lambda(u,\{v_j\}) = (u(u-d))^N \prod_{k=1}^{M} S_1(u-v_k-1/2)S_{-1}(u-v_k-1)$$
$$-((u-1)(u-d))^N \prod_{k=1}^{M} S_1(u-v_k) - (u(u-d+1))^N \prod_{k=1}^{M} S_{-1}(u-v_k-d)\,, \tag{4.13}$$

where $S_l(u) = (u+l/2)/(u-l/2)$ and $d = 3/2$. The regularity condition of $\Lambda(u,\{v_j\})$ in u (Manakov's principle): $Res\Lambda(u,\{v_j\}) = 0$ at $u = v_k$, gives the Bethe equations for the set of the quasimomenta $\{v_j\}$:

$$(S_1(v_k))^N = \prod_{j=1}^{M} S_2(v_k-v_j)S_{-1}(v_k-v_j).$$

Hence the M super-particle eigenstates are parametrized by one set of the quasimomenta, according to the Lie super-algebra of rank one.

Another way to define the form of the reflection equation, different from the factorizability requirement, is related to the covariance properties of the K-matrix: If K satisfies the RE, then the same is true for the transformed $K' = TKT^{-1}$. It follows from (4.5), that

$$R_{12}\,(u,v)\Gamma K_1\,(u)\Gamma R_{21}\,(v,-u)K_2\,(v) \tag{4.14}$$
$$= K_2\,(v)R_{12}\,(u,-v)\Gamma K_1\,(u)\Gamma R_{21}\,(-v,-u).$$

Further generalizations of the YBE and the corresponding RE [37] are related to the colored algebras [59] and/or more complicated commutation relations among entries of the T-matrices and the K-matrices. However the structure of these equations can be easily obtained following the standard ZF algebra pattern. In particular, if the multiplicative factor $\omega(a,b)$ of the color ZF algebra is nondegenerate [59], where $a, b \in \mathcal{A}$, and $\mathcal{A}$ is an abelian grading group, then the graded FRT-relation is

$$R_{12}\,(u,v)\Gamma^{-1}T_1\,(u)\Gamma T_2\,(v) = T_2\,(v)\Gamma^{-1}T_1\,(u)\Gamma R_{12}\,(u,v), \tag{4.15}$$

where $(\Gamma)_{ab;cd} = \omega(a,b)\delta_{ac}\delta_{bd}$, and we identify for simplicity the matrix indices with the grading group elements (which is not always the case as e.g. the Z_2-grading). Now Γ is not necessary unipotent. The corresponding RE follows from the covariance arguments:

$$R_{12}\,(u,v)\Gamma^{-1}K_1\,(u)\Gamma R_{21}\,(v,-u)K_2\,(v) \tag{4.16}$$
$$= K_2\,(v)R_{12}\,(u,-v)\Gamma^{-1}K_1\,(u)\Gamma R_{21}\,(-v,-u).$$

To extract the commuting functionals of the $\mathcal{A}$-graded monodromy matrix T, or the corresponding Sklyanin's matrix $\mathcal{T}$, with the entries as the homogeneous elements of the grading, the $\mathcal{A}$-graded trace [59] has to be used:

$$t(u) = tr_{\mathcal{A}} T(u) = \sum \omega(a,a) T_{aa}(u) \, .$$

This trace can be considered as a particular example of the quantum trace [51] with Γ as the R-matrix. It is also possible to extend these constructions further by using instead of the multiplicative factor $\omega(a,b)$ and/or Γ an appropriate R-matrix. However, the main problem for the moment is to find an interesting example, the solution of which requires the mentioned above equations.

5. Integrable Models with Finite Degrees of Freedom

The intensive development of the YBE and of the theory of quantum groups was strongly influenced by conformal field theory. The latter one, as most of the field theoretical integrable models, has the two dimensional space-time. However, the rich structure of the YB-algebra and RE-algebra, or of the quadratic R-matrix algebras with the spectral parameter dependence, permits to include into this formalism a variety of known integrable models with finite degrees of freedom (e.g. [38-40, 41, 46]) and to find new ones, which are physically interesting systems in the space of three (or more) dimensions.

Let us consider as an example the Kowalewski-Chaplygin-Goryachev top (KCG top). The Hamiltonian of this model is [46]

$$H = \frac{1}{2}(J_1^2 + J_2^2 + 2J_3^2) + c_1 x_1 + c_2 x_2 + c_3(x_1^2 - x_2^2) + c_4 x_1 x_2 + c_5/x_3^2, \quad (5.1)$$

where c_i, $i = 1, \cdots, 5$ are arbitrary constants, and J_i, x_i are the angular momenta and coordinates. This system is integrable, provided the constraint

$$l = \sum_{j=1}^{3} x_j J_j = 0 \quad (5.2)$$

holds.

One gets the famous Kowalewski's top for $c_3 = c_4 = c_5 = 0$. The corresponding auxiliary linear problem (the L-operator) is related to the simplest R-matrix: the Yang solution for $sl(2)$ with $c = \kappa$. The L-operator shows quadratic dependence on the spectral parameter u:

$$L(u) = \begin{pmatrix} y_0 u^2 + y_2 u + y_1 & y_4^+ u + y_6^+ \\ y_4^- u + y_6^- & y_3 \end{pmatrix} = \begin{pmatrix} a & b \\ c & d \end{pmatrix}(u), \quad (5.3)$$

where $y_4^{\pm} = y_4 \pm y_5$ and the same for $y_6^{\pm}$; $y_0, ..., y_7$ are eight dynamical variables of the model.

The corresponding YB-algebra, generated by four entries $a(u), ..., d(u)$ of the L-operator, has its centre generated by the q-determinant of the L-operator (5.3) [5, 3]:

$$\begin{aligned} det_q L(u) &= a(u+i\kappa/2)d(u-i\kappa/2) - b(u+i\kappa/2)c(u-i\kappa/2) \\ &= Q_1 u^2 + Q_2 u - Q_3 + \kappa^2/4Q_1 \quad , \end{aligned} \tag{5.4}$$

where

$$Q_0 = y_0,\ Q_1 = y_0 y_3 - y_4^2 - y_5^2 Q_2 = y_2 y_3 - 2y_4 y_6 - 2y_5 y_7,$$

$$Q_3 = y_6^2 + y_7^2 - \frac{1}{2}\{y_1, y_2\} + \frac{1}{2}\kappa^2 y_0 y_3,\ \{y_i, y_j\} = y_i y_j + y_j y_i.$$

The YB-algebra for the entries of $T(u) \equiv L(u)$ results in quadratic commutation relations for the dynamical variables y_k. The nontrivial problem is to realize the latter ones in terms of physically significant variables. One of the realization is given [7] by the momenta p_i and coordinates q_i, $i = 1, 2$, of the two site Toda lattice, so that the L-operator (5.3) is the product of two elementary ones ($i = 1,\ 2$):

$$L_i(u) = \begin{pmatrix} u - p_i & -e^{q_i} \\ e^{-q_i} & 0 \end{pmatrix} \quad . \tag{5.5}$$

The realization we are looking for is defined by the generators J_k, x_k, $k = 1, 2, 3$, of the Lie algebra e(3), provided that $l = \sum_{k=1}^{3} x_k J_k = 0$ [46]:

$$\begin{array}{ll} y_0 = 1 & y_4 = ibx_1 \\ y_1 = -(J_1^2 + J_2^2 + \frac{1}{4} + 2\alpha/x_3^2) & y_5 = ibx_2 \\ y_2 = -2J_3 & y_6 = -\frac{1}{2}ib\{x_3, J_1\} \\ y_3 = b^2 x_3^2 & y_7 = -\frac{1}{2}ib\{x_3, J_2\} \quad , \end{array} \tag{5.6}$$

where α and b are constants related to the YB-algebra central elements. The integrals of motion are generated by the trace of (5.3),

$$t(u) = trL(u) = a(u) + d(u) = u^2 - 2J_3 u - 2H - \frac{1}{4} \quad , \tag{5.7}$$

where $H = \frac{1}{2}(J_1^2 + J_2^2 - b^2 x_3^2) + \alpha/x_3^2$ is the Hamiltonian of Neumann's system. The solution of this system by the separation of variables (from J_k, x_k to the new ones) in the framework of the QISM is achieved using the **functional Bethe Ansatz** [7, 41]: the introduction of the new variables as the operator roots $\hat{u}_i$, $i = 1, 2$, of the entry $c(u)$ (5.3), and the conjugated variables $\hat{m}_i$ as the values of $a(u)$ and $d(u)$ at these roots. Then the eigenvalue equation for the transfer matrix $t(u)$ (two degrees of freedom for the Newmann's case) is reduced to the one-dimensional problem. To embed the KCG top into such an approach one has to use the RE and the Sklyanin monodromy matrix.

The c-number solutions of the RE with the $sl(2)$ R-matrix have the form

$$K_a(u) = \alpha_a\sigma_0 + u(\beta_a^-\sigma_+ + \beta_a^+\sigma_-), \tag{5.8}$$

where $a = \pm$, $(\beta_-^-, \beta_-^+, \beta_+^-, \beta_+^+) = (1, -\beta_1, \beta_2, -1)$. Using the L-operator (5.3) one can construct according to the general recipe [29] the monodromy matrix

$$\mathcal{T}(u) = L(u)K_-(u - ix/2)\sigma_2 L^T(-u)\sigma_2, \tag{5.9}$$

and the corresponding generating function $\tau(u)$ of the integrals of motion

$$\tau(u) = trK_+(u + ix/2)\mathcal{T}(u) \tag{5.10}$$

The latter one gives rise to the KCG top Hamiltonian (5.1).

Another wide class of models embedded recently into this scheme [49] includes the so called quasi-solvable models; some of them are physically relevant.

Acknowledgement
This work is supported in part by the DGICYT (Spain). The author wishes to thank the hospitality of the Department of Theoretical Physics of Valencia University, J. A. de Azcarraga for useful discussions, H. Grosse and L. Pittner for the kind invitation to the Schladming International Winter School of Theoretical Physics.

References

1. L. D. Faddeev, *Integrable models in* $(1+1)$*-dimensional quantum field theory*, Proceed. Les Houches XXXIX, (eds J.-B. Zuber and R. Stora), North-Holland Publ., 1984, 561-608.
2. L. D. Faddeev and L. A. Takhtajan, Russ. Math. Surveys **34** (1979) 11-68.
3. P. P. Kulish and E. K. Sklyanin, Lect. Notes Phys. **151** (1982) 61-119.
4. H. B. Thacker, Rev. Mod. Phys. **53** (1981) 253-285.
5. A. G. Izergin and V. E. Korepin, Sov. J. Part. Nucl. **13** (1982) 207-223.
6. H. J. de Vega, Int. J. Mod. Phys. **A4** (1989) 2371-2463.
7. E. K. Sklyanin, *Quantum inverse scattering method: selected topics*, in: *Quantum group and quantum integrable systems*, (ed. M.-L. Ge), Singapore; WS, 1992.
8. F. W. Nijhoff and H. W. Capel, *Quantization of integrable mappings*, in: *Geometric and quantum aspects of integrable systems*, (ed. G. F. Helsmincle), Lect. Notes Phys. **424** (1993) 187-211.
9. M. Jimbo (ed.) *Yang-Baxter equation in integrable systems*, Singapore: WS, 1990, 715 pp.
10. E. H. Lieb and D. C. Mattis (eds.), *Mathematical Physics in one dimension*, NY-London: Acad. Press, 1966.
11. R. J. Baxter, *Exactly solved models in statistical mechanics*, NY-London: Acad. Press, 1982.

12. M. Gaudin, *La fonction d'onde de Bethe*, Paris: Masson, 1983.
13. L. D. Faddeev and L. A. Takhtajan, *Hamiltonian methods in the theory of solitons*, Berlin: Springer-Verlag, 1987.
14. F. A. Smirnov, *Form factors in completely integrable models of quantum field theory*, Singapore: WS, 1992.
15. V. E. Korepin, N. M. Bogoliubov and A. G. Izergin, *Quantum inverse scattering method and correlation functions*, Cambridge University Press, 1993.
16. M. Jimbo and T. Miwa, *Algebraic analysis of solvable lattice models*, CBMS/85N, 1994.
17. C. Gómez, M. Ruiz-Altaba and G. Sierra, *Quantum groups in two-dimensional physics*, Cambridge University Press, 1995 (to be published).
18. E. Abdalla, M. C. B. Abdalla and K. Rothe, *Non perturbative methods in two dimensional quantum field theory*, WS, Singapore, 1991.
19. A. Zamolodchikov and Al. Zamolodchikov, Ann. Phys. **120** (1979) 253.
20. C. N. Yang, Phys. Rev. Lett. **19** (1967) 1312.
21. M. Karowski, Nucl. Phys. **B153** (1979)244.
22. I. V. Cherednik, Theor. Math. Phys. **61** (1984) 977.
23. I. V. Cherednik, Int. J. Mod. Phys. **A7**, Suppl. 1B (1992) 707.
24. S. Ghoshal and A. B. Zamolodchikov, Int. J. Mod. Phys. **A9** (1994) 3841.
25. T. Inami and H. Konno, J. Phys. **A27**(1994)211.
26. A. Fring and R. Köberle *Nucl. Phys.* **B421** (1994) 1592,
27. R. Sasaki, *Reflection Bootstrap Equations for Toda Field Theory*, 1993. Preprint YITP/U-93-33.
28. F. C. Alcaraz, M. N. Barber, M. T. Batchelor, R. J. Baxter and G. R. W. Quispel, 1987 *J. Phys. A: Math. Gen.* **20** 6397.
29. E. K. Sklyanin, *J. Phys.* **A21** (1988) 2375
30. V. Pasquier and H. Saleur *Nucl. Phys.* **B330** (1990)523
31. P. P. Kulish and E. K. Sklyanin, *J. Phys.* **A24** (1991) L435.
32. L. Mezincescu, R. I. Nepomechie and V. Rittenberg *Phys. Lett.* **A147** (1990) 70; L. Mezincescu, R. I. Nepomechie, *J. Phys.* **A24** (1991) L17; *Int. J. Mod. Phys.* **A6** (1991)5231, *ibid.* **A7** (1992) 5657.
33. H. J. de Vega and A. González Ruiz *J. Phys.* **A26** (1993) L519.
34. E. K. Sklyanin *Funct. Analy. Apply.* **21** (1987) 164
35. L. Pittner and P. Uray, J. Math. Phys. **36** (1995) 944.
36. M. Jimbo, R. Kedem, T. Kojima, M. Konno and T. Miwa, *XXZ chain with boundary*, Preprint RIMS, hep-th/9411121, 1994.
37. P. P. Kulish, Zap. Nauch. Semin. LOMI, **145** (1985) 140; J. Sov. Math. **35** (1986) 1111; Lett. Math. Phys. **10** (1985) 87.
38. H. M. Babujian and R. Flume, Mod. Phys. Lett. **9**(1994) 2029-2039
39. K. Hikami, P. P. Kulish and M. Wadati, J. Phys. Soc. Japan, **61** (1992) 3071-3081.
40. P. P. Kulish, S. Rauch-Wojciechowski and A. V. Tsiganov, Mod. Phys. Lett., **9** (1994) 2063-2073.
41. E. K. Sklyanin, Comm. Math. Phys. **142** (1992) 123-132.
42. M. Karowski and A. Zapletal, Nucl. Phys. **B419** (1994) 111.
43. H. Grosse, S. Pallua, P. Prester and E. Raschhofer, J. Phys. **A27** (1994) 4761-4771.
44. M. L. Ge and Q. Zhao, Phys. Lett. **A175** (1993) 199.
45. C. K. Lai, J. Math. Phys. **15**(1974)954-956.
46. V. B. Kuznetsov and A. V. Tsiganov, J. Phys. **A22** (1989) L73-L79.
47. E. Ogievetsky, N. Yu. Reshetikhin and P. B. Wiegmann, Nucl. Phys. **B280** (1987) 45
48. V. Bazhanov, A. Schadrikov, Theor. Math. Phys. **73** (1987) 402.

49. A. V. Zabrodin, *Quantum transfer matrices for discrete and continuous quasi-exactly solvable problems.* Preprint HEP-TH-9412116, 1994.
50. P. P. Kulish, Zap. Nauch. Sem. LOMI, **109** (1981) 83; J. Sov. Math. **23** (1984) 2111.
51. P. P. Kulish and R. Sasaki, Prog. Theor. Phys. **89** (1993) 741.
52. E. Cremmer and J.-L. Gervais, Comm. Math. Phys. 144 (1992) 279.
53. P. Martin and V. Rittenberg, Int. J. Mod. Phys. **A7**, Suppl.(1992) 707.
54. M. Karowski, Lect. Notes Phys. **380** (1990) 183.
55. L. Freided and J.-M. Maillet, Phys. Lett. **B262** (1991) 278; ibid. **B263** (1991) 403.
56. M. L. Ge and H. C. Fu, J. Phys. **A25** (1992) L1233
57. S. Lukyanov and Ya. Pugai, *Bosonization of ZF algebras: direction toward deformed Virasoro algebra.* Preprint RU-94-41, 1994.
58. M. Gaudin, J. Phys. France **49** (1988) 1857; Y. Shibukawa and K. Ueno, Lett. Math. Phys. **25** (1992) 239-248; P.P. Kulish and V.D. Lipovsky, Phys. Lett. **A 127** (1988) 413.
59. V. Rittenberg and D. Wyler, Nucl. Phys. **B139** (1978) 189; M. Scheunert, J. Math. Phys. **24** (1983) 2671; D. Gurevich, Leningrad Math. J.,**2** (1991) 801; D. S. McAnally, Lett. Math. Phys. **33** (1995) 249.

Scattering Matrices and Affine Hecke Algebras

V. Pasquier

Service de Physique Théorique de Saclay, F-91191 Gif sur Yvette Cedex, France, Laboratoire de la Direction des Sciences de la Matière du Commissariat à l'Energie Atomique.

These lectures deal with two related topics. The first one is the construction of the scattering matrix for an arbitrary Weyl group. The second topic is the application of the affine Hecke algebra to construct physical models with enlarged symmetries.

1. Introduction

These lectures deal with two related topics.

The first one is the construction of the scattering matrix for an arbitrary Weyl group. The aim is to define a set of commuting operators in terms of generators which obey the Yang-Baxter equation. These operators, which coincide with the usual scattering matrices, form a group isomorphic to the weight lattice of the corresponding Lie group. In the first lecture we interpret the Yang Baxter generators as elementary reflections. This geometrical picture enables us to construct the scattering matrices in a systematic way, and to show that they commute.

In this lecture we have used the language of Coxeter-Weyl groups and cells exposed in [6, 3].

The second topic is the application of the affine Hecke algebra to construct physical models with enlarged symmetries.

The affine Hecke algebra is a deformation of the affine Weyl group. The reflections across the origin are deformed into the none affine generators, of the Hecke algebra, and the translations are deformed into generators, which coincide with the scattering matrices of the first lecture. Using a specific representation of the Yang Baxter generators, which uses the Hecke algebra, we show that the scattering matrices obey the affine Hecke relations. We then derive a representation in a space of polynomials in several variables. This enables us to construct a set of commuting Hamiltonians essentially given by the center of the affine Hecke algebra. These Hamiltonians are Hermitian, and their spectrum can be computed in a simple way. One of their main interests is that they commute with q-deformations of the usual affine Lie algebras defined in terms of quadratic algebras.

The abstract theory of affine Hecke algebras is exposed in [7]. We expand here the point of view of [1, 8]. Parallel arguments are in [5] and some recent results can be found in [4].

2. The Scattering Matrix

Two important tools of solvable models are the transfer matrix and the scattering matrix. In his study of the delta interacting gas, Yang introduced a commuting family of scattering matrices. He obtained the spectrum by diagonalizing simultaneously these matrices. A simple way to prove the commutation relations for the scattering matrices consisted in defining a commuting family of transfer matrices depending of a continuous parameter. By letting the parameter take some special values, he recovered the scattering matrices. This approach is expanded in the book of M.Gaudin [2].

I do not know a generalization of the transfer matrix for an arbitrary Weyl group, and when this can be done, (A,B,C,D cases); one cannot simply obtain the scattering matrix by letting the spectral parameter take special values, except in the A case. On the other hand, the scattering matrix can be easily generalized to arbitrary Weyl groups. If a proper correspondence is used, the scattering matrices are identified with translations in the affine Weyl group. To generalize the construction of the scattering matrices it is useful to give them a geometrical interpretation. It enables to construct commuting operators from a set of generators obeying the (generalized) Yang-Baxter algebra.

In the first part, I recall some known facts about Weyl groups following the presentation of Carter [6]. I modify the presentation of the Weyl group in terms of generators and relations to introduce generators which obey the Yang-Baxter equation. The difference between the two approaches is analogous to the description of a rigid body motion in the moving frame compared to the description in the rest frame.

In the second part, I extend these considerations to the affine case, which enables to obtain the expression of the scattering matrix. The geometrical interpretation is the following: Periodic trajectories inside the fundamental cell associated to a Weyl group are in correspondence with translations in the weight lattice. To these trajectories, we associate operators by taking the time ordered product of the Yang-Baxter operators of the reflections across the walls hit by the trajectory. These operators form a commutative group isomorphic to the weight lattice.

2.1 Systems of Roots and Weyl Groups

Root Systems. Let V be an euclidian space of dimension l. For r a vector of V we denote by w_r the reflection in the hyperplane orthogonal to r.

$$w_r(x) = x - 2\frac{(r,x)}{(r,r)}r. \tag{2.1}$$

A system of roots is defined by a set Φ of non zero vectors spanning V, which obey the following properties:

1) If $r, s \in \Phi$, then $w_r(s) \in \Phi$.
2) If $r, s \in \Phi$, then $2(r,s)/(r,r)$ is a rational integer.
3) If $r, \lambda r \in \Phi$, then $\lambda = \pm 1$.

It will be convenient to denote $r^V = 2r/(r,r)$. The lattice spanned by the r^V is denoted Q^V.

We shall also consider the lattice spanned by the vectors p, such that (p,r) is a rational integer for any root $r \in \Phi$. This lattice denoted P^V contains the coroot lattice Q^V.

One can define a basis of V , Π, which is a subset of Φ and is such that any root of Φ is a linear combination of roots of Π with coefficients which are either all non-negative or all non-positive. Π is called a fundamental system of roots. $\Pi = \{r_1, r_2, ..., r_l\}$, and a root can be written:

$$r = \sum_{i=1}^{l} \lambda_i r_i, \tag{2.2}$$

where either $\lambda_i \geq 0$ for all i, or $\lambda_i \leq 0$ for all i. Accordingly, we say that r is a positive root $r \in \Phi^+$ or a negative root $r \in \Phi^-$. We say that $r > s$ if $r - s$ is a linear combination with coefficients ≥ 0 of elements in Π. Every positive system of roots contains just one fundamental system. Thus there is a one to one correspondence between positive systems and fundamental systems in Φ.

Chambers. For our purpose, it is useful to have a geometric interpretation of a fundamental system. For each root $r \in \Phi$, we denote by H_r the hyperplane orthogonal to r. The set $V - \cup H_r$ is disconnected, its connected components are called chambers. The roots orthogonal to the bounding hyperplanes of a chamber and pointing into the chamber form a fundamental system Π. A chamber C_Π is thus defined as the set of points $x \in V$ such that:

$$(r_i, x) > 0, \forall r_i \in \Pi. \tag{2.3}$$

Moreover, every fundamental system arises in this way from some chamber.

Let C be a chamber, and $\delta(C)$ be its boundary. The bounding hyperplanes of C are called the walls of C, and their intersections with $\delta(C)$ the faces of C. Two chambers which have a face in common are called adjacent.

Weyl Group. The group W generated by the reflections w_r is called the Weyl group of Φ. Each element of W transforms Φ into itself. One can show that the Weyl group is generated by the fundamental reflections w_r with $r \in \Pi$. Given two chambers C and C' (alternatively two fundamental systems Π and Π'), there is a unique element of the Weyl group such that $w(C) = C'$. Let us here give a construction of this element. This presentation will make the Yang-Baxter equation appear naturally in this context.

We consider a sequence $C_0, C_1, ..., C_m$ of chambers with $C_0 = C$, $C_m = C'$, such that two consecutive chambers are adjacent. Each chamber is characterized by a fundamental system Π_i, and there is a unique $r_i \in \Pi_i$ such that $w_{r_i} C_i = C_{i+1}$. We define the element of the Weyl group such that $w(C) = C'$ by:

$$w = w_{r_{m-1}} w_{r_{m-2}} ... w_{r_0}. \tag{2.4}$$

Length of the Weyl Group Elements. The length of the element w, $l(w)$ can be defined in two different ways. The first one being:
a) $l(w)$ is the number of positive roots turned by w into negative roots.

Equivalently, if we denote Φ^+, Φ'^+ the positive roots respectively associated to the fundamental systems of C and C'.:

$$l(w) = |\Phi^+ \cap \Phi'^-|. \tag{2.5}$$

Let $r \in \Pi'$. Consider the word $w_r w$; since w_r changes the sign of only $\pm r$ in Φ', one has in particular:

$$\begin{aligned} l(w_r w) &= l(w) + 1 \text{ if } r \in \Phi^+; \\ l(w_r w) &= l(w) - 1 \text{ if } r \in \Phi^-. \end{aligned} \tag{2.6}$$

Now consider an expression of the form (2.4) to represent the Weyl group element w. From (2.6) it follows that the minimal number of reflections in (2.4) is larger than $l(w)$. Let us assume that the number of terms is strictly larger than $l(w)$. This means that the expression contains a reflection w_{-r} with $r \in \Phi^+$. From the above argument, this is only possible if the reflection w_r has occured earlier. Thus the expression is of the form:

$$w = w_{r_{m-1}} ... w_{-r} X w_r ... w_{r_0}. \tag{2.7}$$

Let us denote by $w_r(X)$ the expression obtained from X by substituting $w_{w_r(p)}$ for w_p in the expression of X. A shorter expression for w is given by:

$$w = w_{r_{m-1}} ... w_r(X) ... w_{r_0}. \tag{2.8}$$

It follows that another definition of the length is:
b) $l(w)$ is the the minimal length of an expression w in the form (2.4).

A word of minimal length is one for which all the r_i in the expression (2.4) of w are in $\Phi^+ \cap \Phi'^-$.

Given two chambers C and C', let us describe the way to construct an element w of minimal length such that $w(C) = C'$. For this we proceed by induction on l. If $l = 0$, $C = C'$ and $w = 1$. If $l \neq 0$, $\Phi'^- \cap \Pi$ is not empty. Let r_0 belong to this set. We put $w_{r_0}(C) = C_1$ and $\Pi_1 = r_0 \Pi$. The set of positive roots defined by Π_1 is obtained from Φ by replacing r_0 by $-r_0$. Hence

$$|\Phi_1^+ \cap \Phi'^-| = l - 1, \tag{2.9}$$

and we can continue the construction replacing C by C_1.
Yang-Baxter Equation. We now give a presentation of the Weyl group W as an abstract group in terms of generators x_r, $r \in \Phi$, using the above construction. Namely, we consider the elements (2.4) with the reflection w_r replaced by x_r, and we impose relations to the x_r in such a way that two elements w_1 and w_2 coincide whenever the Weyl group reflections are equal.

The operators x_r are subject to the following relations:
a) unitarity:

$$x_{-r}x_r = 1. \tag{2.10}$$

b) Generalized Yang-Baxter equation:
Let r and s be two roots in some fundamental system Π, and m_{rs} be the order of $w_r w_s$. Then we have:

$$x_{r_1}x_{r_2}x_{r_3}...x_{r_m} = x_{r_m}x_{r_{m-1}}x_{r_{m-3}}...x_{r_1}, \tag{2.11}$$

where:

$$r_1 = s,\ r_2 = w_s(r),\ r_3 = w_s w_r(s), ...,\ r_m = r.$$

For example, in the case where the root system is A_n, if we take $r = e_1 - e_2$, $s = e_2 - e_3$, then $m_{rs} = 3$, and the above equation writes:

$$x_{e_2-e_3}x_{e_1-e_3}x_{e_1-e_2} = x_{e_1-e_2}x_{e_1-e_3}x_{e_2-e_3}. \tag{2.12}$$

One recognizes the usual Yang-Baxter equation. These equations are generalizations of the Yang-Baxter equation for arbitrary Weyl groups.

2.2 Affine Root Systems and Weyl Groups

Affine-Root Systems. In order to define a scattering matrix, we need to extend the previous considerations to the affine case.

We denote by E_0 an affine vector space of dimension d. For $r \in \Phi$ and $k \in Z$ we define a hyperplan in E_0 by:

$$H_{r,k} = \{x \in E_0 | (r, x) = k\}. \tag{2.13}$$

The orthogonal reflections with respect to the Hyperplanes $H_{r,k}$ generate a group called the affine Weyl group. This group is also the semi-direct product of the Weyl group defined earlier and the translations in the coroot lattice Q^V.

The affine roots are defined by a couple of non-affine root and a rational integer. We denote $\tilde{\Phi}$ the set of affine roots. Each affine root $r + k\delta$ defines the hyperplane $H_{r,k}$ and the corresponding reflection in the affine Weyl group:

$$w_{r+k\delta}(x) = x - ((r, x) - k)r^V. \tag{2.14}$$

A fundamental system $\tilde{\Pi}$ is given by a basis of $\tilde{\Phi}$, such that any root of $\tilde{\Pi}$ is a linear combination of roots of $\tilde{\Pi}$ with coefficients, which are either all positive ($\tilde{\Phi}^+$) or all negative ($\tilde{\Phi}^-$). For example, given a fundamental system Π, if we denote r_m is the largest root of Φ, we obtain a fundamental system $\tilde{\Pi}$ given by:

$$\tilde{\Pi} = \{r | r \in \Pi\} \cup \{-r_m + \delta\}. \tag{2.15}$$

The positive roots are given by:

$$\tilde{\Phi}^+ = \{r + k\delta | r \in \Phi, k > 0\} \cup \{r | r \in \Phi^+\}. \tag{2.16}$$

Cells. The set $E_0 - \cup H_{r,k}$ is disconnected, its connected components are called cells. A cell $C_{\tilde{\Pi}}$, $\tilde{\Pi} = \{r_i + k_i\delta,\ i = 0, ..., r\}$, is characterized by the fundamental system given by the roots, which define its bounding hyperplanes and which point into it:

$$C_{\tilde{\Pi}} = \{x \in E_0\ |(r_i, x) \geq k_i,\ \forall i\}. \tag{2.17}$$

The faces of a cell are defined as earler, and two cells which have a face in common are called adjacent.

*Representation of the Weyl Group Elements by Words.*We denote by E_p the affine vector space centered at some point p. $E_0 = \{0, x\}$, $E_p = \{p, x\}$, with the identification $\{0, x\} = \{p, x - p\}$. To simplify the notations, $\{0, x\}$ is denoted x. In the following, we shall restrict to the spaces E_p, with p in the coweight lattice: $p \in P^V$.

Let us define operators which intertwine E_p and E_{p+a}, and which are equal to one when we use the equivalence relation to identify the two spaces. They simply translate the origin of the affine space. They are defined by:

$$t^a\{p, x\} = \{p + a, x - a\}.$$

We also consider the affine reflections x_r for $r \in \Phi$ acting in E_p as:

$$x_r\{p, x\} = \{p, w_r(x)\}.$$

Inside the parentheses, w_r denotes the Weyl reflection defined earlier. Using the equivalence relation to identify E_p with E_0, the reflection $x_r|_{E_p}$ is identified with $w_{r+(r,p)\delta}$.

Given two cells C and C', there is a unique element w of the affine Weyl group, which maps the first one into the second one. Let us here construct a word representing this element using the elementary transformations x_r and t^a. We consider a sequence of cells $C_0 = C, C_1, ..., C_m = C'$, such that two consecutive cells are adjacent. To simplify the discussion, we fix C to be defined by the fundamental system (2.15). We consider a linear transformation acting in E_0 as follow:

$$S = t^{a_m} x_{r_{m-1}} t^{a_{m-1}} x_{r_{m-2}} ... t^{a_1} x_{r_0} t^{a_0}. \tag{2.18}$$

The reflection x_{r_i} acts in $E_{a_i + a_{i-1} + ...}$ and maps the cell C_i onto the cell C_{i+1}. The origin $a_i + a_{i-1} + ...$ must therefore belong to the common wall of the cell C_i and C_{i+1}. The root r_i is orthogonal to this wall, and we take it to point into the cell C_i. It is clear that S intertwines E_0 with E_p where: $p = a_0 + a_1 + ... + a_m$.

Using the equivalence relation to identify E_0 with E_p, one can view the transformation S in the affine Weyl group. Conversely any element w in the affine Weyl group is characterized by a cell C' and can be represented by a word (2.18). We can always choose $p \in Q^V \cap C'$; in this way w characterizes a unique point $p \in Q^V$ which is a summit of C'. In this case, we can identify the Weyl group element w with the word denoted S_w which represents it.

Here, we need to be more general, and we take $p \in P^V \cap C'$. Unless $P^V = Q^V$, this does not determine a unique point p. In that case, we denote by $S_{w,p}$ the word (2.18) which intertwines E_0 and E_p, and which coincides with the element w in the affine Weyl group. p is a summit of C' and can be written in a unique way as

$$p = q - \gamma, \tag{2.19}$$

where $q \in Q^V \cap C'$.

Scattering Matrix. Let us restrict to the case where C' is obtained from C by a translation :

$$C' = C + \xi. \tag{2.20}$$

In order for C' to define a cell, ξ must be in P^V. We denote by w_ξ the corresponding element in the affine Weyl group. (w_ξ is not a translation unless ξ is in the coroot lattice Q^V.) In the construction of $S_{w_\xi,p}$, one can choose $p = \xi$. We denote by S_ξ the corresponding word $S_\xi = S_{w_\xi,\xi}$.

In this case, γ defined in (2.19) must be a summit of C; therefore, either $\gamma = 0$, or $\gamma \in P^V$ and $(\gamma, r_m) = 1$, where r_m is the largest root of Φ. Such weights are called minuscule weights and are representatives of P^V/Q^V in the weight lattice.

Let us now consider that the sequence of cells C_i is defined up to a translation. We can view S_ξ as intertwinning E_p and $E_{p+\xi}$ for $p \in P^V$ arbitrary Then it is clear that two words S_ξ and $S_{\xi'}$ can be multiplied, and one has:

$$S_{\xi_1} S_{\xi_2} = S_{\xi_1+\xi_2}. \tag{2.21}$$

Thus these linear transformations form a commutative group isomorphic to the translations in the coweight lattice. Note that the above relation is not true for the Weyl group elements: $w_{\xi_1} w_{\xi_2} \neq w_{\xi_1+\xi_2}$, unless ξ_2 is in the root lattice.

Remark. We can be slightly more general and define a semi-group multiplication law on the words $S_{C_1,p_1}^{C_2,p_2}$, which transform C_1, p_1 into C_2, p_2 by defining the product:

$$S_{C_2,p_2}^{C_3,p_3} S_{C_1,p_1}^{C_2,p_2} = S_{C_1,p_1}^{C_3,p_3}. \tag{2.22}$$

Length. We can define the length of a Weyl group element w as the minimal length of S_w in terms of operators x_r. The length is also given by the expression which generalizes (2.5), (2.6) in an obvious way:

$$l(w) = |\tilde{\Phi}^+ \cap \tilde{\Phi}'^-|, \tag{2.23}$$

where $\tilde{\Phi}'^- = w(\tilde{\Phi}^-)$. The same definition applies to the generalized words $S_{w,p}$, since p is not relevant in the definition of the length.

In the case where the word (2.18) is of minimal length, one has:

$$\tilde{\Phi}^+ \cap \tilde{\Phi}'^- = \{r_i + \delta \sum_{j=0}^{i-1} (r_i, a_j)\}. \tag{2.24}$$

Consider the dominant weights defined by:

$$P^V_{\text{dom}} = \{\xi \in P^V | (\xi, r) \geq 0 \,, \forall r \in \Pi\}. \tag{2.25}$$

Let us restrict to the words S_ξ where ξ is a dominant weight: $\xi \in P^V_{\text{dom}}$. In this case it is not difficult to see that:

$$\tilde{\Phi}^+ \cap \tilde{\Phi}'^- = \{r + \delta k | \;\; r \in \Phi^+, \;\; 0 \leq k < (\xi, r)\}, \tag{2.26}$$

and we have:

$$l(w_\xi) = \sum_{r \in \Phi^+} (\xi, r). \tag{2.27}$$

Moreover, it follows from (2.24) that the reduced expression of S_ξ, the operator x_r, occurs exactly (r, ξ) times, and the operator x_{-r} never occurs.

It is also straightforward to construct a word of minimal length by extending the method used in the non-affine case. A geometrical interpretation of this construction is the following: One draws a straight line joining an arbitrary point A in the cell C to its translated $A' = A + \xi$ in C'. Travelling along this geodesic one goes successively through the cells C_i. The word S_ξ is obtained taking the product (from the right to the left) of the reflections across the hyperplanes, which are successively come through by the trajectory. The translations t^a must be inserted in order for the origin of the affine space to belong to the reflecting hyperplane, and one must have $\sum_{i=0}^m a_i = \xi$.
Definition of Words by Generators and Relations. We now proceed as before and replace the reflection operators x_r and the translation operators t^a by abstract operators. We impose relations to these operators so that the two words $S_{w,p}$ and $S_{w',p'}$ are equal when:
a) they both intertwine E_0 with the same affine space E_p.
b) After identifying E_0 with E_p, the two affine Weyl group elements w and w' coincide.

For this we impose the conditions (2.10) and (2.11) to the operators x_r, and for the operators t^a we require:
1) that they form an additive group:

$$t^a t^b = t^{a+b}, \tag{2.28}$$

2) that t^a commutes with x_r, whenever r and a are orthogonal:

$$t^a x_r = x_r t^a, \text{ if } (a, r) = 0. \tag{2.29}$$

The elements S_ξ constructed in that way clearly commute, we call them scattering matrices. Using this construction we have obtained a method to construct a group of commuting operators isomorphic to the coweight lattice.

Example: For example let us consider the case where the Weyl group is A_n. If we take the vector $\xi = -e_i$, the scattering matrix is given by:

$$S_{-e_i} = x_{e_i-e_{i-1}} x_{e_i-e_{i-2}} \dots x_{e_i-e_1} t^{-e_i} x_{e_i-e_n} \dots . x_{e_i-e_{i+1}}. \tag{2.30}$$

In the case where the t^a are equal to 1, these coincide with the scattering matrices considered by Yang.

In the next section, we shall use these transfer matrices to construct the q-analogue of the Dunkl operators, and we shall show that they obey the defining relations of a affine Hecke algebra.

3. The Affine Hecke Algebra

In this section, we use the previous formalism to construct some representations of the affine Hecke algebra. In the first part, we recall some well known results about representations of the Yang-Baxter algebra, using generators which obey the (non-affine) Hecke algebra relations. This gives a certain representation of the scattering matrices. In the second part, we consider a limiting form of these scattering matrices to obtain a representation of the affine Hecke algebra. In the third part, we derive a representation of the Hecke and of the affine Hecke algebra, acting in the group algebra of the weight lattice. We obtain in this way the q-Dunkl operators.

3.1 Representation of the Yang-Baxter Operators

Hecke Algebra. Let us consider a set of operators g_r, indexed by the roots $r \in \Phi$, on which the Weyl group acts in a natural way:

$$w_r g_s = g_{w_r(s)} w_r, \tag{3.1}$$

and which obey the following relations:

a) Braid group relations:

Let r and s be two fundamental roots in Π, and $m_{r,s}$ be the order of $w_r w_s$; then:

$$g_r g_s \dots = g_s g_r \dots, \quad m_{r,s} \text{ terms on either side;} \tag{3.2}$$

b) Hecke Relations:

$$(g_r - q_r)(g_r + q_r^{-1}) = 0, \tag{3.3}$$

where q_r is a complex number which depends on length of the root r.

The subset g_r with $r \in \Pi$ generates the Hecke algebra associated to Φ defined by:

$$g_w g_{w'} = g_{ww'} \ \text{ if } \ l(ww') = l(w) + l(w'). \tag{3.4}$$

Note that we have $g_{w_r} = g_r$ only for $r \in \Pi$.

Group Algebra of the Weight Lattice. The the group algebra of the Weight lattice is denoted P and is called the space of spectral parameters:

$$P = \{\sum c_\lambda e^\lambda | \lambda \in P^V\}, \tag{3.5}$$

with the product given by:

$$e^\mu e^\lambda = e^{\lambda+\mu}. \tag{3.6}$$

The Weyl group is not acting on P: $w_r e^\lambda = e^\lambda w_r$.
Yang-Baxter Generators. It is straightforward to verify that the following operators obey the unitarity and the Yang-Baxter equation (2.10, 2.11):

$$x_r = w_r \frac{e^r g_r - g_r^{-1}}{e^r q_r - q_r^{-1}}. \tag{3.7}$$

Relations Satisfied by the t^a: In order for the operators t^a to obey the relations (2.29), we require that they commute with P and satisfy the following relations:

$$\begin{aligned} wt^a &= t^{w(a)} w, \ \forall w \in W; \\ t^a g_r &= g_r t^a, \text{ if } (r,a) = 0. \end{aligned} \tag{3.8}$$

Remarks:
a) Another way to realize (2.29) consists in having the operators t^a commute with the Weyl group and obey the following relation with P:

$$t^a e^r = e^{(r,a)} e^r t^a. \tag{3.9}$$

Such a realization occurs in the so called quantum Kniznik-Zamolodchikov equation.
b) Let us consider the case of a root system, for which the third condition defining a root system is not satisfied (nonreduced system) and there are proportional roots r and $2r$. Those are the BC systems, for which the roots are: $\pm e_i$, $\pm 2e_i$, $\pm e_i \pm e_j$. In this case one can find an expression for x_r, which interpolates between the two forms of (3.7) for x_r and x_{2r}:

$$x_r = w_r \frac{e^{2r} g_r - g_r^{-1} + e^r \alpha}{e^{2r} q_r - q_r^{-1} + e^r \alpha}. \tag{3.10}$$

This formula coincides with x_{2r} in (3.7) if $\alpha = 0$ and $q_{2r} = q_r$. It coincides with x_r if $\alpha = q_r - q_r^{-1}$. It also satisfies the unitarity condition (2.28) and the Yang-Baxter equation (2.29).

3.2 Affine Hecke Relations

In this section, we consider the Hecke algebra (3.4). We complete it by adding a group of translations isomorphic to the coweight lattice generated by the S_ξ. Moreover, we require that the translations and the generators g_r obey the commutation relations which define an affine-Hecke algebra. In addition to the relations (3.4) satisfied by the g_r, $r \in \Pi$, and the additive group formed by the S_ξ, $\xi \in P^V$, we require that the g_r and the S_ξ obey the relations:

Let $r \in \Pi$ and $\xi \in P^V$; then:

$$\begin{array}{ll} i) & \text{if } (r,\xi)=0,\ g_r S_\xi = S_\xi g_r; \\ ii) & \text{if } (r,\xi)=1,\ g_r S_\xi = S_{w_r(\xi)} g_r^{-1}. \end{array} \tag{3.11}$$

Here we consider a limit of the operators S_ξ constructed in (2.18), which obey these relations. We consider the following limiting form (obtained when we set $e^r = \infty$ for $r > 0$ with respect to a fixed fundamental system Π) of the operators x_r (3.7):

$$\begin{aligned} x_r &= w_r g_r \text{ if } r \in \Phi^+, \\ x_r &= w_r g_r^{-1} \text{ if } r \in \Phi^-. \end{aligned} \tag{3.12}$$

Let us first show these relations in the case where ξ is a dominant weight, $\xi \in P^V_{dom}$. We recall that in this case S_ξ has a reduced expression:

$$S_\xi = t^{a_m} x_{r_{m-1}} t^{a_{m-1}} x_{r_{m-2}} \dots t^{a_1} x_{r_0} t^{a_0}, \tag{3.13}$$

where $r_i \in \Phi^+$, and x_{r_i} occurs (r_i,ξ) times in this expression.

Given a word S and $w \in W$, we denote by $w(S)$ the word obtained by substituting everywhere $t^{w(a)}$ and $x_{w(r)}$ for t^a and x_r in the expression of S. In general, $w(S) \neq wSw^{-1}$, because there can be some x_{r_i} in the expression of S, for which $r_i \in \Phi^+$ and $w(r_i) \in \Phi^-$. The main point of the following proof is to bring us back to a situation where the equality applies.

i) *Proof of the First Relation* (3.11): We consider the two words $x_r S_\xi$ and $w_r(S_\xi)x_r$. Both words intertwine E_0 with E_ξ, and after identifying the two spaces they coincide with the Weyl transformation w defined by its action on the cell C defined by (2.15): $w(C) = w_r(C) + \xi = w_r(C+\xi)$. Thus they are equal, and one has:

$$w_r g_r S_\xi = w_r(S_\xi) w_r g_r. \tag{3.14}$$

Now, since $(\xi, r) = 0$ and S_ξ is reduced, S_ξ contains only operators x_{r_i} with $r_i \neq r$, and $w_r(r_i)$ are all in Φ^+. Thus, $w_r(S_\xi) = w_r S_\xi w_r$. Substituting this equality in (3.14), $i)$ follows.

ii) *Proof of the Second Relation* (3.11): Since $(\xi, r) = 1$, one can write S_ξ in the form: $S_\xi = S' x_r$. S' is a word of length $l(S') = l(S_\xi) - 1$, and x_r does not appear in the expression of S'. Using the geometrical interpretation, one shows that:

$$x_{-r} w_r(S') = S_{w_r(\xi)}. \tag{3.15}$$

Substituting $w_r(S') = w_r S' w_r$, $x_{-r} = g_r^{-1} w_r$, and $S' = S x_{-r}$ in (3.15), relation *ii*) follows.

If ξ is not a dominant weight, we can always write $\xi = \xi_1 - \xi_2$, where ξ_1 and ξ_2 are dominant weights such that: $(\xi_1, r) = 0$ and $(\xi_2, r) = 0$ in case *i*), $(\xi_1, r) = 1$ and $(\xi_2, r) = 0$ in case *ii*). Writing $S_\xi = S_{\xi_1} S_{\xi_2}^{-1}$ we obtain the general result.

3.3 Another Presentation of the Affine-Hecke Algebra

Consider the affine generator given by:

$$g_{-r_m+\delta} = t^{-r_m/2} g_{-r_m} t^{r_m/2}. \tag{3.16}$$

We could have considered the algebra generated by the subset g_r with $r \in \tilde{\Pi}$ defined by the relations (3.2, (3.3), replacing the non-affine root system by the affine root system $\tilde{\Pi}$. Commuting the Weyl group elements w_r to the left of the word $S_{\tilde{w}}$, and reorganizing the translations t^a so as to make only the generators g_r with $r \in \tilde{\Pi}$ appear to the right, we obtain the following expression:

$$S_{\tilde{w}} = w^{-1} g_{\tilde{w}}, \tag{3.17}$$

where $g_{\tilde{w}}$ is in the subalgebra generated by the $g_r, r \in \tilde{\Pi}$, and w is the projection of $\tilde{w}$ in the Weyl group.

In the general case, we obtain the following expression:

$$S_{\tilde{w},p} = t^{-\gamma} w^{-1} g_{\tilde{w}}; \tag{3.18}$$

γ is defined in (2.19). In particular:

$$S_\xi = t^{-\gamma} w_\xi^{-1} g_{\tilde{w}_\xi}; \tag{3.19}$$

in this case γ is a minuscule weight. The left-hand side operator $t^{-\gamma} w_\xi^{-1}$ depends only on γ, the projection of ξ in P^V/Q^V. These operators can be added to the generators g_r with $r \in \tilde{\Pi}$ to give a presentation of the affine Hecke algebra by generators and relations.

3.4 Representation of the Hecke-Algebra in a Polynomial Space

In this section, we construct a representation of the scattering matrix (2.18) in a space of polynomials in several variables. If we use the expression (3.7) for the operators x_r, all we need is a representation of the operators g_r, $r \in \Pi$, such that the Hecke-algebra relations (3.1), (3.2), (3.3) are satisfyed. We also need a representation of the operators t^a satisfying (3.8). Such a representation of the Hecke algebra is known in the mathematic literature (Lusztig, Lascou and Schutzenberger). Our aim here is to deduce it from the Yang-Baxter equation. In doing so, we shall discover two different sets of operators, which obey the affine-Hecke relations: One is given by the S_ξ, the other by the polynomials on which the g_r act.

To construct the representation of the g_r, we recall that the group algebra of the weight lattice is denoted P. There is a natural action of the Weyl group on P given by:

$$se^\lambda = e^{s(\lambda)}s. \tag{3.20}$$

We consider the Hecke algebra (3.4) generated by f^r, $r \in \Pi$. Here the f^r commute with the Weyl group action and the multiplication by e^λ.

We define the operators y_r as follows:

$$y_r = s_r \frac{e^r f_r - f_r^{-1}}{e^r q_r - q_r^{-1}}. \tag{3.21}$$

It is easy to show that the Yang-Baxter equations (2.10), (2.11) are equivalent to the fact, that y_r with $r \in \Pi$ obey the defining relations of the generators of the Weyl group:

$$y_r^2 = 1, \tag{3.22}$$

$$(y_r y_s)^{m_{rs}} = 1, \tag{3.23}$$

where m_{rs} is the order of $w_r w_s$. It suffices for that to write $x_r = w_r z_r$ in (3.7), and to commute the Weyl group elements w_r to the left of the expressions (2.10), (2.11). If one does the same by commuting s_r to the left of ((3.22), (2.23), the identities to verify are the same in both cases.

Let us now factorize the group algebra: $\{\sum c_{\lambda,w,w'} e^\lambda s_w f_{w'}\}$ by the relation $y_r = 1$ to the right. We denote by $\pi(.)$ the projection, which consists in eliminating the reflections s_r to the right of an expression in the quotient. This operation eliminates the s_r in the following way:

$$\pi(...s_r) = \pi(...) \frac{e^r f_r - f_r^{-1}}{e^r q_r - q_r^{-1}}. \tag{3.24}$$

The consistency of this projection is assured by the relations (3.22), (3.23). Alternatively, we can define the projection π as follows:

$$\pi(...(q_r s_r + (q_r - q_r^{-1}) \frac{1}{e^r - 1}(s_r - 1))) = \pi(...)f_r. \tag{3.25}$$

Let us rewrite the above relation as $\pi(...g_r) = \pi(...)f_r$. The consistency of this operation is assured by the fact that the g_r obey the Hecke algebra relations. The expression for g_r is given by:

$$g_r = q_r s_r + (q_r - q_r^{-1}) \frac{1}{e^r - 1}(s_r - 1), \tag{3.26}$$

and the expression (3.12) of x_r is:

$$x_r = q_r + (q_r - q_r^{-1}) \frac{1}{e^{-r} - 1}(1 - s_r). \tag{3.27}$$

One can verify, that this representation of the g_r for $r \in \Pi$ and the translations e^λ obey the defining relations (3.11) of an affine Hecke algebra. In fact, it is easy to see that the relations (3.11) are equivalent to the following relations:

$$g_r Q = s_r(Q) g_r + (q_r - q_r^{-1}) \frac{1}{e^r - 1}(s_r(Q) - Q), \tag{3.28}$$

where Q is in the group algebra of the weight lattice. So the above representation is simply obtained by considering the action of g_r on the group algebra P, which satisfy the affine Hecke relations with g_r, and setting it equal to q_r to the right of an expression.

One important consequence of (3.28) is that any polynomial in P, which is Weyl invariant ($s_r(Q) = Q$ for all $r \in \Pi$), is in the center of the affine Hecke algebra: $[g_r, Q] = 0$ for all $r \in \Pi$.

Finally, in order to obtain a representation of the operators S_ξ, we must give the realization of the operators t^a, which satisfies the relations (3.8). The operators t^a obey the following commutation relations with P:

$$t^a e^\lambda = e^{(\lambda, a)} e^\lambda t^a. \tag{3.29}$$

Remarks: In (3.26), g_r is expressed in terms of the reflection s_r. Alternatively, we can express the reflections s_r in terms of the generators g_r. We obtain an expression very similar to (3.21):

$$s_r = \frac{e^r g_r - g_r^{-1}}{e^r q_r - q_r^{-1}}; \tag{3.30}$$

but now, the Weyl group relations (3.22), (3.23) satisfied by s_r rely on the fact, that e^r and g_r obey the defining relations of the affine Hecke algebra.

In this representation, there are two sets of operators which obey the affine Hecke relations with the operators g_r. One is the group algebra P generated by the spectral parameters e^ξ, and the other is the group algebra generated by the S_ξ computed as in (3.13).

4. Affine Hecke Algebra, Quadratic Algebras and Physical Models

Our aim is to relate the Affine-Hecke relations to the quadratic algebras. In this section we are concerned with specific representations of the quadratic algebras which use the x_r. We show that the spectral parameters, which enter the definition of x_r in (3.7), can be taken to obey the affine-Hecke relations (3.11). By considering the operators which commute with the quadratic algebras, one obtains a commuting set of Hamiltonians which describe physical models.

In the first part, we give a brief description of the quadratic algebras which we need in the following. In the second part, we describe the procedure of quantization of the spectral parameters of the monodromy matrices, which obey the quadratic relations. Finally, in the fourth part, we give a brief description of the physical models which result from this construction.

4.1 Quadratic Algebras

Let us consider the quadratic relations for the matrices T_r:

$$x_{e_a - e_b} T_{e_a} T_{e_b} = T_{e_b} T_{e_a} x_{e_a - e_b}, \tag{4.1}$$

and

$$x_{e_a - e_b} T_{e_a} x_{e_a + e_b} T_{e_b} = T_{e_b} x_{e_a + e_b} T_{e_a} x_{e_a - e_b}. \tag{4.2}$$

Note that these quadratic algebras are obtained from the Yang-Baxter equations (2.11) with $m_{r,s} = 3, 4$, by replacing two of the generators by operators called "monodromy matrices". The monodromy matrix T_{e_a} can be expanded in the parameter e^a called its spectral parameter. A realization of the relations (4.1), (4.2) is respectively given by:

$$T_{e_a} = x_{e_a - e_n} x_{e_a - e_{n-1}} ... x_{e_a - e_1}, \tag{4.3}$$

$$T_{e_a} = x_{e_a - e_n} x_{e_a - e_{n-1}} ... x_{e_a - e_1} x_{e_a} x_{e_a + e_1} ... x_{e_a + e_n}. \tag{4.4}$$

The vectors $e_a, e_b, e_1, ..., e_n$ form an orthogonal basis. We denote V the vector space spanned by the e_i, $i = 1, ..., n$. e_a, e_b are called auxiliary vectors. The relations (4.1), (4.2) result from the Yang-Baxter equations (2.11) satisfied by the x_r. In what follows, we consider the representation (3.7) of the operators x_r in terms of generators f_r of the Hecke algebra and w_r of the Weyl group:

$$x_r = w_r \frac{e^r f_r - f_r^{-1}}{e^r q_r - q_r^{-1}}. \tag{4.5}$$

We recall, that in this formula the reflection w_r acts on the Hecke generator f_r, and that the spectral parameter e^r commutes with f_r and w_r.

4.2 Quadratic Algebras and Affine Hecke Algebras

The Weyl group of A_n (S_n) acts by permuting the vectors e_i of V in the first case (4.1), and the Weyl group of B_n, C_n acts by permuting the vectors and taking their opposite in the second case (4.2). In each case, we consider the fundamental systems given by:

$$\Pi = \{e_1 - e_2, ..., e_{n-1} - e_n\}, \tag{4.6}$$

$$\Pi = \{e_1 - e_2, ..., e_{n-1} - e_n, e_n\}. \tag{4.7}$$

The generators f_r with $r \in \Pi$ generate a Hecke algebra. Let us consider another realization of the Hecke algebra (3.4) generated by g_r with $r \in \Pi$. The g_r commute with the w_r and the f_r.

We factorize the group algebra of the g_r and the f_r by the relation $f_r = g_r$ to the right. We denote $\pi(.)$ the operation which consists in eliminating the generators g_r in the quotient: $\pi(...g_r) = \pi(...)f_r$. We now require, that the quadratic relations (4.3), (4.4) are still satisfied, when one replaces the monodromy matrices T_{e_a} by their projection $\pi(T_{e_a})$. Let us show that the relations are satisfied, if the spectral parameters e^r obey the defining relations (3.11) of the affine Hecke algebra with the generators g_r.

In order for the quadratic relations (4.1), (4.2) to be satisfied after taking the quotient, the condition to be satiesfied is:

$$\pi(g_r T_a) = f_r \pi(T_a) \quad \text{for } r \in \Pi. \tag{4.8}$$

It ensures that $\pi(T_a T_b) = \pi(T_a)\pi(T_b)$, and therefore that the relations (4.1), (4.2) are satified, when one substitutes $\pi(T_a)$ for T_a in them.

To verify (4.8) one commutes the generators g_r, $r \in \Pi$, through T_a, using the affine Hecke relations with the spectral parameters which enter the definition of the x_r. Once g_r has been pushed to the right, one replaces it by f_r, and one commutes f_r back to the left, using the Hecke relations and the fact that that the Weyl reflections w_r act on the f_r. The computation is simplified by the fact that the denominator of T_a is a Weyl invariant polynomial in P. It therefore commutes with the g_r, and one can keep only the numerator of the x_r, ($x_r = w_r(e^r f_r - f_r^{-1})$).

4.3 Physical Models

To simplify the discussion, we restrict ourselves to the A_{n-1} case (4.1). We consider n particles with coordinates θ_j on the unit circle. Each particle carries a spin denoted σ_j. They are described by a wave function $\Psi(z_j)$, which is supposed to be polynomial in the variables $z_j = e^{i\theta_j}$.

There is an action of the Hecke algebra on the coordinates which follows from (3.26):

$$g_{j,j+1} = q s_{j,j+1} + (q - q^{-1}) \frac{z_{j+1}}{z_j - z_{j+1}} (s_{j,j+1} - 1), \tag{4.9}$$

where the permutation $s_{j,k}$ permutes the coordinates z_j and z_k. There is also a representation of the Hecke algebra on the spin variables denoted $f_{j,j+1}$ ($f_{j,j+1}$ is supposed to act on the spin of the particles j and $j+1$).

The operators $t_i = t^{e_i}$ act by shifting the variable θ_i:

$$t_i \Psi(z_1, ..., z_i, ...) = \Psi(z_1, ..., t z_i, ...). \tag{4.10}$$

We call "Physical states" those for which the two Hecke algebras act in the same way:

$$g_{j,j+1}\Psi_{\text{phys}} = f_{j,j+1}\Psi_{\text{phys}}. \tag{4.11}$$

It is clear, that on these states we can replace the action of the operators $g_{j,j+1}$ by $f_{j,j+1}$. The operation which consists in replacing the generators $g_{j,j+1}$ by $f_{j,j+1}$ coincides with the projection $\pi(...g_{j,j+1}) = \pi(...)f_{j,j+1}$ described before. An operator O acting on physical states preserves this space, if $\pi(g_{j,j+1}O) = f_{j,j+1}O$.

Let us denote by S_j the Hecke generators S_{e_j} constructed in (2.30). S_j and $g_{k,k+1}$ obey the relations:

$$\begin{aligned} [g_{k,k+1}, S_j] &= 0 \text{ if } j \neq k, k+1; \\ g_{j,j+1}S_j &= S_{j+1}g_{j,j+1}^{-1}. \end{aligned} \tag{4.12}$$

We consider the following Hamiltonians acting on physical states:

$$H_l = \sum_{1\le i_1<i_2..<i_l\le n} S_{i_1}...S_{i_l}. \tag{4.13}$$

These operators obviously commute with each other. Moreover, since they are symmetric polynomials in the S_j, they commute with $g_{j,j+1}$ $\forall j$. Thus their action preserves the physical states. In the simple case where $f_{j,j+1} = q$ (or $-q^{-1}$) (the scalar case), the projection of these Hamiltoniens by π coincides with the trigonometric models defined by Ruijenhaars. In this case, the Hamiltonians are invariant under permutations of the coordinates, which makes the identification with Ruijenhaars models easy to do. Unfortunately, in the general case, this projection is much more difficult to describe explicitly.

One can also construct the monodromy matrix (4.3), the expression of $x_{e_a-e_i}$ being:

$$x_{e_a-e_i} = s_{a,i}\frac{e^u f_{e_a-e_i} - S_i f_{e_a-e_i}^{-1}}{e^u q - S_i q^{-1}}. \tag{4.14}$$

Here e^u is the spectral parameter of the monodromy matrix. Because of (4.8), the monodromy matrix also preserves the physical states, and it commutes with H_l. This remains true when one replaces the operators by their projection by π. Thus, the algebra defined by the quadratic relations (4.1) is a symmetry algebra for the H_l.

Let us show that the operators S_j constructed in this way can be diagonalized simultaneously and are unitary.

Eigenvalues of the S_j: We show that the operators S_j are represented by triangular matrices. Let us recall the expression for the S_j:

$$S_j = x_{j-1,j}^{-1}...x_{1,j}^{-1}t_j x_{j,n}...x_{j,j+1}, \tag{4.15}$$

where the operator $x_{i,j}$ takes the limiting form for $i < j$:

$$x_{i,j} = q + (q - q^{-1})\frac{z_i}{(z_i - z_j)}(s_{i,j} - 1); \tag{4.16}$$

$x_{i,j}$ commutes with $z_i z_j$ and with z_k for $k \neq i,j$. It acts in the following way on the monomials z_i^m, z_j^m, $m>0$, and 1:

$$\begin{aligned} x_{i,j} z_i^m &= q^{-1} z_i^m - (q - q^1)(z_i^{m-1} z_j + ... + z_i z_j^{m-1}) \text{ for } m > 0, \\ x_{i,j} z_j^m &= q z_j^m + (q - q^{-1})(z_i^m + ... + z_i z_j^{m-1}) \text{ for } m > 0, \\ x_{i,j} z_i^0 &= 0. \end{aligned} \tag{4.17}$$

To a monomial $z_1^{k_1}...z_n^{k_n}$ we associate a partition $|k| = (k_{p_1} \geq k_{p_2} ... \geq k_{p_n})$, where we order the k_j in decreasing order.

Let us consider, which new monomials $z_1^{k'_1}...z_n^{k'_n}$ can appear, when $x_{i,j}$ acts on this monomial. First, all the k'_l for $l \neq i,j$ are equal to k_l. Then, if the partition of the new monomial is different from $|k|$, it can be obtained from $|k|$ by a sequence of squeezing operations:

$$(..., k_i, ..., k_j, ...) \to (..., k_i - 1, ..., k_j + 1, ...) \text{ if } k_i > k_j;$$

$$(..., k_j, ..., k_i, ...) \to (..., k_j - 1, ..., k_i + 1, ...) \text{ if } k_i < k_j.$$

Finally, if the partition of the new monomial is equal to $|k|$, $k'_i = k_i$ and $k'_j = k_j$ if $k_i > k_j$.

Let us define an order of the monomials by saying, that $z_i^{k_i}$ is larger than $z_i^{k'_i}$, if either $|k'|$ is obtained from $|k|$ by a sequence of squeezing operations, or $|k| = |k'|$ and $k'_i - k_i > 0 \; \forall i$. It follows from the above analysis, that the action of S_j on a monomial produces only monomials, which are smaller with respect to this order. Thus the eigenvalues of the operators S_j are given by the diagonal elements in the monomial basis.

Given the partition $|k| = (k_1, ..., k_n)$, the eigenvalues corresponding to the monomials associated to it are all obtained by permutations of the multiplet:

$$(S_j) = (t^{k_j} q^{n+1-2j}). \tag{4.18}$$

If we set $q = t^{\beta/2}$ with β a real parameter, the operators S_j have the physical interpretation of exponentials of momentum operators: $S_j = t^{K_j}$.

$$K_j = k_j + \beta(\frac{n+1}{2} - j)$$

obey a generalized Pauli principle, since they must be β apart from each other.

Scalar Product. A scalar product can be defined [5], so that the operators $g_{j,j+1}$ and S_j are unitary: $g_{j,j+1}^+ = g_{j,j+1}^{-1}$, $S_j^+ = S_j^{-1}$. For $q = t^{k/2}$, with k an integer, the scalar product is given by:

$$< \Psi_1, \Psi_2 >= \int \prod_{i=1}^{n} d\theta_i \; \overline{\Psi}_1(z_j) \Psi_2(z_j) C(z_j), \tag{4.19}$$

where the bar symbol stands for the complex conjugation. q, t, z_j are supposed to be complex numbers of modulus 1, and the integration over θ_j keeps the coefficient of z^0 of the integrand.

The measure $C(z_k)$ is given by:

$$C(z_j) = \prod_{i<j} \prod_{l=-k}^{k-1} (\sqrt{t^l z_i/z_j} - \sqrt{t^{-l} z_j/z_i}). \tag{4.20}$$

In the case of the Ruijenhaars models, many properties of the wave functions Ψ can be obtained, for example their norms. We refer to [4] for a more complete analysis of their properties.

References

1. D.Bernard, M.Gaudin, F.D.M.Haldane and V.Pasquier, J.Phys. **A 26** (1993) 5219.
2. M.Gaudin, *La fonction d'onde de Bethe*, Edition du CEA, Masson 1983.
3. N.Bourbaki, *Groupes et algebres de Lie*, ch 4,5,6.
4. I.G. Macdonald, *Seminaire N.Bourbaki*, expose 797,1995.
5. I. Cherednik, C.M.P 169, **441**, (1995) 461.
6. R.W.Carter, *Simple groups of Lie type* (1972).
7. G.Lusztig, J. Amer. Math. Soc. 2, **599**, (1989) 635.
8. V.Pasquier, Lecture Notes in Physics, **436** (1993) 36 .

Some Algebraic and Analytic Structures in Integrable Systems

Nicolai Reshetikhin

Department of Mathematics, University of California, Berkeley, CA 94720, USA

1. Introduction

These notes are focused on certain algebraic and analytic aspects of the theory of classical and quantum integrable systems.

As is well known by now, the theory of classical integrable systems is naturally related to the geometry and representation theory of Poisson-Lie groups and the corresponding Lie bialgebras. At the same time quantum integrable systems are related to the representation theory of Hopf algebras.

Perhaps the most interesting analytical problems appear in the theory of infinite dimensional integrable systems. In particular, in questions related to the behavior of finite dimensional systems in the infinite dimensional limit. Integrable systems demonstrate a rich symmetry in comparison to nonintegrable ones, which allows us in many cases to compute the spectrum, and in some cases the correlation functions, for infinite dimensional systems.

In Section 2 some basic facts about classical integrable systems, Lie bialgebras, Poisson-Lie groups and their role in the theory of integrable systems are reviewed. The special role of integrable classical field theories is emphasized, and a classical integrable system describing a spin chain is presented as an example of an integrable model related to the Lie group $\widehat{SL_2}$. Only Hamiltonian classical field theories are mentioned. The Lagrangian formulation, its relation to the Hamiltonian formulation, and problems related to degenerate Lagrangians and gauge theories are not discussed, since it seems that up to now there are no sufficiently interesting examples of integrable systems of this type.

Section 3 contains a general discussion of quantum integrable systems, the role Hopf algebras play in the theory, and, as an example, it contains the quantization of the classical spin chain described in Section 2. It is shown how to obtain the generalized Heisenberg spin chains in this way.

Section 4 is devoted to a discussion of the thermodynamic limit in the example of the Heisenberg spin chain. This is where the most interesting analytical part of the theory appears. In this limit the finite dimensional Hilbert space of the spin chain on the finite lattice becomes a nonseparable Hilbert space. Its decomposition into the direct integral of the separable spaces is discussed.

The conclusion contains a summary of some of the problems discussed in the text.

2. Algebraic Structures in Classical Integrable Systems

In this section some basic facts about classical integrable systems and their corresponding algebraic structures will be reviewed.

2.1

The notion of an integrable system is classical mechanics in well established and goes back to Liouville's theorem. It tells us that if, on an n-dimensional symplectic manifold $\mathcal{M}$, there exist n functionally independent functions $I_1, \ldots, I_N$, which commute with respect to the Poisson bracket on $\mathcal{M}$, then the following holds:

- The "level surfaces"
$$\mathcal{M}(c_1, \ldots, c_n) = \{x \in \mathcal{M}, I_i(x) = c_i\} \tag{2.1}$$
are invariant with respect to the flow of any Hamiltonian $H = F(I_1, \ldots, I_N)$.
- For every such Hamiltonian and every level surface $\mathcal{M}(c_1, \ldots, c_n)$, there exists an affine coordinate system $(p_1, \ldots, p_n)$, in which the Hamiltonian flow generated by H is linear: p_i =constant.
- The coordinate system $(p_1, \ldots, p_n)$ on $\mathcal{M}(c_1, \ldots, c_n)$ can be completed to a canonical coordinate system (p_i, q_j) in every sufficiently small neighborhood of $\mathcal{M}(c_1, \ldots, c_n)$. These coordinates are called action-angle variables, and in some cases these coordinates are global.

Let L and M be two $N \times N$ matrices, which are also functions on the phase space. Such a pair is called a Lax pair, if the equation
$$\frac{dL}{dt} = [L, M]$$
is equivalent to the equations of motion for the elements of L, and if the matrix elements of L form a complete independent set of functions on $\mathcal{M}$.

If a Hamiltonian system admits a Lax representation, and if f is a function on $N \times N$ matrices, which is invariant with respect to conjugations, then the function $I_f = f(L)$ is an integral of the motion.

Classical Hamiltonian field theories represent an important class of Hamiltonian dynamical systems. Mathematically, they are characterized by the following data:

- a world space (a topological space);
- a target space (a smooth manifold);
- a symplectic structure on the space of fields (fields are maps from a world space to a target space);
- a Hamiltonian which is a smooth function on the space of fields.

The most interesting cases are when the world space is either a manifold with a metric on it, or a "discrete space" which is usually a cellular complex (a collection of 0-dimensional cells (vertices), 1-dimensional cells (edges), 2-dimensional cells (plaquettes), etc., glued together in one complex). If translational symmetry is required in the discrete case, the world space is a lattice. The corresponding field theory in this case is called lattice field theory (although the same terminology is frequently used for any field theory with a discrete space). Here we will restrict ourselves to a discussion of spin chains only, in which case the world space is the set of vertices of a cyclic graph of order N. It can be also regarded as a cellular complex with N 0-cells and N 1-cells, such that the boundary of every 1-cell consists of two distinct 0-cells, and every 0-cell is a boundary of two distinct 1-cells. Infinite chains are the limit $N \to \infty$ of this.

A classical Hamiltonian field theory is called integrable, if it is an integrable Hamiltonian system.

Now let us focus on the algebraic aspects of integrable classical Hamiltonian systems.

2.2

After a long period of experimentation with the Hamiltonian interpretations of Lax pairs, and of discovering new patterns in the theory of classical integrable systems, the role of Lie bialgebras and Poisson-Lie groups in the construction of practically all known integrable systems has been established.

Let us recall the notion of a Lie bialgebra. A pair consisting of a Lie algebra $\mathfrak{g}$ and a map $\delta : \mathfrak{g} \to \mathfrak{g} \wedge \mathfrak{g}$ is called a *Lie bialgebra* [1], if

– δ is a 1-cocycle, i.e.

$$\delta([x, y]) = [x, \delta(y)] + [\delta(x), y], \tag{2.2}$$

where $[x, a \wedge b] \equiv [x, a] \wedge b + a \wedge [x, b]$,

– δ satisfies the co-Jacobi identity $(\delta \wedge id)\delta = 0$.

If $(\mathfrak{g}, \delta)$ is a Lie bialgebra and $\mathfrak{g}^*$ is a vector space dual to $\mathfrak{g}$ with the pairing $\langle \ldots \rangle$, the following operations determine the Lie bialgebra structure on $\mathfrak{g}^*$:

$$\langle [x, y]_*, a\rangle = \langle x \wedge y, \delta(a)\rangle,$$
$$\langle \delta_*(x), a \wedge b\rangle = \langle x, [a, b]\rangle,$$

where $\langle x \wedge y, b \wedge c\rangle = \langle x, b\rangle\langle y, c\rangle - \langle x, c\rangle\langle y, b\rangle$.

Thus, we can consider any Lie bialgebra as a pair of Lie algebras $\mathfrak{g}$ and $\mathfrak{g}^*$ with the compatibility condition (2.2).

A Lie bialgebra $(\mathfrak{g}, \delta)$ is called *quasitriangular*, if there is an element $r \in \mathfrak{g} \otimes \mathfrak{g}$ such that

1. $I = r + \sigma(r)$ is a $\mathfrak{g}$ -invariant symmetric tensor in $\mathfrak{g} \otimes \mathfrak{g}$;

2. $\delta(a) = -ad_a(r)$ where $ad_a(x \otimes y) = [a,x] \otimes y + x \otimes [a,y]$;
3. $(\delta \otimes id)(r) = [r_{13}, r_{23}]$ and $(id \otimes r)(r) = [r_{13}, r_{12}]$.

Here and below we use the following notation for the elements of the tensor product of Lie algebras: If $r = \sum_i r^i \otimes r_i$, then $[r_{12}, r_{23}] = \sum_{ij} r^i \otimes [r_i, r^j] \otimes r_j$,etc..

One of the corollaries of this definition is that r satisfies the classical Yang-Baxter equation

$$[r_{12}, r_{13}] + [r_{12}, r_{23}] + [r_{13}, r_{23}] = 0. \tag{2.3}$$

If the element $I = r + \sigma(r)$ in $\mathfrak{g} \otimes \mathfrak{g}$ determines a non-degenerate bilinear form on $\mathfrak{g}^*$, the Lie bialgebra $\mathfrak{g}$ is called *factorizable.* There exists a double construction of Lie bialgebras known also as the Manin triple, which produces a factorizable Lie bialgebra for any given Lie bialgebra.

The next important notion is a Poisson-Lie group. A Lie group G is called a *Poisson-Lie group*, if G is a Poisson manifold and the multiplication $G \times G \to G$ is a Poisson map.

It is known that, if $\mathfrak{g}$ is a Lie bialgebra, it can be "exponentiated" to a unique connected, simply connected, Poisson-Lie group. Since every Lie bialgebra comes with its dual counterpart, it determines two Poisson-Lie groups, G and G^*. The duality between these two Poisson manifolds can be described directly in terms of dual pairs of Poisson manifolds [2], for details see [5].

Let G be a Lie group with the quasitriangular Lie bialgebra $\mathfrak{g}$, and let $\pi : G \to End(V)$ be a finite dimensional representation of G. The matrix elements of π relative to some basis in V satisfy the following Poisson brackets:

$$\{\pi \otimes \pi\} = [r^{\pi\pi}, \pi \otimes \pi], \tag{2.4}$$

where $r^{\pi\pi} = (\pi \otimes \pi)(r)$. In cases when functions π generate algebras of functions, the brackets (2.4) determine the Poisson algebra structure.

If G is a Poisson-Lie group, and if the corresponding Lie bialgebra is factorizable, then the Poisson-Lie group G is called factorizable. This particular class of Poisson-Lie groups plays an important role in the theory of classical integrable systems.

If $\mathfrak{g}$ is a quasitriangular Lie bialgebra, the element r in $\mathfrak{g} \otimes \mathfrak{g}$ determines two maps, $r_\pm$, from $\mathfrak{g}^*$ to $\mathfrak{g}$, which are homomorphisms of Lie bialgebras [3]. The images of these maps in $\mathfrak{g}$ are Lie subbialgebras $\mathfrak{g}_\pm$ in $\mathfrak{g}$. The map from $\mathfrak{g}^*$ to $\mathfrak{g}$ determined by the element I is an isomorphism of vector spaces, if $\mathfrak{g}$ is a factorizable Lie algebra. The "exponentiation" of the maps $r_\pm$ provides two Poisson-Lie subgroups $G_\pm$ in G. The remarkable property of factorizable Poisson-Lie groups is, that (on a certain dense open subset of G) the factorization

$$g = g_+(g_-)^{-1} \tag{2.5}$$

exists and is unique. For simplicity we will assume here, that this factorization exists and is unique everywhere on G.

In the most interesting examples related to loop groups, the factorization problem is equivalent to Riemann-Hilbert factorization of matrix valued functions. In many cases this problem can be solved in terms of theta-functions (see, for example, [4] and references therein).

2.3

Let us formulate the theorem, which describes how to construct integrable classical systems using factorizable Poisson-Lie groups, and how to integrate the equations of motion of such systems in term of the factorization problem.

Let G be a factorizable Poisson-Lie group, and $\theta : G \to G$ be a Poisson-Lie automorphism of it (a Poisson map which is an automorphism of the group G).

Denote by $I_\theta(G) \hookrightarrow C(G)$ the subalgebra of functions on G invariant under twisted conjugations:

$$h \mapsto gh\theta(g^{-1}), \tag{2.6}$$

and let $\pi \to \mathrm{End}(V)$ be a finite-dimensional representation of the group G. By L_π we denote the element of $\mathrm{End}(V) \otimes C(G)$, which takes the value $L_\pi(g) = \pi(g)$ on the element $g \in G$.

Theorem 1.[3]

1. The algebra $I_\theta(G)$ is a commutative Poisson subalgebra of the Poisson algebra $C(G)$.
2. The Hamiltonian system on G generated by $H \in I_\theta(G)$ has the following Lax representation:

$$\frac{dL}{dt}(t) = L(t)M_\theta(g(t)) - M(g(t))L(t),$$

where

$$L_\pi(t) = L_\pi(g(t)), \quad M(g) = -\pi(r_+(\nabla H(g))), \quad M_\theta(g) = M(\theta(g)).$$

3. Let $g_\pm(t)$ be the solution to the following factorization problem:

$$\exp(t\nabla H(g)) = g_+(t)g_-(t)^{-1}.$$

Then

$$g(t) = g_\pm(t)^{-1}g\theta(g_\pm(t))$$

is the trajectory of the Hamiltonian system on G generated by H with $g(0) = g$. Here we assumed an identification between $\mathfrak{g}$ and $\mathfrak{g}^*$ via the form $r + \sigma(r)$.

For the proof of this theorem see [3, 5]. It has a long history which dates back to the works [49, 50] (see [4] and references therein).

This theorem describes how to construct certain Hamiltonian flows on factorizable Poisson-Lie groups, and how to integrate these flows using the factorization method. In order to have integrable systems in the sense of Liouville's theorem, one has to restrict the dynamics described by the theorem above to a symplectic leaf of the Poisson manifold G.

Consider the example of the integrable classical Hamiltonian field theories known as spin chains which are described by theorem 1.

2.4

Let G be a factorizable Poisson-Lie group. Consider the Poisson-Lie group $G_N := G^{\times N}$ with the product structure of a group and of a Poisson manifold:

$$(g_N, \ldots, g_1)(h_N, \ldots, h_1) = (g_N h_N, \ldots, g_1 h_1),$$

$$\{f_N \otimes \ldots \otimes f_1, k_N \otimes \ldots \otimes k_1\} = \sum_{i=1}^{N} f_N k_N \otimes \ldots \otimes \{f_i, k_i\} \otimes \ldots \otimes f_1 k_1 .$$

Here, we used the following notation:

$$(f_N \otimes \ldots \otimes f_1)(g_N, \ldots, g_1) = f_N(g_N) \ldots f_1(g_1) .$$

Clearly G_N is a factorizable Lie group:

$$(g_N, \ldots, g_1) = ((g_N)_+, \ldots, (g_1)_+)((g_N)_-, \ldots, (g_1)_-)^{-1} .$$

The map $\theta : G_N \to G_N$,

$$\theta((g_N, \ldots, g_1)) = (g_{N-1}, \ldots, g_1, g_N),$$

is an automorphism of the Poisson-Lie group G_N.

Remark 1.

We choose this particular automorphism of G_N in order to have a system with a specific physical interpretation. We want to construct a system which describes an interacting chain of atoms. In principle, one can choose any permutation in $G^{\times N}$ (or any other Poisson-Lie automorphism of G_N) as the θ-automorphism.

Conjugation by an element $g \in G_N$ twisted by the θ defined above acts as follows:

$$h = (h_N, \ldots, h_1) \mapsto (g_N h_N g_{N-1}^{-1}, g_{N-1} h_{N-1} g_{N-2}^{-1}, \ldots, g_1 h_1 g_N^{-1}) .$$

If $f \in C^\infty(G)$ is a function invariant under conjugation by elements in G, the function $f_\theta \in C^\infty(G_N)$, defined as

$$f_\theta(g_N, \ldots, g_1) = f(g_N \ldots g_1),$$

is invariant under twisted conjugations and therefore belongs to $I_\theta(G)$.

The equations of motion corresponding to the Hamiltonian $H \in I_\theta(G)$ have the following Lax representation:

$$\frac{dL_i}{dt} = L_i M_{i-1} - M_i L_i \ .$$

Here $M_i = \pi(r_+(\nabla_i H(g)))$, where

$$\xi(\nabla_i H(g)) = \frac{d}{ds} H(g_N, \ldots, e^{\xi s} g_i, \ldots, g_1)|_{s=0} \ .$$

If H is given by the function f above, we have

$$M_i = \sum_{\mu\nu=1}^{\dim \mathfrak{G}} \pi(I_\nu)\pi_\nu f(g_i g_{i-1} \cdots g_1 g_N \cdots g_{i+1}) r^{\mu\nu},$$

where $\{I_\mu\}$ is a basis in $\mathfrak{g}$, $\pi_\mu(g) = \frac{d}{ds} f(e^{sI_\mu} g)|_{s=0}$, and $r^{\mu\nu}$ are coordinates of the r-matrix.

The integrable systems obtained in this way are integrable spin chains. They are classical field theories with the world space being the set of vertices of a cyclic graph of order N.

An important notion in classical field theory is the notion of local fields. For theories with a discrete world space, local fields are functions on the world space with values in the classical algebra of observables. If F is such a function, then $\{F(x), F(y)\} = 0$ for all x and y separated by some neighborhood of a given size. The size of the separation interval may be different for different local fields.

In the case of spin chains the function $F^{(k)} : \{1, \ldots, N\} \to C(G^{\times N})$, such that

$$F^{(k)}(i)(g_1, \ldots, g_N) = F(g_i, \ldots, g_{i+k}), \tag{2.7}$$

where $F : G^{\times k} \to \mathbf{R}$, is an example of a a local field. It is clear that $\{F^{(k)}(x), F^{(k)}(y)\} = 0$ for all x and y separated by more then k edges of the graph.

Some of the most interesting Hamiltonians from the of view of classical field theories are Hamiltonians with local densities (also called local Hamiltonians). These Hamiltonians have the following form:

$$H = \sum_x H(x). \tag{2.8}$$

Here the sum is taken over the whole world space (assuming this makes sense), and $H(k)$ is a local field. If the model is a spin chain and the local field $\{H(x)\}$ is of the type (2.7), the Hamiltonian is said to describe the local interactions of $k+1$ nearest neighbors. When $k = 1$, it is called a nearest neighbor interaction.

The classical Heisenberg spin chain, known also as the discrete Landau-Lifshitz model, is an example of a spin chain with nearest neighbor interactions. The phase space of the model is $S^{2\otimes N}$. The Hamiltonian of the model is

$$H = \sum_{x=1}^{N} \sum_{\alpha=1,2,3} S_x^\alpha S_{x+1}^\alpha .$$

This model is not integrable, but the following simple modification gives an integrable Hamiltonian with nearest neighbor interactions [9]:

$$H = \sum_{x=1}^{N} \log(1 + \sum_{\alpha=1,2,3} S_x^\alpha S_{x+1}^\alpha). \tag{2.9}$$

In the next section we will describe a 1-dimensional family of integrable Hamiltonians on $S^{2\otimes N}$ with nearest neighbor interactions, which generalizes (2.9).

2.5

Let G be the (complex) loop group $\widehat{SL}_2$. There are two natural Lie bialgebra structures on its Lie algebra, which correspond to the homogeneous and principal gradations on the Lie algebra $\widehat{sl}_2$ [6]. These two Lie bialgebras are factorizable and correspond to the rational and trigonometric r-matrices,, respectively [7, 8].

Consider the Poisson-Lie group structure on $\widehat{SL}_2$ corresponding to the principal gradation. It admits the following description. Let $r(z)$ be the following matrix acting in the tensor product $\mathbf{C}^{2\otimes 2}$:

$$r(z) = \frac{z + z^{-1}}{z - z^{-1}} \; \sigma^3 \otimes \sigma^3 + \frac{2}{z - z^{-1}} \; (\sigma^+ \otimes \sigma^- + \sigma^- \otimes \sigma^+).$$

Here $r(z)$ is regarded as a power series in z, and the σ's are Pauli matrices. Let $\widehat{SL}_2$ be the formal Laurent power series in z with coefficients which are 2×2 matrices with the determinant equal to 1. The coefficients of this power series, $g_{ij}(z)$, generate the algebra $C(\widehat{SL}_2)$ of algebraic functions on the Lie group.

The following bracket between coordinate functions on $\widehat{SL}_2$ determines the Poisson-Lie algebra structure on the Lie group, corresponding to the principal gradations on the Lie algebra $\widehat{sl(2)}$:

$$\{g(z) \otimes g(w)\} = -i[r(z/w), g(z) \otimes g(w)]. \tag{2.10}$$

Here the matrix on the left side consists of the Poisson brackets between matrix elements of $g(z)$ and $g(w)$.

Notice that such a description of the Poisson-Lie group structure on $\widehat{SL}_2$ is an example of the structure described in (2.4). Here the two-dimensional evaluation representation $\pi(z)$ (see, for example, [30]) plays the role of π, and $r(z/w) = (\pi(z) \otimes \pi(w))(r)$.

Consider the following real form of the complex Poisson manifold $\widehat{SL}_2$:

$$g(z)^* = g(z^{-1})^{-1}.$$

Here the conjugation $*$ maps z to z^{-1}, and the equality is of formal power series in z^{-1}. Let us call this real form $\widehat{SU}_2$.

Let a be a nonzero complex number. Consider the 4-dimensional submanifold in $\widehat{SL}_2$ formed by matrices

$$g(z) = L(z/a)det(L(z/a))^{-1/2}, \tag{2.11}$$

where

$$L(z) = \begin{pmatrix} X_0 + \frac{(z+z^{-1})X_3}{2(z-z^{-1})} & \frac{X_1 - iX_2}{2(z-z^{-1})} \\ \frac{X_1+iX_2}{2(z-z^{-1})} & X_0 - \frac{(z+z^{-1})X_3}{2(z-z^{-1})} \end{pmatrix}, \tag{2.12}$$

where everything is regarded as power series in z.

This submanifold is a Poisson submanifold, if and only if the coordinate functions X_a satisfy the following Poisson brackets [9]:

$$\{X_a, X_b\} = X_0X_c, \quad \{X_3, X_0\} = 0, \{X_1, X_0\} = -X_2X_3, \quad \{X_2, X_0\} = X_1X_3. \tag{2.13}$$

Here a, b, c are some cyclic permutation of $1, 2, 3$. There are two Casimir functions on this submanifold:

$$c_0 = X_1^2 + X_2^2 + X_3^2, \quad c_1 = X_0^2 - \frac{1}{4}X_3^2.$$

Consider the real submanifold $\mathcal{L}_R$ with $c_0 = 4\sinh(R/2)^2$, $c_1 = 1$, and with X_a real. It is quite clear from the definition of $\mathcal{L}_R$ that, as a Poisson manifold, it is homeomorphic to the two-dimensional sphere S^2 with the standard symplectic structure (for details see [10, 9]).

The formula (2.11) with $|a| = 1$ describes the manifold $\mathcal{L}_R$ as a symplectic leaf in the real form $\widehat{SU}_2$.

Since the brackets (2.10) determine the Poisson-Lie group structure on $\widehat{SL}_2$, the matrix product of the elements (2.12),

$$T(z) = L_N(z/a_N)L_{N-1}(z/a_{N-1})\dots L_1(z/a_1)\prod_{j=1}^{N} \det\,(L_j(z/a_j))^{-1/2}, \tag{2.14}$$

also determines a Poisson submanifold in $\widehat{SL}_2$. Fix the symplectic leaves of the real forms of $\widehat{SL}_2$ described above for each factor in this product. Then the formula (2.14) gives the embedding of $\mathcal{L}_{R_N} \times \dots \times \mathcal{L}_{R_1}$ as a symplectic leaf in $\widehat{GL}_2$. Here, $|a_i| = 1$. This manifold is a $2N$-dimensional symplectic manifold homeomorphic to the product of N copies of S^2, with the symplectic structure being a linear combination of standard symplectices on S^2.

Consider the trace function on $\mathcal{L}_{R_N} \times \ldots \times \mathcal{L}_{R_1}$:

$$t(z) = \mathrm{tr}\ (T(z)). \tag{2.15}$$

Part (i) of Theorem 1 implies that the functions $t(z)$ form a commutative family (with respect to the Poisson brackets) of smooth functions on $\mathcal{L}_{R_N} \times \ldots \times \mathcal{L}_{R_1}$:

$$\{t(z), t(z')\} = 0.$$

The coefficients of the Taylor expansion of this function provide N independent functions on a $2N$ dimensional symplectic manifold, which commute with respect to the Poisson brackets. Therefore we can choose any function of these Taylor coefficients as the Hamiltonian of some completely integrable system.

Remark 2.

If d is a nonzero complex number, the "d-weighted trace"

$$t(z; d) = dT_{11}(z) + d^{-1}T_{22}(z)$$

also generates a commutative Poisson subalgebra. Such functions on $\widehat{SL}_2$ are invariant with respect to conjugations "twisted by d" [3]. They provide examples of spin chains with quasiperiodic boundary conditions.

The choice of the Hamiltonian of a model is usually determined by physical arguments. In this particular case one can construct a Hamiltonian, which describes only nearest neighbor interactions of classical spins. Let us focus on the real form corresponding to the manifold $\mathcal{L}_R$. First notice that

$$L_{ij}(e^{R/2}) = \alpha_i\beta_j, L_{ij}(e^{-R/2}) = \alpha_j{}'\beta_i{}', \tag{2.16}$$

where α and β are some two dimensional vectors, and $x_1' = -ix_2$, $x_2' = ix_1$.

Consider the function (2.15), where $a_i = 1$, and $R_i = R$ for some R and all $i = 1 \ldots N$. It follows from the identities (2.16) that

$$\tilde{t}(e^{R/2}) = \prod_{j=1}^{N}(\alpha_{j+1}, \beta_j), \tilde{t}(e^{-R/2}) = \prod_{j=1}^{N}(\alpha_j, \beta_{j+1}), \tag{2.17}$$

where $(\alpha, \beta) = \alpha_1\beta_1 + \alpha_2\beta_2$. Here and below $\tilde{t}(z)$ denote the function $t(z)$ without the factors involving the determinant.

The following function on $\mathcal{L}_R^{\times N}$ provides a Hamiltonian, which describes a nearest neighbor interaction in the spin chain, and which can be regarded as a discrete integrable approximation for the Landau-Lifshitz system:

$$H_1 = \log\tilde{t}(e^{R/2}) + \log\tilde{t}(e^{-R/2}) - N\log 2 = \sum_{n=1}^{N}\log\{X_0^nX_0^{n+1} \tag{2.18}$$
$$+(X_1^nX_1^{n+1} + X_2^nX_2^{n+1} + \cosh^2(R/2)X_3^nX_3^{n+1})/4\sinh^2(R/2)\}.$$

More generally, for very integer $k = 1, 2, \ldots, N$, the functions

$$H_k = \frac{d^k}{du^k}(\log \tilde{t}(e^u) + \log \tilde{t}(e^{-u}))|_{u=R/2}$$

can be represented in the form

$$H_k = \sum_{n=1}^{N} H^{(k)}(X^n, \ldots, X^{n+k+1}),$$

where $H^{(k)}$ is a function on $\mathcal{L}_R^{\times k}$. Such functions are called higher local Hamiltonians, since they describe systems with interactions between k nearest neighbors.

Integrable systems with the Hamiltonians described above admit a quite explicit description of the action-angle variables [11]. For finite N, the solutions to the equations of motion for the Hamiltonian (2.18) can be described in terms of theta-functions. It would be interesting to trace how they degenerate in the limit $N \to \infty$, *i.e.* for macroscopically large chains.

2.6

Hamiltonian flows on symplectic manifolds describe the evolution of smooth Hamiltonian dynamical systems. The iteration of maps of symplectic manifolds, which preserve the symplectic structure, describe evolutions of discrete Hamiltonian systems.

Discrete Hamiltonian systems have been studied intensively in recent years. The paper [12] and the references therein provide a good overview on this subject.

As in the case of smooth Hamiltonian evolutions, there is a natural notion of integrability for discrete evolutions. The discrete evolution $\phi : \mathcal{M} \to \mathcal{M}$ on a $2n$-dimensional symplectic manifold $\mathcal{M}$ is called integrable, if there exist n smooth functions on $\mathcal{M}$, which are invariant with respect to this map, and which commute with respect to the Poisson bracket.

Remarkably, this requirement is a very strong restriction on the map ϕ. Even in case when $\mathcal{M} = \mathbf{R}^2$ (then every smooth evolution is integrable), not every map, which preserves the standard symplectic structure $dp \wedge dq$, is integrable. For example, consider the map:

$$\phi(p, q) = (p + V(q), q).$$

Here p and q are coordinate functions, and $V(q)$ is a smooth potential function.

The case $V(q) = -4k \sin(q)$, were k is a positive real, has been studied in detail and is one of the examples of a nonintegrable map well known in KAM theory. This map is sometimes called the "standard map".

Let us consider two examples of integrable maps.

1. Let us return to the previous example, but choose
$$V(q) = 4\arg(1 + k\exp(-iq)).$$
This gives an example of an integrable map with the following integrals of the motion:
$$H = 2(\cos q + \cos(p+q)) + k\cos(p+2q) + k^{-1}\cos p.$$
This example was described in [13] as a particular case of the discrete sine-Gordon model [13, 14, 15].
2. The second example is a discrete version of the affine Toda system. Consider the manifold $\mathbf{R}^{\times NL}$ with coordinates $q_{i,n}$. Consider the Poisson algebra structure on the algebra of functions on $\mathbf{R}^{\times NL}$, with the following Poisson bracket between the coordinate functions:
$$\begin{aligned} \{q_{i,n}, q_{j,m}\} &= 0, \quad n-m = 0 \ (\mathrm{mod}\ 2), \\ \{q_{i,n}, q_{j,m}\} &= \Omega(i-j, \frac{n-m-1}{2}), \quad n-m = 1(\mathrm{mod}\ 2), \end{aligned} \tag{2.19}$$
where
$$\begin{aligned} \Omega(i,n) &= 2\delta_L(n)(\delta_N(i) - \delta_N(i+1)) \\ &+ 2\delta_L(n+1)(\delta_N(i) - \delta_N(i-1)). \end{aligned} \tag{2.20}$$
Here $\delta_N(i) = 1$ if $i = 0(\mathrm{mod} N)$, and $\delta_N(i) = 1$ otherwise.
Consider the map ϕ and the sequences of points in $\mathbf{R}^{\times NL}$ defined as follows:
$$\begin{aligned} q_{i,2n}(2t+1) &= \phi^t(q_{i,2n}), \\ q_{i,2n+1}(2t) &= \phi^t(q_{i,2n+1}). \end{aligned}$$
$$\begin{aligned} &q_{i,n}(t+1) + q_{i,n}(t-1) - q_{i,n+1}(t) - q_{i,n-1}(t) \\ &= \log\left\{\frac{1+k^2\exp q_{i-1,n+1}(t)}{1+k^2\exp q_{i,n+1}(t)}\,\frac{1+k^2\exp q_{i+1,n-1}(t)}{1+k^2\exp q_{i,n-1}(t)}\right\}. \end{aligned}$$
One can show that this map, restricted to the symplectic leaves of the Poisson manifold described above, determines some integrable discrete dynamics. This discrete Hamiltonian system can be regarded as a discrete integrable approximation for the affine Toda field theory. The discrete evolution equations for $q_{i,n}(t)$ and their solutions in terms of determinants were first written in [29]. Its solutions were interpreted in terms of vertex operators in [31]. The Hamiltonian formulation and its quantization is due to [16].
For numerous reasons, the following variable known as the τ-function is important:

$$\frac{\tau(x,y+1)\tau(x,y-1)}{\tau(x+1,y)\tau(x-1,y)} = k^2 \exp(q_{y,x-y}) .$$

One can resolve this system for τ in such a way that the time evolution of τ-coordinates is determined by the bilinear equation

$$\begin{aligned} &\tau(x,y,t+1)\tau(x,y,t-1) \\ &= \tau(x+1,y,t)\tau(x-1,y,t) + \tau(x,y+1,t)\tau(x,y-1,t). \end{aligned} \tag{2.21}$$

In the case $N = 2$, which corresponds to the discrete version of the sinh-Gordon theory [14, 13], we can choose $\tau(x,y+1)\tau(x,y-1) = 1$, and the system (2.21) becomes

$$\tau(x,t+1)\tau(x,t-1) = \tau(x+1,t)\tau(x-1,t) + 1 . \tag{2.22}$$

We will see how this equation appears in a completely different context in chapter 4.

Both of these examples and some others can be obtained from the iteration of a certain map of Poisson-Lie groups known as a factorization map. There is also an analog of Theorem 1 for these maps, which solves the equations of motion in terms of the factorization problem on a Poisson-Lie group. For details see [17, 18]. The first examples of such dynamics were constructed in [32].

3. Algebraic Structures in Quantum Integrable Systems

3.1

A quantization of a Hamiltonian system on a symplectic manifold $\mathcal{M}$ with a Hamiltonian H involves the following:

– The construction of a quantization of the algebra of observables, i.e. the construction of a family A_h of associative algebras (this family gives the quantum algebra of observables), which is a deformation of the Poisson algebra of functions on the symplectic manifold $\mathcal{M}$ (the classical algebra of observables). The deformation parameter is usually denoted by h. The point $h = 0$ corresponds to the classical algebra of observables. At this point the algebra becomes commutative with the Poisson bracket

$$\{a,b\} = i \lim_{h\to 0} (\tilde{a}\tilde{b} - \tilde{b}\tilde{a})/h.$$

Here $\tilde{a}, \tilde{b}$ are elements of A_h, which are equal to a, b respectively when $h = 0$, and we assume that A_h is identified with A_0 as a complex vector space. The algebra A_0 is the complexified algebra of smooth functions on $\mathcal{M}$. For each h the algebra A_h is provided with the antilinear antiinvolution$*$, which acts as a complex conjugation on A_0. Physical observables are $*$-invariant (selfadjoint) elements of A_h. For details about quantization of Poisson algebras see the review [35] and references therein.

- For a given classical Hamiltonian H, define a selfadjoint element of the quantum algebra of observables that is equal to H when $h = 0$, which plays the role of the quantum Hamiltonian.

This is a very simplistic account of a quantization, but it brings us to the question: given a classical integrable system, which quantization of this system should we call an integrable one, and what are the characteristic properties of quantum integrable systems?

Consider the case of quantum mechanics on a smooth manifold: we have a smooth real n-dimensional manifold $\mathcal{N}$. Its cotangent bundle $T^*\mathcal{N}$ is an n-dimensional symplectic manifold. Assume that we have an integrable Hamiltonian system on $T^*\mathcal{N}$ with Hamiltonians, which are polynomial functions in the "cotangent directions". The algebra of differential operators on $\mathcal{N}$ provides the quantization A_h of the algebra of functions on $T^*\mathcal{N}$, which are polynomials in the "cotangent directions". Quantized Hamiltonians are differential operators on $\mathcal{M}$, with the symbols of these differential operators at $h = 0$ being equal to the corresponding classical Hamiltonians. The following is a well known statement:

A quantum system, whose Hamiltonian is a differential operator on an n-dimensional smooth manifold $\mathcal{M}$, is called integrable, if there exist n independent commuting differential operators on $\mathcal{M}$, which commute with the Hamiltonian.

These differential operators are called quantum integrals of the motion.

More generally, *if a quantum system is a quantization of a classical integrable system on a $2n$-dimensional symplectic manifold, it is called integrable, if the Hamiltonian of the system commutes with n algebraically independent, mutually commuting elements (quantum integrals) of the algebra of observables.*

Quantum systems with Hamiltonians which are differential operators on smooth manifolds, but which do not have enough quantum integrals, describe what is known as quantum chaos [19].

3.2

As demonstrated in chapter 2, Poisson-Lie groups play a fundamental role in theory of classical integrable systems. They describe the algebraic nature of practically all known integrable systems. Hopf algebras play a similar role for the quantum case.

The basic notions about Hopf algebras can be found in [20]. Here we recall some of them.

- The algebra A (over $\mathbf{C}$) is called a Hopf algebra, if the following conditions hold.
  1. It has a comultiplication $\Delta : A \to A \otimes A$, which is a homomorphism of algebras.

2. The comultiplication is cocommutative:
$$(\Delta \otimes id) \circ \Delta = (id \otimes \Delta) \circ \Delta.$$
3. There exists a counit $\epsilon : A \to \mathbf{C}$ with the property
$$(\epsilon \otimes id) \circ \Delta = (id \otimes \epsilon) \circ \Delta = id.$$
4. There exists an antiautomorphism of A, $S : A \to A$, called the antipode, which satisfies the identities
$$\sum_i S(a^i)a_i = \epsilon(a) = \sum_i a^i S(a_i),$$
where we denoted $\Delta(a) = \sum_i a^i \otimes a_i$.

– A Hopf algebra A is called quasitriangular, if there exists an invertible element $R \in A^{\otimes 2}$ such that
$$\begin{aligned} \Delta'(a) &= R \cdot \Delta(a) \cdot R^{-1}, \\ (\Delta \otimes id)(R) &= R_{13}R_{23}, \\ (id \otimes \Delta)(R) &= R_{13}R_{12}. \end{aligned}$$
Here $\Delta' = \sigma \circ \Delta(a)$, and σ is the permutation map in $A \otimes A$.
The remarkable property of quasitriangular Hopf algebras is that the element R, known as the universal R-matrix, satisfies the Yang-Baxter relation:
$$R_{12}R_{13}R_{23} = R_{23}R_{13}R_{12} \ .$$
– Another important notion is duality between Hopf algebras. A pair of Hopf algebras A and A^* (over k) is called a dual pair, if there exists a bilinear non-degenerate map
$$(\ ,\ ) : A^* \otimes A \to k,$$
such that
$$\begin{aligned} (ab, c) &= (a \otimes b, \Delta(c)) \\ (a, bc) &= (\Delta_*(a), b \otimes c). \end{aligned}$$
Here Δ is the comultiplication for A, Δ_* is the comultiplication for A^*, and $(a \otimes b, c \otimes d) = (a, c)(b, c)$.

– The algebra A^{*cop} is called co-opposite to A^*, if it coincides with A^* as an algebra, but has the comultiplication $\Delta'_*(a) = \sigma \circ \Delta_*(a)$.

– Let A and A^{*cop} be a dual pair of Hopf algebras. Define the maps $R_\pm$: $A^{*cop} \to A$ as follows:
$$R_+(x) = \sum (x, R^{(1)})R^{(2)}, \tag{3.1}$$
$$R_-(x) = \sum R^{(1)}(x, S(R^{(2)})), \tag{3.2}$$
where we used the notation $R = \sum R^{(1)} \otimes R^{(2)}$.

- The quasitriangular Hopf algebra A is called factorizable, if every element $a \in A$ can be represented uniquely in the form

$$a = \sum_i a_+^i S^{-1}(a_-^i),$$

where $\sum_i a_+^i \otimes a_-^i = (R_+ \otimes R_-)\Delta'_*(x)$ for some $x \in A^{*\mathrm{cop}}$. Equivalently, the map I,

$$I : A^{*\mathrm{cop}} \xrightarrow{\Delta'_*} A^{*\mathrm{cop}} \otimes A^{*\mathrm{cop}} \xrightarrow{R_+\otimes R_-} A \otimes A \xrightarrow{m(id\otimes S^{-1})} A,$$

where Δ'_* is the comultiplication for $A^{*\mathrm{cop}}$, m is the multiplication for A, and S is the antipode for A, is an isomorphism of vector spaces.

For a given dual pair A, A^*, the double of A, $\mathcal{D}(A, A^*)$, is the Hopf algebra defined as follows:

- $\mathcal{D}(A, A^*)$ is isomorphic as a coalgebra to the tensor product of coalgebras $A^{*\mathrm{cop}} \otimes A$;
- the isomorphism mentioned above can be extended to inclusions of A and $A^{*\mathrm{cop}}$ as Hopf subalgebras in $\mathcal{D}(A, A^*)$;
- the algebra structure on $\mathcal{D}(A, A^*)$ is determined uniquely by the extra requirement

$$\Delta(a) = R\Delta(a)R^{-1},$$

where $R = \sum_i (1 \otimes e_i) \otimes (e^i \otimes 1)$ is the universal R-matrix for the double. Here $\{e_i\}$ is a basis in A and $\{e^i\}$ is the dual basis in A^*.

The double of a Hopf algebra is a factorizable Hopf algebra. From its definition follows the isomorphism of algebras

$$\mathcal{D}(A, A^*)^* \simeq A' \otimes A^*,$$

where A' is the algebra with the opposite comultiplication to A.

Let $\{e_i\}$ be a linear basis in A, and $\{e^i\}$ the dual basis in $A^{*\mathrm{cop}}$. Denote by $\{f_i\}$ the basis in A' dual to $\{e^i\}$, and by $\{f^i\}$ the dual basis in $A^{*\mathrm{cop}}$. Consider the following elements of $\otimes\mathcal{D}(A, A^*) \otimes \mathcal{D}(A, A^*)^*$:

$$T^{(+)} = \sum_i e^i \otimes f_i, \quad T^{(-)} = \sum_i S(e^i) \otimes f_i,$$

$$T = T^{(+)}T^{(-)}.$$

Consider the tensor product $\mathcal{D}(A, A^*) \otimes \mathcal{D}(A, A^*) \otimes \mathcal{D}(A, A^*)^*$, and the elements $T_2^{(\pm)} = 1 \otimes T^{(-)}$, $T_1^{(\pm)} = (\sigma \otimes id)_2^{(\pm)}$, where σ is the permutation map. We have the following relations:

$$\begin{aligned}
(R \otimes 1)T_1^{(\pm)}T_2^{(\pm)} &= T_2^{(\pm)}T_1^{(\pm)}(R \otimes 1), \\
T_1^{(\pm)}T_2^{(\pm)} &= T_2^{(\pm)}T_1^{(\pm)}, \\
(R \otimes 1)T_1T_2 &= T_2T_1(R \otimes 1).
\end{aligned} \tag{3.3}$$

Another set of relations describes the action of the antipode on T, $T^{(\pm)}$:

$$T(S\otimes id)T = 1, T^{(\pm)}(S\otimes id)T^{(\pm)} = 1.$$

Let End(V) be a finite dimensional representation of the double. Given a basis in V, consider the matrix $T^{\pi} = (\pi\otimes id)T$. It is clear that the elements of this matrix generate some Hopf subalgebra in the double.

In some cases this subalgebra $D^{\pi}(A, A^*)$ coincides with the whole algebra. The relations

$$(R^{\pi\pi}\otimes 1)T_1^{\pi}T_2^{\pi} = T_2^{\pi}T_1^{\pi}(R^{\pi\pi}\otimes 1)$$

follow from (3.3), $R^{\pi\pi} = (\pi\otimes\pi)(R)$, together with, perhaps, some others, because the determining relations can be regarded as the determining relations for the algebra $D^{\pi}(A, A^*)$.

For each finite dimensional representation π of $D(A, A^*)$, consider the following elements of $D(A, A^*)^*$:

$$t_{\pi} = \mathrm{tr}\otimes id(T^{\pi}),$$

where tr is the quantum trace, see, for example, [51]. It follows immediately from the identity (3.3), that for any pair of finite dimensional representations π and π' we have

$$[t_{\pi}, t_{\pi'}] = 0.$$

Remark 3.
The elements t_{π} also posess the following remarkable properties, which allow us to consider them as characters of finite dimensional representations of $D(A, A^*)$: If $\pi : D\to \mathrm{End}(V)$ has a subrepresentation $\phi : D\to \mathrm{End}(W)$, and if $\psi : D\to \mathrm{End}(W/V)$ is the corresponding quotient representation, then

$$t_{\phi} - t_{\pi} + t_{\psi} = 0,$$
$$t_{\psi\otimes\phi} = t_{\psi}t_{\phi}.$$

The elements t_{π} generate a commutative subalgebra, which gives the quantum integrals of the motion.

Let V be an A-module. Then the action of A on V can be extended to the action of $D(A, A^*)$. Then V can be regarded as both a $D(A, A^*)$-module and a $D(A, A^*)^*$-module (since $D(A, A^*)^* \simeq A'\otimes A^*$, and for every Hopf algebra B with the counit ε we have a one-dimensional representation $a\in B$, $a\mapsto \varepsilon(a)$).

Let G be a factorizable Poisson-Lie group; then the algebra of functions on G is a Poisson-Hopf algebra (a commutative Hopf algebra, which is also a Poisson algebra, and where the comultiplication is a map of Hopf algebras). Denote by $C_h(G)$ its quantization as a Hopf algebra, and by $C_h(G^*)$ the quantization of the algebra of functions on the dual Poisson-Lie group. These algebras are generally called quantum groups [1].

As for the Poisson-Lie groups, there exists the notion of factorizable Hopf algebras. Moreover, if a Poison-Lie group G is factorizable, then there exists a factorizable Hopf algebra quantization of the algebra of functions on G^*. This follows from the work of [33].

The quantum analog of Theorem 1 was announced recently in [34]. It also exists for the discrete quantum evolutions related to the factorization map [17]. Whether such theorems can be used in the quantum case to construct explicit solutions to the Heisenberg equations of motion, is still an open question. In particular, one would expect, that certain devices similar to the ones which exist in the classical case (the τ- function, for example) should have their quantum counterparts.

Finally, we note that the construction of the commutative subalgebras of quantum groups described above is inspired by families of commuting transfer matrices in the 8-vertex model [22] and is an essential part of the quantum inverse scattering method [23]. The details can be found in [24, 23, 17].

3.3

One of the most natural ways to define $C_q(\widehat{SL}_2)$ is as the Hopf algebra dual to $U_q(\widehat{sl}_2)$, generated by the matrix elements of evaluation representations of $U_q(\widehat{sl}_2)$. Denote by $\pi_z^{(\ell)} : U_q(\widehat{sl}_2) \to \mathrm{End}(V^{(\ell)})$ the irreducible $(\ell+1)$-dimensional evaluation representation. The algebra $U_q(\widehat{sl}_2)$ (or, more precisely, a certain completion of it) is a factorizable Hopf algebra with the universal R-matrix $R \in U_q(\widehat{sl}_2) \widehat{\otimes} U_q(\widehat{sl}_2)$. The evaluation of R on the tensor product $\pi_z^{(\ell)} \otimes \pi_w^{(m)}$ gives the linear operator $R^{(\ell,m)}(z/w)$ acting in the product $V^{(\ell)} \otimes V^{(m)}$, which is also a meromorphic function of z/w. For the details see [30] and references therein. It is known that all finite dimensional representations of $U_q(\widehat{sl}_2)$ can be obtained from the two-dimensional one, $\pi_z^{(1)}$, by certain standard operations. This is why the elements of the 2×2 matrix $\mathcal{T}(z) = (\pi_z^{(1)} \otimes \mathrm{id})(\mathcal{T})$ generate the algebra $C_q(\widehat{SL}_2)$. When defining $\mathcal{T}(z)$ in this way, one has to be careful with convergence. Let us simply assume that $C_q(\widehat{SL}_2)$ is in the category of modules, where $\mathcal{T}(z)$ is a formal Laurent power series in z.

Here is the formal definition of $C_q(\widehat{SL}_2)$ in terms of generators and relations. Let q be a nonzero complex number. The algebra $C_q(\widehat{SL}_2)$ is a complex algebra generated by elements the $T_{ij}^{(k)}$, where $i, j = 1, 2$, and $k \in \mathbf{Z}$. Consider the matrix $\mathcal{T}(z)$, which is the generating function for the elements $T_{ij}^{(k)}$:

$$\mathcal{T}_{ij}(z) = \sum_{k \in \mathbf{Z}} T_{ij}^{(k)} z^k.$$

The determining relations in $C_q(\widehat{SL}_2)$ can be written as the following matrix identity with entries in $C_q(\widehat{SL}_2)$:

$$\begin{aligned} R(z)\mathcal{T}(zw) \otimes \mathcal{T}(w) &= (1 \otimes \mathcal{T}(w))(\mathcal{T}(zw) \otimes 1)R(z), \qquad (3.4) \\ \mathcal{T}(qz)_{11}\mathcal{T}(z)_{22} &- \mathcal{T}(qz)_{12}\mathcal{T}(z)_{21} = 1, \end{aligned}$$

where the tensor product is taken over $C_q(\widehat{GL}_2)$. The matrix $R(z)$ acts in $\mathbf{C}^2 \otimes \mathbf{C}^2$ and has the following structure in the standard basis:

$$R(z) = f(z) \begin{pmatrix} 1 & 0 & 0 & 0 \\ 0 & \frac{z-z^{-1}}{zq-z^{-1}q^{-1}} & \frac{(q-q^{-1})}{zq-z^{-1}q^{-1}} & 0 \\ 0 & \frac{q-q^{-1}}{zq-z^{-1}q^{-1}} & \frac{z-z^{-1}}{zq-z^{-1}q^{-1}} & 0 \\ 0 & 0 & 0 & 1 \end{pmatrix}. \tag{3.5}$$

This matrix is the result of the evaluation of the universal R-matrix for $U_q(\widehat{sl}_2)$ on the tensor product of two-dimensional evaluation representations:

$$R(z/w) = (\pi_z^{(1)} \otimes \pi_w^{(1)})(R).$$

The scalar factor $f(z)$ has the following form:

$$f(z) = q^{-1/2} \frac{(z^2q^2; q^4)_\infty^2}{(z^2; q^4)_\infty (z^2q^4; q^4)_\infty},$$

where

$$(x;p)_\infty = \prod_{k=1}^{\infty} (1 - xp^n).$$

The Hopf algebra structure on $C_q(\widehat{SL}_2)$ is determined by the following action of the antipode on the generators:

$$\Delta(\mathcal{T}_{ij}(z)) = \sum_{k=1,2} \mathcal{T}_{ik}(z) \otimes \mathcal{T}_{kj}(z). \tag{3.6}$$

The right side here, as well as in the second relation in (3.4), is understood as a product of Laurent power series. The antipode is determined by the relation

$$\mathcal{T}(z)(S \otimes id\mathcal{T}(z) = 1.$$

Let d be a nonzero complex number. It follows from the identity (3.4), that the power series

$$t_1(z;d) = d\mathcal{T}_{11}(z) + d^{-1}\mathcal{T}_{22}(z) \tag{3.7}$$

generates a commutative subalgebra in $C_q(\widehat{SL}_2)$.

Let $\pi_z^{(\ell)} : U_q(\widehat{sl}_2) \to \mathrm{End}(V^{(\ell)})$ be an $(\ell+1)$-dimensional irreducible representation of $U_q(\widehat{sl}_2)$. Define the $(\ell+1)\times(\ell+1)$ matrix $\mathcal{T}^{(\ell)}(z) = (\pi_z^{(\ell)} \otimes \mathrm{id})(T)$, where T is the canonical element in $C_q(\widehat{SL}_2) \otimes U_q(\widehat{sl}_2)$. The matrix $\mathcal{T}^{(\ell)}(z)$ also can be obtained from $\mathcal{T}(z)$ via the fusion procedure [52, 53].

Define

$$t_\ell(z) = (\mathrm{id} \otimes tr_{V^{(\ell)}})\mathcal{T}^{(\ell)}(z)\ .$$

The fusion relations for $\mathcal{T}(z)$ [53] imply the following recursive relations for $t_\ell(z)$ [53]:

$$t_1(z)t_\ell(zq) = t_{\ell+1}(z) + t_{\ell-1}(zq^2),$$

which can be solved in terms of determinants [54]:

$$t_\ell(z) = \det \begin{pmatrix} t_1(z) & 1 & & 0 \\ 1 & t_1(zq) & \ddots & \\ & \ddots & \ddots & 1 \\ 0 & & 1 & t_1(zq^{\ell-1}) \end{pmatrix}$$

The remarkable fact is that elements $\{t_\ell(z)\}$ also satisfy another set of relations, which also follow from the fusion relations:

$$t_\ell(zq^{\frac{1}{2}})t_\ell(zq^{-\frac{1}{2}}) = t_{\ell+1}(z)t_{\ell-1}(z) + 1. \tag{3.8}$$

3.4

In the construction of $C_q(\widehat{SL}_2)$ modules we will use the algebra $C_q(SL_2^*)$, which is a complex algebra generated by the elements e, f, k with the determining relations

$$\begin{aligned} ke &= qek, \quad kf = q^{-1}fk, \\ ef - fe &= (q - q^{-1})(k^2 - k^{-2}). \end{aligned}$$

Remark 4.
We will not use the fact, that $C_q(SL_2^*)$ is a Hopf algebra with the comultiplication $\Delta : C_q(SL_2^*) \to C_q(SL_2^*)^{\otimes 2}$ defined on the generators as follows:

$$\begin{aligned} \Delta(k) &= k \otimes k, \Delta(e) = k \otimes e + e \otimes k^{-1}, \\ \Delta(f) &= k \otimes f + f \otimes k^{-1}. \end{aligned}$$

It is clear that the algebra $C_q(SL_2^*)$ is isomorphic to the quantized universal enveloping algebra $U_q(sl_2)$ when $q^2 \neq 1$, via the map

$$e \to (q - q^{-1})e', f \to (q - q^{-1})f', k \to k',$$

where e', f', k' are the generators of $U_q(sl_2)$.

The element

$$c = fe + k^2 q + k^{-2}q^{-1} - 2 \tag{3.9}$$

generates the center of the algebra $C_q(SL_2^*)$. Denote by $A_q(R)$ the quotient algebra of $C_q(SL_2^*)$ over the ideal generated by the element $c - 4\sinh^2(R/2 - h/2)$.

The algebra $A_q(R)$ is the quantization of the Poisson algebra $C(\mathcal{L}_R)$ described in section 2.5. The homomorphism of Poisson algebras $A_1(R) \to C(\mathcal{L}_R)$ is given by the following map:

$$\begin{aligned} X_0 &\mapsto (k + k^{-1})/2, X_3 \mapsto k - k^{-1}, \\ X_1 &\mapsto e + f, X_2 \mapsto i(f - e). \end{aligned}$$

The algebra $A_q(R)$ can be represented by the following action on a function of one complex variable:

$$
\begin{aligned}
eg(z) &= z^2 \frac{g(q^{1/2}z) - tg(q^{-1/2}z)}{q - q^{-1}}, \\
fg(z) &= -z^{-2} \frac{g(q^{1/2}z) - tg(q^{-1/2}z)}{q - q^{-1}}, \\
kg(z) &= g(q^{1/2}z),
\end{aligned}
$$

where $t = \exp R$.

Fix $S \in \mathbf{Z}_+/2$. It is easy to see that the space $V^{(2S)}$ of Laurent polynomials of degree $\leq 2S$ is an invariant subspace for the action of $A_q(2Sh)$. For generic q this gives an irreducible $2S+1$ dimensional representation of the algebra $C_q(SL_2^*)$. There is a natural basis in $V^{(2S)}$, in which the action of the elements k is diagonal (this is a basis formed by monomials in z). We will identify $V^{(2S)}$ with $\mathbf{C}^{2S+1}$ assuming this choice of basis. Notice that, since $C_q(SL_2^*)$ is isomorphic to $U_q(sl_2)$, the space $V^{(l)}$ can be regarded as an irreducible finite dimensional module over both these algebras.

3.5

Let $a \in \mathbf{C}^*$. It is easy to see from the definition of $C_q(\widehat{SL}_2)$ given above, that the map $\hat{\phi}_a : C_q(\widehat{SL}_2) \to C_q(SL_2^*)$, acting on the generating functions $T_{ij}(z)$ as

$$
\hat{\phi}_a(\mathcal{T}(z)) = g(z) \begin{pmatrix} qz/ak - a/zk^{-1} & f \\ e & qz/ak^{-1} - a/zk \end{pmatrix},
$$

is a homomorphism of algebras. Below we will use the notation $\mathcal{L}(z/a)$ for this matrix. The function $g(z)$ is equal to the following infinite product:

$$
g(z) = -z \frac{(t^{-1}q^3z^2; q^4)_\infty (tq^5z^2; q^4)_\infty}{(t^{-1}qz^2; q^4)_\infty (tq^3z^2; q^4)_\infty},
$$

where the element t "parameterizes" the central element c:

$$
c = tq + t^{-1}q^{-1}.
$$

Remark 5.
The homomorphism $\hat{\phi}_a$ can be regarded as the composition of the factorization map $R_+ : C_q(\widehat{SL}_2)) \to U_q(\widehat{b}_+)$ and the evaluation homomorphism $ev_z : U_q(\widehat{b}_+) \to U_q(sl_2)$ (using the isomorphism between $U_q(sl_2)$ and $C_q(SL_2^*)$). Note also that there are evaluation maps from $U_q(\widehat{sl}_2)$ to $U_q(sl_2)$. This is why both algebras $U_q(\widehat{sl}_2)$ and $C_q(\widehat{SL}_2)$ act on $U_q(sl_2)$-modules, and in particular on the irreducible finite dimensional modules $V^{(l)}$.

Consider the composition of the quotient homomorphism $C_q(SL_2^*) \to A_q(R)$ and the homomorphism $\hat{\phi}_a$. This gives a homomorphism of algebras

$\psi_a : C_q(\widehat{GL_2}) \to A_q(R)$. This homomorphism is a quantization of the map $C(\widehat{GL_2}) \to C(\mathcal{L}_R)$. The following diagram is commutative:

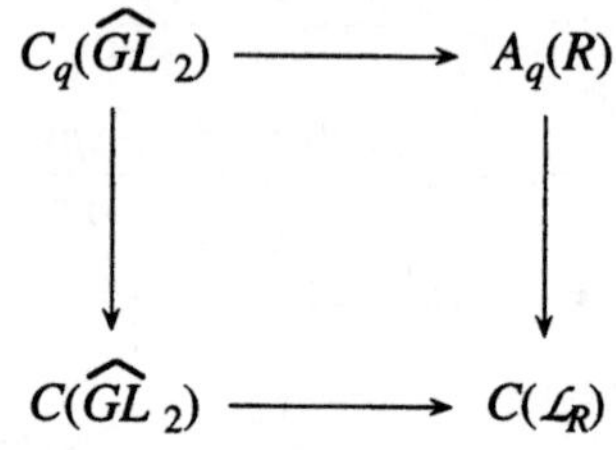

The composition of the tensor product of the maps ψ_a and the iterated comultiplication gives the homomorphism $C_q(SL_2^*) \to A_q(R_N)\otimes \ldots \otimes A_q(R_1)$, which maps the matrix $T_{ij}(z)$ to the matrix

$$T^{(R_1,\ldots,R_N)}(z\alpha_1,\ldots,a_N) = \mathcal{L}_N(z/a_N)\ldots\mathcal{L}_1(z/a_1), \qquad (3.10)$$

where we have taken the matrix product of the matrices $\mathcal{L}(z)$ as 2×2 matrices. The n-th factor in (3.10) acts on the n-th factor of $A_q(R_N)\otimes\ldots\otimes A_q(R_1)$.

If the parameters R_i in (3.10) are chosen from $h\mathbf{Z}_+$, we can restrict the action of (3.10) to the tensor product $\mathbf{C}^2\otimes V^{(2S_N)}\otimes\ldots\otimes V^{(2S_1)}$. The weighted trace of (3.10) over the auxiliary 2-dimensional space restricted to the action on $V^{(2S_N)}\otimes\ldots\otimes V^{(2S_1)}$,

$$t_1(z) = dT_{11}(z) + d^{-1}T_{22}(z),$$

gives commuting families of linear operators in this vector space.

As a result we obtained the transfer matrix of the inhomogeneous higher spin generalization of the Heisenberg spin chain with quasiperiodic boundary conditions. The case $d=1$ correspond to the periodical boundary conditions.

The homogeneous Heisenberg model of spin S corresponds to the choice $S_N = \ldots = S_1 = S$ and $a_N = \ldots = a_1 = 1$. We will denote corresponding transfer matrices as $t_l^{(2S)}(u)$.

In the case $s = \frac{1}{2}$, the transfer matrix $t_1(z)$ can be written as

$$t_1^{(1)}(z) = tr_0(R_{0N}(z)\ldots R_{01}(z)), \qquad (3.11)$$

where $R(z)$ is the matrix (3.5). All the factors in the product act in $(\mathbf{C}^2)^{\otimes(N+1)}$. The subindices show, in which of the factors in this tensor product the elements acts nontrivially (elements of the tensor product are enumerated from left to right from 0 to N).

The matrix (3.11) is the transfer matrix of the 6-vertex model [39, 22]. It is also a generating function for local spin Hamiltonians:

$$H_1 = \frac{d}{du}\log(t_1^{(1)}(u))|_{u=0} = \sum_{n=1}^{N}(\sigma_n^x\sigma_{n+1}^y + \sigma_n^y\sigma_{n+1}^y + \Delta\sigma_n^z\sigma_{n+1}^z), \qquad (3.12)$$

$$H_k = (\frac{d}{du})^k \log t_1^{(1)}(u)|_{u=0} = \sum_{n=1}^{N} H^{(k)}(\sigma_n^\alpha, \dots, \sigma_{n+k}^\alpha). \tag{3.13}$$

Here $\sigma^x, \sigma^y, \sigma^z$ are the Pauli matrices:

$$\sigma^x = \begin{pmatrix} 0 & 1 \\ 1 & 0 \end{pmatrix}, \quad \sigma^y = \begin{pmatrix} 0 & i \\ -i & 0 \end{pmatrix}, \quad \sigma^z = \begin{pmatrix} 1 & 0 \\ 0 & -1 \end{pmatrix}.$$

The subindices in (3.12) (3.13) show, in which factors in $(\mathbf{C}^2)^{\otimes N}$ the matrices act nontrivially.

Similarly, if we take coefficients of $t_{2S}^{(2S)}(u)$, they will provide local Hamiltonians acting in $(\mathbf{C}^{2S+1})^{\otimes N}$:

$$H_k = (\frac{d}{du})^k t_{2S}^{(2S)}(u)|_{u=0} = \sum_{n=1}^{N} H^{(k)}_{n,\dots,n+k}\ .$$

Here $H^{(k)}$ is the matrix acting in $(\mathbf{C}^{2S+1})^{\otimes k+1}$. The subindices show how this matrix acts in $(\mathbf{C}^{2S+1})^{\otimes N}$.

Remark 6.

As noted above, the modules over the algebra $C_q(\widehat{SL}_2)$, which were used in the construction of these spin Hamiltonians, are also $U_q(\widehat{sl}_2)$ modules with respect to the diagonal action on the evaluation modules.

3.6

The spectrum of the elements $t_\ell(z)$ evaluated on the tensor product $V^{(2S_N)} \otimes \dots \otimes V^{(2S_1)}$ can be constructed using the algebraic Bethe ansatz. The details can be found in [55, 24, 53].

Let $x_1, \dots, x_n$ be a solution to the following system of algebraic equations:

$$d^2 \prod_{j=1}^{N} \frac{x_\alpha t_j - a_j}{x_\alpha a_j - t_j} = \prod_{\beta \neq \alpha}^{n} \frac{x_\alpha q^2 - x_\beta}{x_\alpha - x_\beta q^2}. \tag{3.14}$$

Denote by Ω_0 the tensor product of highest weight vectors in $V^{(2S_N)} \otimes \dots \otimes V^{(2S_1)}$:

$$\Omega_0 = v_{S_N}^{(0)} \otimes \dots \otimes v_{S_1}^{(0)}.$$

Then the vector

$$\Omega(x_1, \dots, x_n) = T_{21}(x_1) \dots T_{21}(x_n)\Omega_0 \tag{3.15}$$

is the common eigenvector of operators $t_\ell(z)$ with the following eigenvalue:

$$\begin{aligned} & t_\ell(z)\Omega(x_1, \dots, x_n) \\ & = \{\Lambda(z)\Lambda(zq)\dots\Lambda(zq^{\ell-2})\Lambda(zq^{\ell-1}) \\ & +\Lambda(z)\Lambda(zq)\dots\Lambda(zq^{\ell-2})\Lambda(zq^{\ell})^{-1} \\ & +\Lambda(z)\Lambda(zq)\dots\Lambda(zq^{\ell-1})^{-1}\Lambda(zq^{\ell})^{-1} + \dots + \\ & +\Lambda(zq)^{-1}\Lambda(zq^2)^{-1}\dots\Lambda(zq^{\ell})^{-1}\}\Omega(x_1, \dots, x_n). \end{aligned}$$

Here

$$\Lambda(z) = d\prod_{j=1}^{N} g_{t_j}(z/a_j)(za_j^{-1}t_j^{\frac{1}{2}} - za_jt_j^{-\frac{1}{2}})\prod_{\alpha=1}^{n}\frac{zx_\alpha^{-\frac{1}{2}} - z^{-1}x_\alpha^{\frac{1}{2}}}{zqx_\alpha^{-\frac{1}{2}} - z^{-1}x_\alpha^{\frac{1}{2}}q^{-1}}.$$

One can show that for generic $a_1, \ldots, a_N$ these formulas describe all eigenvectors of $t_1(z)$, and that the spectrum is simple. For particular values of $a_1, \ldots, a_N$ degeneracies are possible, and in this case there are eigenvectors different from (3.15). However, it is expected that their number is "thermodynamically small" (see the last section).

Since the transfer matrices commute, these equations also gives the spectrum of all local Hamiltonians described in section 3.5.

4. The Thermodynamic Limit

4.1

Quantum field theories are special quantum systems. Roughly speaking they correspond to a pair: a quantum system and a world space, as in classical Hamiltonian field theories.

As in any quantum system, observables are selfadjoint elements of the algebra of observables. The notion of world space determines the notion of fields in quantum field theory. When the world space is a manifold, fields are usually regarded as maps from the space of smooth locally supported functions on the world space to the algebra of observables. When the world space is discrete (a space lattice), the fields can be regarded as functions on the world space with values in the algebra of observables.

The topological structure on the world space provides the notion of mutually local fields (local operators) in quantum field theory. For example, when the world space is discrete (a cellular complex) and fields are functions of vertices (of 0-cells), two fields F_1 and F_2 are mutually local, if $[F_1(x), F_2(y)] = 0$, when the points x and y are separated by a neighborhood of given size. In quantum field theories, where the world space is a smooth manifold, locality means that the commutator is a distribution supported at $x = y$. The details can be found in any textbook on quantum field theory (see, for example, [26, 27]).

One of the fundamental principles of field theory is locality. This principle dominates classical mechanics because of the fact that the equations of motion in Newtonian mechanics are second order differential equations. In classical field theories on smooth manifolds, locality appears in a similar way: the equations of motion (classical or quantum) should not include derivatives of order higher than two. This principle is equivalent to the "additivity" principle for Euclidean quantum field theories: when we have two manifolds with a common boundary, then the partition function on the union of these two manifolds is the sum of the product of partition functions corresponding to each individual manifold, taken over all boundary states. In this form locality is one of the determining principles of the field theoretical formulation of topological field theory [56].

An important class of observables are elements with local densities. The local Hamiltonians (3.13) are examples of such elements.

The magnetic chain described in chapter 3 is an example of a quantum field theory on a finite cyclic graph of order N. It quantizes the classical spin chain described in section 2.

Theories with discrete world spaces appear in different contexts. One of the most natural ways to remove ultraviolet divergences in relativistic quantum field theory is to consider lattice regularizations. Many models in condensed matter physics are quantum field theories with a discrete world space (such as crystals). Spin chains are example of such models.

Field theories in "infinite world spaces" are the most interesting from the physical point of view, because they may exhibit phase transitions. The limit when the volume of the world space increases to infinity has to be taken with certain care. The resulting theory has an infinite dimensional phase space in the classical space, and non-separable space of states in the quantum case. A simplistic account of the notion of the *thermodynamic limit* is presented in the next section.

4.2

Let $H^{(\mathcal{M})}$ be the Hamiltonian of a quantum field theoretical model with a world space $\mathcal{M}$ of finite size (either its volume if $\mathcal{M}$ is compact, or the number of vertices if $\mathcal{M}$ is a graph). The partition function of the system is the trace

$$Z^{(\mathcal{M})}(\beta) = \mathrm{Trace}_{\mathcal{A}}(e^{-\beta H^{(\mathcal{M})}}),$$

where $\mathrm{Trace}_{\mathcal{A}}$ is a trace over the algebra of observables (a positive, cyclic invariant functional, normalized as $\mathrm{Trace}_{\mathcal{A}}(P) = 1$ for a selfadjoint 1-dimensional projector P).

A model with the Hamiltonian $H^{(\mathcal{M})}$ is said to have a thermodynamic limit, if, as $\mathrm{vol}(\mathcal{M}) \to \infty$, the partition function has an asymptotic behavior of the form

$$Z^{(\mathcal{M})}(\beta) = \exp(-\mathrm{vol}(\mathcal{M})\ \beta(F(\beta) + o(1))).$$

This kind of asymptotic behavior usually occurs for Hamiltonians with local densities (provided there no pathologies). The next leading term depends on the boundary conditions. The function $F(\beta)$ is the density of the free energy and "does not depend on $\mathcal{M}$".

If $A^{(\alpha)}(x)$ are local fields, one expects that there exist limits of the correlation functions of the form:

$$\langle A^{(1)}(x_1,\tau_1)\dots A^{(k)}(x_k,\tau_k)\rangle_T = \lim_{\mathrm{vol}(\mathcal{M})\to\inf} Z^{(\mathcal{M})}(\beta)^{-1}\times$$
$$\mathrm{Trace}_{\mathcal{A}}(e^{-(\beta-\tau_1)H}A^{(1)}(x_1)e^{-(\tau_1-\tau_2)H}\dots e^{-(\tau_{k-1}-\tau_k)H}A^{(k)}(x_k)e^{-\tau_k H}).$$

This limit determines the linear functional $\langle\dots\rangle_T$ on the limiting algebra of observables. This functional in the limit of an "infinite size world space" is the state of thermodynamic equilibrium at temperature $T=\beta^{-1}$. For more details about thermodynamic limits in local quantum field theories and statistical mechanics see, for example, [26, 27, 28].

When the elements $I_1,\dots,I_k$ mutually commute and commute with the Hamiltonian H, one can construct the state of thermodynamic equilibrium with fixed average values of I_i. To do this we have to take the linear combination

$$H_{\mu_1,\dots,\mu_k} = H + \sum_{l=1}^{k}\mu_l I_l,$$

where μ_l is the "chemical potential" corresponding to the quantity I_l. Assuming the existence of a lower bound for the Hamiltonian $H_{\mu_1,\dots,\mu_k}$, we can define the partition function $Z^{(\mathcal{M})}(\beta,\mu_1,\dots,\mu_k)$. The average value of I_i in this state is

$$\langle I_i\rangle_{\mu_1,\dots,\mu_k,T} = \frac{\partial}{\partial\mu_i}\log Z^{(\mathcal{M})}(\beta,\mu_1,\dots,\mu_k).$$

If the operators I_i have local densities and there no pathologies in the thermodynamic limit, then

$$\langle I_i\rangle_{\mu_1,\dots,\mu_k,T} = \mathrm{vol}(\mathcal{M})(c_i(\beta;\mu_1,\dots,\mu_k)+o(1)),$$

where the densities of the average values are functions of the chemical potentials and the temperature.

For each value of chemical potentials $\mu_1,\dots,\mu_k$ this gives a state of thermodynamic equilibrium $\langle\dots\rangle_{\mu_1,\dots,\mu_k,T}$ on the limiting algebra of observables.

4.3

Now let us return to integrable models of quantum field theory. For simplicity let us focus on the spin-1/2 anisotropic Heisenberg chain described in section 3.

Let H_N be the Hamiltonian of the Heisenberg spin chain of length N. Denote by $H_N^{(k)}$, $k=1,\dots,N-1$ the higher local Hamiltonians (3.13) with $H_N^{(1)}=H_N$.

As $N\to\infty$ the total space of states $(\mathbf{C}^2)^{\otimes N}$ becomes an infinite dimensional non separable Hilbert space. The following is an outline of how to describe the separable components of this space.

It is important to note that there exist the following bounds for the self-adjoint operators $H_N^{(k)}$:

$$m_k N \le H_N^{(k)} \le M_k N, \tag{4.1}$$

where m_k and M_k, which do not depend on N, are lower and upper bounds for the local densities $H_N^{(k)}(n)$ of the higher Hamiltonians.

Consider an arbitrary linear combination of local Hamiltonians

$$H^{(N)}_{\mu_1,\dots,\mu_N} = \sum_{l=1}^{N} \mu_l H_N^{(l)}.$$

The the partition function is

$$Z(\beta, \{\mu_1, \dots, \mu_N\}) = \mathrm{Tr}_{(\mathbf{C}^2)^{\otimes N}}(e^{-\beta H_\mu^{(N)}}).$$

From the estimates (4.1) we have

$$\begin{aligned} 2^N \exp\left\{-\beta N \sum_{\ell=1}^{N} M_\ell(s)\mu_\ell\right\} \quad & < Z(\beta, \{\mu, \dots, \mu_N\}) \\ & < 2^N \exp\left\{-\beta N \sum_{\ell=1}^{N} m_\ell(s)\mu_\ell\right\} . \end{aligned}$$

These inequalities imply that, if there exists a limit

$$F_\mu(\beta) = -\lim_{N\to\infty} \frac{1}{\beta N} \log(Z(\beta, \{\mu\}),$$

then

$$\sum_{\ell\ge1} m_\ell(s)\mu_\ell - T\log 2 \le F_\mu(\beta) \le \sum_{\ell\ge1} M_\ell(s)\mu_\ell - T\log 2 . \tag{4.2}$$

It is clear now that $\{\mu_\ell\}$ should satisfy some convergence properties. We will require stronger restriction on $\{\mu\}$. We assume that the sequence $\{\mu_1, \dots, \mu_N, \dots\}$ is such that starting from some N it vanishes. This ensures the locality of the limiting Hamiltonian. It also means that we consider the variety $\mathbf{R}^\infty$ of all possible chemical potentials $\{\mu_\ell\}$ as the inductive limit:

$$\dots \hookrightarrow \mathbf{R}^N \hookrightarrow \mathbf{R}^{N+1} \hookrightarrow \mathbf{R}^{N+2} \hookrightarrow \dots,$$

where $\mathbf{R}^{\mathbf{N}} \hookrightarrow \mathbf{R}^{N+1}$ as $x \mapsto (x, 0)$.

There exists a technique of using the Bethe equations (3.14) to compute the free energy of the system. It goes back to the works by Yang and Yang [40]. For more details see [36, 24] and references therein.

The equations (3.14) for the homogeneous spin chain of spin $\frac{1}{2}$ is

$$\left(\frac{\sin(\frac{v_\alpha+i\eta}{2})}{\sin(\frac{v_\alpha-i\eta}{2})}\right)^N = \prod_{\beta\ne\alpha} \frac{\sin(\frac{v_\alpha-v_\beta+2i\eta}{2})}{\sin(\frac{v_\alpha-v_\beta-2i\eta}{2})} ,$$

where we used the variables $x_\alpha = \exp(v_\alpha)$, $q = \exp(-\eta)$.

In the limit $N \to \infty$, the solutions to (3.14) typically form strings of the form

$$v_{\alpha,m} + i(m+1-2k)\eta \ , \qquad k = 1, 2, \dots, m, \tag{4.3}$$

where $v_{\alpha,m}$ are the real centers of the string. The word "typically" here means that the number of "nontypical" solutions is negligible compared to the number of "typical" ones. How many solutions are typical or nontypical as $N \to \infty$ is not precisely known. This is why assertion (4.3) is usually called the string conjecture. See [36, 37, 38] and references therein for more information.

In states with a finite density of m-strings as $N \to \infty$, the real centers of the m-strings $\{v_{\alpha,m}\}$ have density $\rho_m(v)$. The densities $\{\rho_m(v)\}$ can be determined by minimizing the free energy in the thermodynamic limit. The result is the following formula for $F_\mu(\beta)$ [41, 42]:

$$F_\mu(\beta) = E_0 - T \int_{-\pi}^{+\pi} S(v) \log(1 + e^{\beta\varepsilon_1(v)}) dv, \tag{4.4}$$

where

$$S(v) = \frac{1}{2\pi} \sum_{n \in \mathbf{Z}} \frac{e^{-inv}}{2ch(n\eta)}, \qquad E_0 = \frac{1}{2\pi} \sum_{n \in \mathbf{Z}} \frac{e^{-|n|\eta} \hat{\mu}[n]}{2\cosh(n\eta)} \ .$$

Here $\hat{\mu}[n] = \sum_{l \geq 0} \mu_l (-in)^l$, and the function $\varepsilon_1(v)$ is the solution of the following system of nonlinear integral equations:

$$\begin{aligned} \delta_{m,1}\varepsilon^{(0)}(v) &= \varepsilon_m(v) - \int_{-\pi}^{+\pi} S(v-u) T \log((1 + e^{\beta\varepsilon_{m+1}(u)}) \\ &\times \ (1 + e^{\beta\varepsilon_{m-1}(u)}))du \ , \end{aligned}$$

with an extra condition $\lim_{m \to \infty} \varepsilon_m = 0$. The function $\varepsilon^{(0)}(v)$ has the following form:

$$\varepsilon^{(0)}(v) = \sum_{\ell \geq 0} \mu_\ell \frac{d^\ell}{dv^\ell} S(v).$$

There is a remarkable relation between this system of nonlinear integral equations and the fusion system (2.22). Using the identity

$$S(v + i\eta - i0) + S(v - i\eta + i0) = \delta_{2\pi}(v),$$

where $\delta_{2\pi}(v)$ is a 2π periodic δ-function, and noting that $S(v)$ is a meromorphic function of v, it is easily shown that the system (4.5) is equivalent to the system

$$\varepsilon_m(v + i\eta) + \varepsilon_m(v - i\eta) = T \log((1 + e^{\beta\varepsilon_{m-1}(v)})(1 + e^{\beta\varepsilon_{m+1}(v)})) \ , \tag{4.5}$$

with the condition $\lim_{m \to \infty} \varepsilon_m = 0$.

Introduce the variables $\tau_m(v)$ such that

$$\tau_{m+1}(u)\tau_{m-1}(u) = e^{\beta\varepsilon_m(v)}.$$

It is easy to check that $\tau_m(v)$ satisfies the recursion relation

$$\tau_m(v+i\eta)\tau_m(v-i\eta) = \tau_{m+1}(v)\tau_{m-1}(v) + 1,$$

which coincides with the fusion relation (2.22).

Although this fact is well recognized by now and had been used to generalize the fusion system to arbitrary Lie algebras [46], its full explanation is not yet clear.

4.4

When $T \to 0$, the state of the thermodynamic equilibrium becomes a pure state corresponding to the ground state Ω_μ of the Hamiltonian $H_\mu^{(\infty)}$:

$$\langle A\rangle_{\mu,T} \to (\Omega_\mu, A\Omega_\mu)\ .$$

The ground state Ω_μ can also be obtained in the limit $N \to \infty$ at $T = 0$ by filling the Dirac sea. Here is a brief description of this procedure.

For each N consider the ground state $\Omega_\mu^{(N)}$ of $H_\mu^{(N)}$. It corresponds to the solution $\{v_\alpha^{(0)}\}$. The numbers $\{v_\alpha^{(0)}\}$ can be interpreted as the parameterization of pseudoparticle momenta, and the state

$$\Omega_\mu^{(N)} = \frac{\Omega(v_1^{(0)}, \ldots, v_n^{(0)})}{\sqrt{(\Omega(\{v_\alpha^{(0)}\}), \Omega(\{v_\alpha^{(0)}\}))}}$$

can be interpreted as a state where all negative energy levels are filled.

In the limit $N \to \infty$ the numbers $\{v_\alpha^{(0)}\}$ are distributed along the real line with some density $\rho_1(\alpha)$. This density may be identically zero on certain intervals. We will call these intervals Dirac islands, and the intervals where $\rho_1(\alpha) \not\equiv 0$ are called Dirac seas. Suppose we have a finite number of Dirac seas $(B_1^+, B_1^-), \ldots, (B_n^+, B_n^-)$. All numbers $B_\alpha^\pm$ here are increasing from left to right. This is the "microscopic structure" of the ground state Ω_μ.

The excitations over the ground states can be of the following types:

- Holes in the Dirac seas (corresponding to "removing" one member from the collection $\{v_\alpha^{(0)}\}$);
- particles in the Dirac islands (corresponding to "adding" one real number to the collection $\{v_\alpha^{(0)}\}$);
- m-strings, $m > 1$ (corresponding to adding one m-string solution to the collection $\{v_\alpha^{(0)}\}$).

As a result, the Fock space which has a vacuum Ω_μ has the following structure:

$$\mathcal{H}(\{B_i^+, B_i^-\}_{i=1}^k) = \bigoplus_{\substack{N_h \geq 0 \\ N_p \geq 0 \\ N_{str} \geq 0}} \bigoplus_{\substack{n_1+\ldots+n_k = N_h \\ m_1+\ldots+m_k = N_p}} \bigotimes_{j=1}^{k} L_2^{\text{symm}} \Big(I_1^{+\times n_1} \times \ldots I_k^{+\times n_k} \times I_1^{-\times m_1} \times \ldots I_k^{-\times m_k} \times S^{1^{N_{str}}} \Big)$$

Here we used the notation $I_l^+ = (B_l^+, B_l^-), I_l^- = (B_{l-1}^+, B_l^-)$, where $B_0^+ = -\pi$ and $B_{k+1}^- = \pi$. The symbol "symm" means a certain symmetrization procedure which we will not discuss here (see, for example, [31] for a discussion of the aniferromagnetic ground state).

The energies of holes, particles, and strings can be found from the zero temperature limit of the functions $\varepsilon_m(v)$:

1. $\varepsilon_{\text{hole}}(v) = \varepsilon_1(v),\ \ v \in$ Dirac island;
2. $\varepsilon_{\text{part}}(v) = -\varepsilon_1(v),\ \ v \in$ Dirac sea;
3. $\varepsilon_{m-\text{str}}(v) = \varepsilon_m(v),\ \ m \geq 2,\ \ v \in [-\pi, \pi]$.

Here $\varepsilon_m(v) = \varepsilon_m(v)|_{T=0}$ is a solution of the linearized system

$$\delta_{m,1}\varepsilon^{(0)}(v) = \varepsilon_m(v) - \int_{-\pi}^{\pi} S(v-u)(\varepsilon_{m+1}(u) + \varepsilon_{m-1}(u))du,$$

where again we assume $\lim_{m\to\infty} \varepsilon_m = 0$. The relation between $B_i^\pm$ and μ can be written as

$$\varepsilon_1(B_i^\pm) = 0\ .$$

Now we can formulate the asymptotic decomposition of $(\mathbf{C}^2)^{\otimes N}$ in the limit $N \to \infty$ into the direct integral of separable Hilbert spaces:

$$(\mathbf{C}^2)^{\otimes N} \to \bigoplus_{k\geq 0} \int_{[-\pi,\pi]^{\times 2k}}^{\oplus} \mathcal{H}(\{B_i^+, B_i^-\}_{i=1}^k). \tag{4.6}$$

In this integral we assume the B's are ordered as above.

Remark 7.
This decomposition is similar to the description of quasiperiodic solutions to classical integrable field theories in terms of Baker-Akhiezer functions [43].
Remark 8.
The decomposition (4.6) has a simple quasiclassical meaning. Each term in it corresponds to the quantization of an invariant torus, which is a level surface

for the integrals of motion. In infinite dimensional local field theories there are natural stratifications of infinite dimensional level surfaces, if one takes into account the degree of local Hamiltonians.

The antiferromagnetic ground state corresponds to $\mathcal{H}([-\pi,\pi])$ and is a measure zero point in the decomposition (4.6). However, it has some exceptional properties. For example, it seems to be the unique ground state where the string conjecture (4.3) for solutions of the Bethe equations fails. Also, in this case there is a very nice representation-theoretic description of the Hilbert space (see [31] and references therein). It turns out that there is an isomorphism of $U_q(\widehat{sl}_2)$-modules (the action of $U_q(\widehat{sl}_2)$ in $(\mathbf{C}^2)^{\otimes N}$ is described in §3).

$$\mathcal{H}([-\pi,\pi]) \simeq (V_{\Lambda_0} \oplus V_{\Lambda_1})^* \widehat{\otimes} (V_{\Lambda_0} \oplus V_{\Lambda_1}) , \tag{4.7}$$

where V_{Λ_0} and V_{Λ_1} are two irreducible level 1 highest weight modules over $U_q(\widehat{sl}_2)$, and $\widehat{\otimes}$ is a certain completion. In this case we have a structure of the Hilbert space different from (4.6):

$$\mathcal{H}([-\pi,\pi]) \simeq \oplus_{n\geq 0} L_2^{\text{symm}}(S^{1\times n}, \mathbf{C}^{2\otimes n}).$$

Here "symm" means the symmetrization with respect to a certain action of the affine Hecke algebra. The isomorphism (4.7) has some important consequences and, in particular, allows us to compute explicitly certain physical quantities [31]. The "microscopic" structure of this isomorphism was studied in [44, 45] using q-wedge products.

As $T \to \infty$ the inequalities (4.2) imply:

$$F(\beta) \to T \log 2 + O(1) .$$

The physical meaning of this is that the entropy contribution to the free energy dominates at high temperatures.

This result can also be easily obtained from formula (4.4). Such asymptotics of the free energy computed on the basis of the string conjecture (4.3) is an important argument in favor of this conjecture and of the "thermodynamic completeness" of the Bethe vectors.

5. Conclusion

We mention here some of the problems which remain open in this subject.

1. As it was mentioned in the text, the remarkable coincidence of the system (4.5) for the functions $\varepsilon_m(v)$ and the fusion relation (2.22) yet lacks a clear explanation. This correspondence is known as the correspondence between the Y-system (4.5) and T-system (2.22) (see [46] for generalizations to other Lie algebras). For some related important results in statistical mechanics and quantum field theory see [47] and [48].

2. It would be very interesting to investigate the "bundle of ground states" $\Omega(\{B_i^+, B_i^-\})$ from the point of view of the representation theory of $U_q(\widehat{sl}_2)$ and the infinite affine Hecke algebra $H_q(\infty)$. Another question in this direction is to investigate possible phase transitions, when the coefficients μ_l in the Hamiltonian are varied.
3. The coincidence of the fusion system and discrete Toda system is also awaiting a clear explanation.
4. It would be nice to have a good explanation of what the quantum discrete τ-function is. Some progress in this direction has already been made. In [16] the quantization of Hirota bilinear equations has been derived, where a pair consisting of an algebra and its automorphism was found, such that the iteration of this automorphism applied to generators gives a sequence of elements of the algebra which satisfy the Hirota equation (appropriately ordered). This automorphism comes from the quantum factorization map [17]. Another approach, based on the Pluecker relations for quantum groups, has been suggested in [60].
5. One can argue that the quantization of the bilinear Hirota equations can be relevant to the recent in the study of q-W algebras [58, 59] and the center of quantized universal enveloping algebras at the critical level [57].

These are just some of the questions. There are other directions which were not mentioned in this text. For example, we did not discuss integrable relativistic field theories, integrable perturbations of conformal field theory, and integrable systems corresponding to quantum groups at roots of 1. Also, the problem of the computation of correlation functions and other characteristics of the systems still remains one of the most interesting open questions.

Acknowledgments
I would like to thank to V. Bazhanov, E. Frenkel, M. Jimbo, R. Kashaev, R. Kedem, T. Miwa, H. Saleur, P. Wiegmann and A. Zamolodchikov for valuable stimulating discussions on different aspects mentioned in these notes. Also, special thanks to the organizers of the school for the hospitality, and for the patience while this manuscript was completed.

References

1. V. Drinfeld, *Quantum groups*, Proceedings ICM–86, (Berkeley), vol. 1, 798–820.
2. A.Weinstein, *Local structure of Poisson manifolds*, J. Diff. Geom., **18** 3 (1983), 523-557.
3. M.Semenov-Tian-Shanskii,*Dressing transformations and Poisson group actions*, Pub. Res. Inst. Math. Sci. Kyoto Univ. **21**, no. 6 (1985), 1237–1260.
4. A.Reiman, M.Semenov-Tian-Shanskii, *Group theoretical methods in Integrable systems*, in: *Integrable systems VII* (V.Arnold and S. Novikov, eds.)Encyclopaedia of Math Sciences v.16, Springer 1994.

5. M.Semenov-Tian-Shanskii, *Lectures on r-matrices, Poisson-Lie groups and integrable systems*, in: *Lectures on Integrable Systems* (in memory of J.L.Verdier, O.Babelon, P.Cartier,Y.Kossmann-Schwarzbach, eds), World Scientific,1994, pp.269-318.
6. V.Kac, *Infinite dimensional Lie algebras*, Birkhäuser, Boston, 1983.
7. P.Kulish, E.Sklyanin, *Integrable Quantum Field Teories*, Lecture Notes in Physics, **151** 61-119, Berlin: Springer-Verlag, 1982.
8. A.Belavin, V.Drinfeld, *About solutions to the classical Yang-Baxter equations for simple Lie algebras* , Funct. Anal. and Its Appl., **16**, 3 (1982), 1-29.
9. E.Sklyanin, *On some algebraic structures related to the Yang-Baxter equation*, Funct. Anal. and Its Appl., **16**,4 (1982),27-34.
10. L.Faddeev and L.Takhtajan, *Hamiltonian Methods in the Theory of Solitons*, Springer-Verlag, 1987.
11. E.Sklyanin, *On the Poisson structure of the periodic clasical XYZ chain*, J. Sov. Math, **46** (1989), 1664.
12. A.Veselov, *Integrable maps*, Uspekhi Mat. Nauk. **46**:5 (1991), 3–45.
13. A.Bobenko, N.Kutz and U.Pinkal, *The discrete quantum pendulum*, Physics Letters A **177** (1993), 399–404.
14. L.Faddeev and A.Volkov, *Quantum inverse scattering method on a space-time lattice*, Theor. and Math. Phys., **92** (1992) 207-214.
15. V.Bazhanov, A.Bobenko and N.Reshetikhin, *Quantum discrete sine-Gordon model at roots of 1: integrable quantum system on the integrable classical background*, Comm. Math. Phys., to be published.
16. R.Kashaev and N.Reshetikhin, *Affine Toda field theory as a 3-dimensional integrable system*, preprint, July 1995.
17. N.Reshetikhin, *Integarble discrete systems*, lectures given at "Enrico Fermi school of physics", Varenna, July 1994.
18. P.Deift, Li L.C., Tomei C., *Loop groups, integrable systems, and rank 2 extensions.*, Memoirs of the AMS, **479** (1991).
19. M.Guttzwiller, *Chaos in Classical and Quantum Mechanics*, Springer-Verlag, 1990.
20. S.Montgomery, *Hopf algebras and their actions on rings*, Regional Conference Series in Mathematics, n.82, 1993.
21. L.Faddeev, N.Reshetikhin and L.Takhtajan, *Quantization of Lie groups and Lie algebras*, Algebra and Analysis, **1** (1989) 178-201.
22. R.Baxter *Exactly solved models in statistical mechanics*, Graduate Texts in Math., **60**, Springer-Verlag, 1989, 2nd edition.
23. L.Faddeev, *Integrable models in $1+1$-dimensional quantum field theory*. Pages 563–608, in: *Recent Advances in Field Theory and Statistical Mechanics*, J.B.Zuber and R.Stora, editors, Elsevier Science Publishers, Amsterdam, 1984.
24. N. Bogolubov, A. Izergin and V. Korepin, *Quantum Inverse Scattering Method and Correlation Functions*, Cambridge University Press, 1993.
25. V.Chari and A.Pressley, *A guide to quantum groups*, Cambridge University Press, 1994.
26. C.Izykson and J.-B.Zuber, *Quantium Field Theory*, McGraw Hill, New York (1980).
27. J.Glimm and A. Jaffe, *Quantum Physics*, Springer, New York (1981).
28. C.Itzycson and J.-M.Drouffe, *Statistical Field Theory*, Cambridge Monographs on Mathematical Physics.
29. R. Hirota, *Discrete analogue od a generalized Toda Equation*, J. Phys. Soc. Japan, **50** (1981) 3785.
30. I.Frenkel and N. Reshetikhin *Quantum Affine Algebras and Holonomic Difference Equations*, Commun. Math. Phys. **146** (1992) 1-60.

31. M.Jimbo and T.Miwa, *Algebraic Analysis of Solvable Lattice Models*, Publications of RIMS, Infinite Analysis Lecture Notes, No 8, Kyoto 1994.
32. J.Moser and A.Veselov *Discrete versions of some classical integrable systems and factorization of matrix polynomials*, Preprint, ETH, Zurich, 1989.
33. P.Etingof and D.Kazhdan, *Quantization of Lie bialgebras* preprint, 1995.
34. B.Jurco and M.Schleiker, *Quantized Lax equations and their solutions*, preprint 1995.
35. A.Weinstein, *Deformation quantization*, it Séminaire Bourbaki, 46 ème année, 1993–94, no. 789.
36. M.Gaudin, *La function d'onde de Bethe*, Masson; Paris 1983.
37. *On exceptional solutions of the Bethe ansatz equations*, Theor. Mat. Phys. **69** (1987) 1071.
38. C.Destry and J.Lowenstein, Nucl. Physics, **205** (1982), 2985.
39. E.H.Lieb and F.Y.Wu, in: *Phase Transitions and Critical Phenomena*, Vol.1, eds. C.Domb, M.S.Green, Academic Press, London 1972.
40. C.N.Yang and C.P.Yang *Thermodynamics of a one-dimensional system of bosons with repulsive delta-function interactions*, J. Math. Phys., **10** (1967) 1115-1122.
41. M.Takahashi and M. Suzuki, *One-dimensional anisotropic model at finite temperature*, Progress of Theoretical Physics **48**, 6B (1972), 2187-2209.
42. M.Gaudin, *Thermodynamics of the Heisenberg-Ising Ring for $\Delta \geq 1$*, Physical Review Letters **26** 21 (1971) 1301-1305.
43. I.Krichever, *Methos of algebraic geometry in the theory of nonlinear equations*, Russian Math. Surv. **32** 6 (1977) 185-213.
44. E.Stern *Semi-Infinite Wedges and Vertex Operators*, International Mathematiccs Research Notices, (1995) 201-220.
45. M.Kashiwara, T.Miwa and E.Stern *Decomposition of q- deformed Fock spaces*, preprint 1995.
46. A.Kuniba, T.Nakanishi and J.Suzuki, *Functional Relations in Solvable Lattice Models*, part I and II, preprints 1993.
47. A.Klumper and P.Pierce, *Conformal weights of RSOS lattice models and their fusion hierarchies*, Physica **A 183** (1992) 304-350.
48. V.Bazhanov, S.Lukianov and A.Zamolodchikov, *Integrable Structure of Conformal Field Theory, Quantum KdV Theory and Thermodynamical Bethe Anstz*, preprint 1994.
49. M.Adler *On the trace functional for formal pseudodifferential operators and the symplectic structure for the KdV type equations*, Invent. Math. **50** (1979) 267-317.
50. B.Kostant *The solution to a generalized periodic Toda lattice and representation theory*, Adv. Math. **34** (1979) 195-338.
51. N.Reshetikhin *Quasitriangle Hopf algebras and invariants of tangles*, Algebra and Analysis **1** (1989) 491-513.
52. P.Kulish, N.Reshetikhin and E.Sklianin *Yang-Baxter equation and representation theory*, Lett. Math. Phys. **5** (1981) 393-403.
53. A.Kirillov and N.Reshetikhin *Exact solution of the integrable XXZ Heisenberg model with arbitrary spin*, J.Physics **A 20** (1987) 1565-1585.
54. V.Bazhanov and N.Reshetikhin *Restricted solid-on solid models connected with simply-laced algebras and conformal field theory*, J.Phys. **A 23** (1990) 1477-1492.
55. L.Faddev and L.Takhtajan, *The quantum inverse scattering method of the inverse problem and the Heisenberg XYZ model*, Russian Math Surveys **34** 5 (1979) 11-68.
56. E.Witten *Topological Quantum Field Theory*, Comm. Math. Phys. **117** (1988) 353-386.

57. E.Frenkel and N.Reshetikhin *Quantum Affine Algebras and deformations of the Virasoro and W-algebras*, preprint 1995.
58. J.Shiraishi, H.Kubo, H.Awata and S.Odake *A quantum deformation of the Virasoro algebra and Macdonald symmetric functions*, preprint 1995.
59. B.Feigin and E.Frenkel *Quantum W-algebras and elliptic algebras*, preprint 1995.
60. S.Kharchev, S.Khoroshkin and D.Lebedev *Intertwining operators and Hirota bilinear equations*, preprint 1995.

Monodromy Representation of the Mapping Class Group $\mathcal{B}_n$ for the su_2 Knizhnik-Zamolodchikov Equation

Yassen S. Stanev[1] and Ivan T. Todorov[2]

[1] Dipartimento di Fisica, Universita di Roma, 'Tor Vergata', I-00133 Roma, Italy
[2] Institute for Nuclear Research, Tsarigradsko Chaussée 72, BG-1784, Sofia, Bulgaria

1. Introduction

The study of a Fuchsian type differential equation in the complex domain has a substantial algebraic part: the characterization of the (finitely generated) monodromy group acting in the space of its solutions. The problem has its roots in a 19th century study of the monodromy of the hypergeometric equation. Understanding its algebraic nature, however, required a development of the mathematical physics of the 1980's: the dual concepts of a quantum group and a QUE algebra [9, 10, 11, 27], [28, 13]. The topic of the present lecture was first studied (under virtually the same heading) by Tsuchiya and Kanie [50] and triggered a considerable activity, which can be traced back from recent reviews - see, e.g. [15]. In spite of vigorous efforts the subject continues to attract attention, and we still face some intriguing open problems.

The present exposition is based on work of the authors (and of P. Furlan and L. Hadjiivanov) [19, 40, 41, 48], in which we construct and exploit an indecomposable representation of $\mathcal{B}_n$ in a (finite dimensional) space $\mathcal{L}_n$ of regular solutions of the KZ equation, [30] (that do not respect, in general, the physical fusion rules). We indicate the main steps of the argument showing that the monodromy of KZ amplitudes belongs to $GL(\mathcal{L}_n, Z(q))$: the general linear group over the ring of *cyclotomic integers*, that is, polynomials in q (where $q^h = -1$) with integer coefficients. This observation prepares the ground for an application to the Schwarz (finite monodromy) problem for the su_2 KZ equation [42] reviewed in [47].

To understand the results covered in this paper one does not have to know much about *2-dimensional* (2D) *(rational) conformal field theory* ((R)CFT). In order to accommodate such readers, we start in Sec. 2, essentially, with a study of the KZ equation, which, supplemented by Möbius invariance and factorizability of solutions, reflects all CFT postulates. Here instead we shall summarize the quantum field theoretical (QFT) ingredients, that yield the properties of conformal blocks listed in Sec. 2. (We assume some background knowledge in the subject - within the content of, say, the first 3 Sects. of [18].)

In the axiomatic approach to 2D CFT we are following here, one starts with (the vacuum representation of) an algebra generated by local currents (including the stress energy tensor as a composite field) and then construct its local (charged) field (positive energy) representations. Such an approach is manageable because of the simplicity of the *local (observable) algebra*: it splits into chiral parts, say a and $\bar{a}$, generated by current components depending on a single light-cone variable; as a consequence of locality left and right movers mutually commute. In the analytic compact (z) picture (the physical values, $z = e^{i(x^0-x^1)}$, lying on the unit circle) the level k current algebra corresponding to a simple compact Lie group G is characterized by the local *commutation relations* (CR)

$$[J^1(z_1), J^2(z_2)] = [C_{12}, J^1(z_1)]\delta(z_{12}) - kC_{12}\delta'(z_{12}). \tag{1.1}$$

Here $z_{12} = z_1 - z_2$, the δ-function on the circle is defined to satisfy

$$\oint_{|z_2|=|z_1|} \delta(z_{12}) f(z_2) \frac{dz_2}{2\pi i} = f(z_1);$$

if $\{t^a\}$ is a basis in the fundamental representation of g, and $\{J^a\}$ is the corresponding basis of current, then J is a matrix in the fundamental representation given by $J = \eta_{ab} J^a t^b$, where

$$\eta^{ab} = tr(t^{1a}t^{2b}), \eta^{as}\eta_{sb} = \delta^a_b, \tag{1.2}$$

and we are using the abbreviate tensor product notation

$$J^1 = J \otimes \mathbf{1}, J^2 = \mathbf{1} \otimes J. \tag{1.3}$$

In general, $C_{12}(\Lambda_1, \Lambda_2)$ is the Casimir invariant in the tensor product of two irreducible representations (IRs) of weights Λ_1 and Λ_2:

$$C_{12}(\Lambda_1, \Lambda_2) = \eta_{ab} t^{1a}_{\Lambda_1} t^{2b}_{\Lambda_2}. \tag{1.4}$$

In C_{12} appearing in (1.1) both Λ_1 and Λ_2 coincide with the fundamental representation.

The chiral ('right movers') component of the stress energy tensor T is given by the Sugawara formula[1]

$$T(z) = \frac{1}{2h} : tr(J(z))^2 :, \quad h = k + \check{g}, \tag{1.5}$$

where the normal product of two currents can be defined in a standard fashion in terms of their mode expansion (each current component appearing as a free field) - see [44, 45], and h is the *height* or shifted level, $\check{g}$ being the dual Coxeter number (see, e.g. [29]); $\check{g}(su_n) = n$. It is the height that characterizes

[1] We use this, by now conventional, name without going into the intricate history of this relation, which appears to go back to Kronig's energy operator in the neutrino theory of light [31].

(in a sense to be elaborated in these lectures) the resulting RCFT; we shall therefore denote the above chiral current algebra by $a_h(g)$. In quantum field theory one is interested in positive energy Hilbert space representations. If we introduce the Fourier-Laurent expansion of basic local chiral conserved currents,

$$J(z) = \sum_{n \in Z} J_n z^{-n-1}, \quad T(z) = \sum_{n \in Z} L_n z^{-n-2}, \tag{1.6}$$

then we can formulate the following sharpened version of energy positivity. Each superselection sector of the chiral theory, that carries by definition a *factorial representation* (i.e. a finite multiple of an IR) of $a_h(g)$, has a *minimal energy* subspace $\mathcal{H}_{\Lambda,\Delta}$, such that

$$J_n \mathcal{H}_{\Lambda,\Delta} = L_n \mathcal{H}_{\Lambda,\Delta} = 0 \text{ for } n = 1, 2, \ldots, \tag{1.7}$$

$\mathcal{H}_{\Lambda,\Delta}$ is a finite dimensional irreducible G module and is an eigensubspace of the chiral conformal energy operator L_0, so that

$$\{\eta_{ab} J_0^a J_0^b - C_2(\Lambda)\}\mathcal{H}_{\Lambda,\Delta} = 0 = (L_0 - \Delta)\mathcal{H}_{\Lambda,\Delta}. \tag{1.8}$$

(1.5) then implies that the *conformal dimension* Δ is a function of the weight Λ of g:

$$\Delta = \Delta(\Lambda), \quad 2h\Delta(\Lambda) = C_2(\Lambda). \tag{1.9}$$

Fortunately, the unitary positive energy (called by mathematicians highest weight) representations of $a_h(g)$ have been described by the time physicists felt the need to use them (for an up-to-date review by the chief contributor to the field, see [29]; a physicist's oriented version can be found in [25]). They are given by the height h and the highest weight Λ. Remarkably, for a given h (which fixes the vacuum current algebra model) there is only a finite number of admissible Λ. For $g = su_2$ the weight $2I$ does not exceed the level, $2I \leq k (= 1, 2, \ldots)$ (the dimension of the corresponding IR of SU_2 being smaller than the height, $2I + 1 < h = k + 2$). Note that the admissible set always contains the trivial *vacuum* representation for which $\Lambda = 0 = \Delta$, and $\mathcal{H}_{00}$ is spanned by a single state $|0>$ that is (g and) *Möbius invariant*: (1.6b) is to be supplemented by

$$L_{-1}|0> (= J_0|0> = L_0|0>) = 0. \tag{1.10}$$

The existence of only a *finite number of superselection sectors* (labeled in the above example by the isospin I) is a characteristic feature of a *rational* CFT.

The fact, that space-time is not simply connected (the compactified light ray being a circle), is at the origin of the appearance of 'multivalued fields', which are conveniently described by *chiral vertex operators* (CVO) [50]. In the case of the $a_h(su_2)$ chiral algebra, to which we specialize, a CVO $V_I(z)$ that intertwines the vacuum sector with a sector of isospin I, one needs, in general, two more labels to specify the operator: the isospins I_i of the initial state (the domain) and I_f of the final state (the target). In particular, the monodromy of $V_{II_i}^{I_f}(z)$ is given by

$$V_{II_i}^{I_f}(e^{2\pi i}z) = \exp\{2\pi i(\Delta(\mathbf{I_f}) - \Delta(\mathbf{I}) - \Delta(\mathbf{I_i}))\} V_{II_i}^{I_f}(z), \tag{1.11}$$

where $\Delta(I)$ is the *conformal dimension* which satisfies, according to (1.7),

$$h\Delta(I) = I(I+1) \quad \text{as } C_2(2I) = 2I(I+1)). \tag{1.12}$$

(The factor $e^{2\pi i}$ in the argument of V is a shorthand for an analytic continuation along a circle of radius $|z|$ in the positive direction.) Single valued Euclidean scalar fields $\varphi_I(z, \bar{z})$ are viewed as bilinear combinations of $V_I(z)$ and $\bar{V}_I(\bar{z})$ with matching additional labels. A (primary) CVO $V_I = V_{II_i}^{I_f}$ satisfies two local conditions (independent of the labels (I_f, I_i) which will be, therefore, omitted): CR with the stress energy tensor

$$[T(z_1), V_I(z_2)] = \delta(z_{12}) \frac{\partial}{\partial z_2} V_I(z_2) - \Delta(I)\delta'(z_{12}) V_I(z_2), \tag{1.13}$$

which reflects the property of T to generate local reparametrizations, and with the current,

$$[J^a(z_1), V_I(z_2)] = \delta(z_{12}) V_I(z_2) t_I^a, \tag{1.14}$$

displaying the fact that J generates local gauge transformations. We shall make the internal symmetry properties of V_I more explicit by introducing a polynomial realization of the finite dimensional IRs of SU_2. (The idea for such a realization, that goes back at least to Hermann Weyl, has been repeatedly revived in various contexts, see e.g. [1, 52].) Regarding V_I as a polynomial of degree $2I$ in an internal space variable ζ (viewing it, in other words, as an *SU_2 coherent state operator*), we can substitute the matrices t_I^a in (1.11) by first order differential operators in ζ:

$$[J(z_1, \zeta_1), V_I(z_2, \zeta_2)] = -\delta(z_{12})\zeta_{12}(\zeta_{12}\partial_{\zeta_2} + 2I) V_I(z_2, \zeta_2), \tag{1.15}$$

where $\zeta_{12} = \zeta_1 - \zeta_2$,

$$J(z, \zeta) = J^-(z) + 2\zeta J^3(z) - \zeta^2 J^+(z) \quad (J^\pm = J^1 \pm iJ^2). \tag{1.16}$$

The current CR (1.1) can be obtained from (1.12) for $I = 1$ by adding the (central extension) Schwinger term:

$$[J(z_1, \zeta_1), J(z_2, \zeta_2)] = -\zeta_{12}(\zeta_{12}\partial_{\zeta_2} + 2) J(z_2, \zeta_2)\delta(z_{12}) - k\zeta_{12}^2 \delta'(z_{12}). \tag{1.17}$$

Compatibility between (1.10), (1.12) and the quadratic expression (1.5) for T in terms of J implies the relation (1.9) for the conformal weights as well as the operator KZ equation

$$h\frac{\partial}{\partial z} V_I(z, \zeta) =: (I V_I(z, \zeta)\partial_\zeta J(z, \zeta) - J(z, \zeta)\partial_\zeta V_I(z, \zeta)) : . \tag{1.18}$$

(A derivation of (1.15) in these lines, which also applies to the $U(1)$ Thirring model,[2] was worked out in parallel with the KZ paper in [44, 45].) The action of V_I on the vacuum state is specified without additional labels (like I_f) and extends the *field-vector correspondence* (known for local chiral operators):

$$V_I(z,\zeta)|0> = e^{(zL_{-1}+J_0^+\zeta)}V_I(0,0)|)0> \ (J_0^+ = J_0^1 + iJ_0^2), \tag{1.19}$$

$V_I(0,0)|0>$ being a *lowest weight* vector of $a_h(su_2)$,

$$V_I(0,0)|0> = |-I> \quad (J_0^-|-I> = (J_0^3 + I)|-I> = 0). \tag{1.20}$$

The exponent in ζ is actually a polynomial of degree $2I$, since $(J_0^+)^{2I+1}|-I> = 0$.

2. KZ Equation for n-Point Functions. Regular Basis of Solutions for $n = 4$

A significant part of the analysis of this section can be carried through for a current algebra model based on an arbitrary simple compact Lie group. Given, however, that the characteristics of the mapping class group we are aiming at have only been understood in the case of su_2, we decided to stick to this simple case from the very beginning. The advantage of such an approach is that we can keep the discussion elementary and concrete all along.

Let $\mathcal{V}_i = \mathcal{V}(I_i), i = 1, \ldots, n$, be irreducible $2I_i + 1$ dimensional SU_2 modules (of isospin I_i), such that the space of n-fold invariant tensors is nonempty:

$$\mathcal{J}(I_1, \ldots, I_n) = Inv \otimes_{i=1}^n \mathcal{V}_i, \quad d_{\mathcal{J}} = \dim \mathcal{J} > 0. \tag{2.1}$$

(This means that $\sum I_i \in N$, and that none of the isospins exceeds the sum of the others.) For each ordered set of isospins $(I_1, \ldots, I_n)$ $(2I_i \in N)$ satisfying (2.1) we define a $d_{\mathcal{J}}$ dimensional vector space $\mathcal{L}_n = \mathcal{L}(I_1, \ldots, I_n)$ of holomorphic functions $w_n = w(z_1, I_1; \ldots; z_n, I_n)$, called *(conformal) n-point blocks* with values in $\mathcal{J}(I_1, \ldots, I_n)$, (single valued) analytic in (a complex neighbourhood of) the convex cone

$$K_n : z_{ii+1} = z_i - z_{i+1} > 0, i = 1, \ldots, n-1, \tag{2.2}$$

and satisfying the following two conditions:

---

[2] As we are told by solid state physicists, the first $2D$ CFT, the Thirring model [43], also has a (non local) predecessor in Tomonaga's paper on Bloch's sound waves method [49], which has long remained unnoticed by field theorists.

2.1 Möbius Invariance

The correlation function w_n is translation invariant, it only depends on the coordinate differences z_{ij}; it is covariant under dilations

$$z_i \to \rho z_i, \rho > 0 : \rho^{\Delta_1+\ldots+\Delta_n} w(\rho z_{1,\ldots,}\rho z_n) = w(z_1, \ldots, z_n), \qquad (2.3)$$

where $\Delta_i = \Delta(I_i)$ is given by (1.9) (and we have skipped the isospin arguments); it is covariant under infinitesimal special conformal transformations $z \to z(1+\varepsilon z)^{-1}$, $\varepsilon \to 0$ implying the differential equation

$$\sum_{i=1}^{n} z_i(z_i \frac{\partial}{\partial z_i} + 2\Delta_i) w(z_1, I_1; \ldots; z_n, I_n) = 0. \qquad (2.4)$$

2.2 KZ Equation for Conformal Blocks

The correlation functions satisfy the system of differential equations

$$\left(h\frac{\partial}{\partial z_i} + \sum j = 1\ j \neq i^n \frac{C_{ij}}{z_{ij}} \right) w(z_1, I_1; \ldots; z_n, I_n) = 0; \qquad (2.5)$$

here $C_{ij} = C_{ij}(I_i, I_j)$ are the Casimir invariants (1.4) in the tensor product $\mathcal{V}_i \otimes \mathcal{V}_j$ (with Λ_i specialized to $2I_i$).

If we use the polynomial realization of the IRs of SU_2 (with Ward identities for CVO derived from (1.12)), then C_{ij} appear as second order partial differential operators:

$$C_{ij} = 2I_iI_j + 2\zeta_{ij}(I_i\partial_j - I_j\partial_i) - \zeta_{ij}^2\partial_i\partial_j \quad (\partial_i = \partial_{\zeta_i}, \zeta_{ij} = \zeta_i - \zeta_j). \qquad (2.6)$$

Both properties are implied by the field theoretical assumptions of Sec. 1. MI is a consequence of reparametrization covariance (1.10) and of the Möbius ($sl(2,R)$) invariance of the vacuum (1.6). KZE follows from (1.15) and (1.12). For someone who is not inclined to enter the intricacies of a $2D$ CFT, the above formulation could as well serve as a starting point for the subsequent discussion.

We shall derive (in Secs. 3 and 4) the following implications of the definition of $\mathcal{L}_n$.

2.3 Monodromy Representations of the Exchange Algebra

The elements w_n of $\mathcal{L}_n$ satisfying the above requirements admit a (path dependent, multivalued) analytic continuation to the product of complex planes minus the diagonal

$$\mathcal{Z}_n = C^n \backslash diag = \{z_i \in C, z_i \neq z_j \text{ for } i \neq j\}. \qquad (2.7)$$

Thus $\mathcal{L}_n$ can be reinterpreted as a section in a *holomorphic vector bundle* $\mathcal{F}_n$ (see [14]) with a base space $\mathcal{Z}_n$ and a $d_{\mathcal{J}}$ dimensional fibre spanned by analytic continuations of (su_2 invariant) conformal blocks w_n. It carries a representation of the *braid (or exchange) algebra* generated by the *exchange operators*

$$\begin{aligned} B_i &= B_i^{I_1,\dots,I_n} : \mathcal{L}(I_1,\dots,I_i,I_{i+1},\dots,I_n) \\ &\to \mathcal{L}(I_1,\dots,I_{i+1},I_i,\dots,I_n), \quad i=1,\dots,n-1, \end{aligned} \tag{2.8}$$

which we proceed to define. Consider the analytic continuations of each $w_n \in \mathcal{L}_n$ along a pair of paths $\mathcal{C}_i^{\pm}$, that exchange two neighbouring arguments z_i and z_{i+1} in positive/negative direction:

$$\mathcal{C}_i^{\pm}: \quad (z_i)z_{i+1} \to \frac{1}{2}(z_i + z_{i+1}) + \frac{1}{2}z_{ii+1} - z_{ii+1}e^{\pm i\pi t}, \quad 0 \le t \le 1. \tag{2.9}$$

They give rise to a pair of exchange operators B_i (2.8) and $\bar{B}_i$, that intertwine the same two spaces and are inverse to each other:

$$\bar{B}_i^{I_1,\dots,I_{i+1},I_i,\dots,I_n} B_i^{I_1,\dots,I_i,I_{i+1},\dots,I_n} = 1. \tag{2.10}$$

For a basis $\{w_n^{\nu}\}$ of real analytic functions in the domain (2.2), the matrix $\bar{B}_i$ is complex conjugate to B_i.

2.4 Braid Invariant Hermitean Form

If $\{w^{\nu}, \nu = 1,\dots,d_J\}$ is a basis in $\mathcal{L}_n$, then there exists a sesquilinear form $Q = (Q_{\mu\nu}^{I_1\dots I_n})$ such that

$$\begin{aligned} G_n &= G_n(z_1,\bar{z}_1,I_1;\dots;z_n,\bar{z}_n,I_n) \\ &= \bar{w}^{\mu}(\bar{z}_1,I_1;\dots,\bar{z}_n,I_n)Q_{\mu\nu}w^{\nu}(z_1,I_1;\dots;z_n,I_n) \end{aligned} \tag{2.11}$$

is invariant under the exchange of any pair of neighbouring arguments:

$$\begin{aligned} &B_i^{*I_1\dots I_{i+1}I_i\dots I_n} Q^{I_1\dots I_{i+1}I_i\dots I_n} B_i^{I_1\dots I_i I_{i+1}\dots I_n} \\ &= Q^{I_1\dots I_i I_{i+1}\dots I_n} \quad (i=1,\dots,n-1). \end{aligned} \tag{2.12}$$

(In particular, for equal isospins, Q would be braid invariant in the usual sense.) Q can be chosen positive semidefinite and normalized in such a way, that G_n would coincide with the Euclidean $2D$ correlation function

$$G_n =< 0|\varphi_1(z_1,\bar{z}_1)\dots\varphi_n(z_n,\bar{z}_n)|0 > . \tag{2.13}$$

The relative normalizations of G_n for different n's (and different sets of isospins) are constrained by small distance factorization properties that can be summarized as follows.

2.5 2D Operator Product Expansions

For each pair of neighbouring arguments $i, i+1$ there is a minimal isospin J,

$$|I_i - I_{i+1}| \leq J = J(I_i, I_{i+1}) \leq I_i + I_{i+1}, \tag{2.14}$$

for which both the 3-point function $< 0|\varphi_i \varphi_{i+1} \varphi_J|0 >$ and the $(n-1)$-point function

$$G_{n-1}^{(J)} = \; < 0|\varphi_1(z_1, \bar{z}_1) \dots \varphi_{i-1}(z_{i-1}, \bar{z}_{i-1}) \varphi_J(z_i, \bar{z}_i) \\ \varphi_{i+2}(z_{i+2}, \bar{z}_{i+2}) \dots \varphi_n(z_n, \bar{z}_n)|0 > \tag{2.15}$$

do not vanish; then the small $z_{i,i+1}$ behaviour of G_n is given by

$$\lim_{z_{i+1} \to z_i} z_{i,i+1}^{\Delta_i + \Delta_{i+1} - \Delta(J)} G_n = C_{I_i I_{i+1}}^J G_{n-1}^{(J)}, \tag{2.16}$$

where

$$< 0| \prod_{i=1}^{3} \varphi_i(z_i, \bar{z}_i; \zeta_i, \bar{\zeta}_i)|0 > \\ = C_{I_1 I_2 I_3} \prod_{i<j} |z_{ij}|^{-2\Delta_{ij}} |\zeta_{ij}|^{2I^{(ij)}}, \tag{2.17}$$

$$\Delta_{ij} = \Delta_i + \Delta_j - \Delta_k, \quad I^{(ij)} = I_i + I_j - I_k, \tag{2.18}$$

$(i, j, k) =$ Perm $(1, 2, 3)$ and

$$C_{I_1 I_2}^J C_{JJ0} = C_{I_1 I_2 J}. \tag{2.19}$$

Eq. (2.16) reflects the idea that the OPE $\varphi_i(z_i, \bar{z}_i)\varphi_{i+1}(z_{i+1}, \bar{z}_{i+1})$ splits into a finite sum of conformal families (labeled by isospins), the leading contribution to the J family being $C_{I_i I_{i+1}}^J z_{i,i+1}^{\Delta(J) - \Delta_i - \Delta_{i+1}} \varphi_J(z_i, \bar{z}_i)$ - see [2].

2.6 Geometric Interpretation

We can interpret the operator in the left hand side of the KZ equation

$$\nabla^i = \frac{\partial}{\partial z_i} + A^i, A^i = \sum_{j \neq i} r_{ij}(z_{ij}), r_{ij}(u) = \frac{C_{ij}}{u}, \tag{2.20}$$

as a covariant derivative, which defines a *flat connection* in the Friedan-Shenker bundle $\mathcal{F}_n$. Indeed, the corresponding connection 1-form has *zero curvature*:

$$A = A^i dz_i \Rightarrow dA + A \wedge A = 0. \tag{2.21}$$

Actually, both dA and $A \wedge A$ vanish: the first due to the symmetry of C_{ij}, the second as a consequence of the *classical Yang-Baxter equation*:

$$[r_{12}(u), r_{23}(v)] + [r_{12}(u), r_{13}(u+v)] + [r_{13}(u+v), r_{23}(v)] = 0$$

for r_{ij} defined in (2.19). In verifying (2.21) we use (1.4) to prove

$$[C_{12}, C_{23}] = [C_{13}, C_{12}] = [C_{23}, C_{13}] = f_{abc} t_{\Lambda_1}^{1a} t_{\Lambda_2}^{2b} t_{\Lambda_3}^{3c}, \tag{2.22}$$

f_{abc} being the structure constants of g ($[t_\Lambda^a, t_\Lambda^b] = f_c^{ab} t_\Lambda^c$), $f_{abc} = \eta_{aa'} \eta_{bb'} f_c^{a'b'}$.

The interest in this remark stems from the fact that a flat connection on a holomorphic vector bundle defines a (holonomy) representation of the fundamental group of the base space.

2.7 Möbius Invariant 4-Point Amplitudes

We now specialize to $n = 4, g = su_2$, and to pairwise equal isospins, setting

$$\begin{aligned} I_4 &= \min_{1 \le i \le 4} I_i = I, \quad I_1 + I_2 + I_3 - I_4 = 2I', \\ |I_{ij}| &= I_{k4} \text{ for } (i,j,k) = Perm(1,2,3), \quad I_{ij} = I_i - I_j. \end{aligned} \tag{2.23}$$

We shall single out the configuration

$$I_1 = I_4 = I (\le) I_2 = I_3 = I', \tag{2.24}$$

which simplifies formulae and is appropriate for imposing positivity on 4-point blocks. Möbius invariance allows us to reduce Eq. (2.5) to a system of ordinary differential equations. As a first step we use SU_2 invariance and Eq. (2.4) to write

$$w(z_1, \zeta_1, I; z_2, \zeta_2, I'; z_3, \zeta_3, I'; z_4 \zeta_4 I) = p_{II'}(z_{ij}, \zeta_{ij}) F(\eta; \xi_1, \xi_2),$$

where $p_{II'}$ is a homogeneous prefactor:

$$\begin{aligned} p_{II'} &= \frac{\zeta_{23}^{2I_-}}{z_{14}^{2\Delta} z_{23}^{2\Delta} \eta^{2\Delta(I,I')}}, \quad I_- = I' - Ih\Delta - I(I+1), \\ h\Delta' &= I'(I'+1), \quad h\Delta(I,I') = I(I'+1). \end{aligned} \tag{2.25}$$

F is a homogeneous polynomial of degree $2I$ in the pair (ξ_1, ξ_2):

$$F(\eta; \xi_1, \xi_2) = (\xi_1 + \xi_2)^{2I} f(\eta, \xi), \xi_1 = \zeta_{12}\zeta_{34}, \quad \xi_2 = \zeta_{14}\zeta_{23}, \quad \xi = \frac{\xi_1}{\xi_1 + \xi_2}, \tag{2.26}$$

η and ξ being the Möbius invariant cross ratios

$$\eta = \frac{z_{12} z_{34}}{z_{13} z_{24}} \left(= 1 - \frac{z_{14} z_{23}}{z_{13} z_{24}} \right), \quad \xi = \frac{\zeta_{12}\zeta_{34}}{\zeta_{13}\zeta_{24}}. \tag{2.27}$$

In fixing the η dependence of $p_{II'}$, we have required that F has a finite (non-zero) limit for $\eta \to 0$ and $\eta \to 1$. Inserting (2.24) in the KZ equation (2.5) (say, for $i = 2$, using $\{C_{12} + C_{23} + C_{24} + 2I_2(I_2+1)\} w_4 = 0$), we obtain

$$\left(h\frac{\partial}{\partial\eta} - \frac{\tilde{C}_{12}}{\eta} + \frac{\tilde{C}_{23}}{1-\eta} \right) F(\eta;\xi_1,\xi_2) = 0, \tag{2.28}$$

where $\tilde{C}_{ij}$ differ C_{ij} (2.6) by constant shifts:

$$\begin{aligned} \tilde{C}_{ij} &= C_{ij} + I_i(I_i+1) + I_j(I_j+1) - I_{k4}(I_{k4}+1) \\ &= C_{ij} + (I_i + I_j - |I_{ij}|)(I'+1), (i,j,k) = \ Perm(1,2,3), \end{aligned} \tag{2.29}$$

or for the configuration (2.23b)

$$\tilde{C}_{12} = C_{12} + 2I(I'+1), \quad \tilde{C}_{23} = C_{23} + 2I'(I'+1). \tag{2.30}$$

Taking the homogeneity of F (2.26) into account, we can express $\tilde{C}_{12}$ as a differential operator in ξ_2, and $\tilde{C}_{23}$ as a differential operator in ξ_1:

$$\tilde{C}_{12} = (\xi_1+\xi_2)\xi_2\frac{\partial^2}{\partial\xi_2^2} + [(2I_- +1)\xi_1 + 2(I_- +1)\xi_2]\frac{\partial}{\partial\xi_2}, \tag{2.31}$$

$$\tilde{C}_{23} = (\xi_1+\xi_2)\xi_1\frac{\partial^2}{\partial\xi_1^2} + (2\xi_1+\xi_2)\frac{\partial}{\partial\xi_2}. \tag{2.32}$$

The space $\mathcal{J}_4 = \mathcal{J}(I_1,\ldots,I_4)$ of 4-point SU_2 invariants has dimension $d_{\mathcal{J}} = 2I+1$. A natural basis in it in our analytic picture is given by

$$J_l = \zeta_{12}^{I_{14}+I_{23}}\zeta_{23}^{I_{34}-I_{12}}\zeta_{13}^{I_{12}+I_{34}}\tilde{J}_l (= \zeta_{23}^{2I_-}\tilde{J}_l \text{ for } I_{34} = I_{21} = I_-, I_{14} = I_{23} = 0), \tag{2.33}$$

$$\tilde{J}_l \equiv \tilde{J}_l^{(2I)}(\xi_1,\xi_2) = \xi_1^{2I-l}\xi_2^l, \quad l = 0,\ldots,2I. \tag{2.34}$$

Clearly, $\tilde{C}_{12}$ does not increase l, while $\tilde{C}_{23}$ does not decrease it:

$$\begin{aligned} \tilde{C}_{12}\tilde{J}_l &= l\{(2I_- + l + 1)\tilde{J}_l + (2I_- + l)\tilde{J}_{l-1}\}, \\ \tilde{C}_{23}\tilde{J}_l &= l(2I-l)\{(2I-l+1)\tilde{J}_l + (2I-l)\tilde{J}_{l+1}\}. \end{aligned} \tag{2.35}$$

We expand F in terms of $\tilde{J}_l$ setting

$$F(\eta;\xi_1,\xi_2) = \sum_{l=0}^{2I} (\xi_2\eta)^l [\xi_1(1-\eta)]^{2I-l} f_l(\eta),$$

and insert in (2.28). The resulting system of $2I+1$ ordinary differential equations for f_l has $2I+1$ linearly independent solutions $f_l^\lambda, \lambda = 0,1,\ldots,2I$, which are given for $0<\eta<1$ by the integral representation [40, 41]

$$\begin{aligned} f_l^\lambda(\eta) = \quad & N_\lambda \int_0^\eta dt_1 \int_0^{t_1} dt_2 \ldots \int_0^{t_{\lambda-1}} dt_\lambda \int_\eta^1 dt_{\lambda+1} \\ & \int_{t_{\lambda+1}}^1 dt_{\lambda+2} \ldots \int_{t_{2I-1}}^1 dt_{2I} P_l(t_i,\eta), \end{aligned} \tag{2.36}$$

where

$$P_l = \prod_{i=1}^{2I} t_i^{\frac{1}{h}(1+2I_-)}(1-t_i)^{\frac{1}{h}}|\eta - t_i|^{\frac{1}{h}-1}\prod_{i<j}|t_{ij}|^{\frac{2}{h}}$$

$$\sum_{\sigma}\frac{1}{l!(2I-l)!}\prod_{s=1}^{l} t_{i_s}^{-1}\prod_{r=l+1}^{2I}(1-t_{i_r})^{-1}, \tag{2.37}$$

the sum extending over all permutations σ: $(1,\ldots,2I) \to (i_1,\ldots,i_{2I})$ (the numerical factor cancels the number of permutations that do not change the integrand); the constant $N_\lambda = N_\lambda(I,I')$ is independent of l.

3. Braid Group Statistics and Quantum Symmetry. Exchange Matrices for 4-Point Blocks

We now come to a remarkable property of chiral CFT: multivaluedness of conformal blocks yields braid group statistics and quantum group symmetry. Let us consider this assertion in the context of general QFT. Multivalued correlation functions in complexified space-time are certainly not newcomers in field theory. It is, however, a theorem for any local QFT, say in $D = 4$ space-time dimensions, that Wightman functions are single valued analytic in a real domain of space-like separated arguments (see, e.g. [BLOT 90]). This is a reflection of basic principles like microcausality and the fact that space-like separated points form a simply connected manifold in the product of Minkowski spaces for $\mathcal{D} \geq 3$. It follows that for 3 and higher space-time dimensions fields should satisfy permutation group statistics, and the superselection structure is associated to a compact gauge symmetry group [8]. By contrast, primary CVO in a $2D$ CFT have nontrivial monodromy and obey braid group exchange relations for causally disjoint arguments. As a consequence their symmetry can only be described by a 'quantum group' - in general, by a weak quasi Hopf algebra [33]. If one gives up Wightman positivity for chiral n-point blocks (and associated unitarity condition for exchange operators), then the quantum symmetry can be described [40, 41, 48] by a standard (Hopf) QUE algebra. As noted in Section 2, the existence of a positive-semidefinite braid invariant Hermitean form (BIHF) allows to recover positivity for $2D$ physical correlators.

3.1 Exchange Algebra

We shall display the matrices B_1 and B_2 in the basis (2.24) (2.33) (2.34) of 4-point blocks, where we are using the abbreviate notation

$$B_i^{I_1 I_2 I_3 I_4} = B_i \text{ for } (I_1 I_2 I_3 I_4) = (II'I'I), \tag{3.1}$$

and will demonstrate that B_3 is expressed in terms of B_1 (for equal isospins we have $B_1 = B_3$).

Proposition 3.1. If we write $B_i : w^\lambda \to B^\lambda_{i\mu} w^\mu$, then the exchange matrix B_1 is upper triangular:

$$B^\lambda_i \mu = \frac{N_\lambda}{N_\mu} (-1)^{2I-\mu} q^{I_-(I_-+2\lambda+1)+\mu(\lambda+1)-I(I+1)-I'(I'+1)} \mu\lambda, q = e^{i\frac{\pi}{h}}, \quad (3.2a)$$

while B_2 is lower triangular:

$$B^\lambda_2 \mu = \frac{N_\lambda}{N_\mu} (-1)^\mu q^{2I-\mu(2I-\lambda+1)-2I'(I'+1)} 2I - \mu 2I - \lambda; \quad (3.2b)$$

B_2 is similar to $B_1^{I'I'II}$:

$$B_2^{II'I'I} = F^{I'I'II} B_1^{I'I'II} F^{II'I'I} \quad (F^{I'I'II} F^{II'I'I} = 1), \quad (3.3)$$

where $F^{I_1,\ldots I_4}$ is a multiple of the operator exchanging the arguments 1 and 3:

$$\begin{aligned} & B_1^{I_2I_3I_1I_4} B_2^{I_2I_1I_3I_4} B_1^{I_1I_2I_3I_4} = B_2^{I_3I_1I_2I_4} B_1^{I_1I_3I_2I_4} B_2^{I_1I_2I_3I_4} \\ = & (-1)^{I_1+I_2+I_3-I_4} q^{I_4(I_4+1)-I_1(I_1+1)-I_2(I_2+1)-I_3(I_3+1)} F^{I_1I_2I_3I_4}. \quad (3.4) \end{aligned}$$

If the ratio $\frac{N_\lambda}{N_\mu}$ of normalization constants is assumed to be independent of the order of isospins, then so are the matrix elements of the involutive matrix F which is antidiagonal:

$$(F^{I_1\ldots I_4})^\lambda \mu \equiv F^\lambda \mu = \frac{N_\lambda}{N_\mu} \delta^{2I}_{\lambda+\mu} (\lambda, \mu = 0, \ldots, 2I = 2\min(I_1, \ldots, I_4)). \quad (3.5)$$

We have used in (3.2) the q-binomial coefficients

$$nm = \frac{[n]}{[m][n-m]!}, [n]! = [n][n-1]!, [0]! = 1, [n] = \frac{q^n - \bar{q}^n}{q - \bar{q}} (\bar{q} = q^{-1} = q^{2h-1}). \quad (3.6)$$

The proof ([40, 41, 48]) uses analytic continuation from the real domain (2.2) as well as the transformation law for the SU_2 invariants (2.31) under permutation:

$$(I_1, \zeta_1) \leftrightarrow (I_2, \zeta_2) \Rightarrow J_l \to (-1)^{2I-l} \sum_{l'=0}^{l} ll' J'_l, \quad (3.7a)$$

$$(I_2, \zeta_2) \leftrightarrow (I_3, \zeta_3) \Rightarrow J_l \to (-1)^{2I_-+l} \sum_{l'=l}^{2I} (2)I - l2I - l' J'_l. \quad (3.7b)$$

(Note that in deriving (3.7) one needs the general expression (2.31a) for arbitrary isospins.)

The family of normalizations for which the N-dependent factor in $F^\lambda \mu$ (3.5) disappears, $N_\lambda = N_{2I} - \lambda$, contains the value

$$N_\lambda = N(I_1,\ldots,I_4)2I\lambda^{-1} \leftrightarrow \frac{N_\lambda}{N_\mu} = \frac{[\lambda]![2I-\lambda]!}{[\mu][2I-\mu]} \quad (I = \min_i I_i), \tag{3.8}$$

for which

$$(B_1^{I_1I_2I_3I_4})^\lambda{}_\mu = (-1)^{I_{14}+I_{23}+2I-\mu}$$
$$q^{(I_{34}+\mu)(I_{34}+\lambda+1)+I_{12}(\mu-\lambda)-I_1(I_1+1)-I_2(I_2+1)} \begin{bmatrix} 2I-\lambda \\ \mu-\lambda \end{bmatrix}. \tag{3.9}$$

The q-binomial coefficient vanishes outside the range of indices (e.g., for $\mu > 2I$). The choice $N_\lambda/N_\mu = 1$ (of the same family) was made in [40, 41]. Physically relevant quantities (such as the eigenvalues of the braid matrices) are independent of N_λ.

The first equation (3.4) has the form of a (parameter free) *quantum Yang-Baxter equation.* For equal isospins it reduces to the basic third degree relation for braid group generators. The exchange matrices obey additional relations (like the involutivity of F in (3.3)), that give rise to a projective representation of the *mapping class group* $\mathcal{B}_4$ *of the 2-sphere with 4 punctures.* In order to write them down we introduce the maps

$$B(I_1|I_2I_3I_4) = B_3^{I_2I_3I_1I_4} B_2^{I_2I_1I_3I_4} B_1^{I_1I_2I_3I_4} : \mathcal{J}^{I_1I_2I_3I_4} \to \mathcal{J}^{I_2I_3I_4I_1}, \tag{3.10a}$$

$$B(I_2I_3I_4|I_1) = B_1^{I_2I_1I_3I_4} B_2^{I_2I_3I_1I_4} B_3^{I_4I_1I_2I_3} : \mathcal{J}^{I_4I_1I_2I_3} \to \mathcal{J}^{I_1I_2I_3I_4}. \tag{3.10b}$$

The additional relations characterizing $\mathcal{B}_4$ have the form

$$\begin{aligned} B(I_2I_3I_4|I_1)B(I_1|I_2I_3I_4) &= \bar{q}^{4I_1(I_1+1)}, \\ B(I_4|I_1I_2I_3)B(I_1I_2I_3|I_4) &= \bar{q}^{4I_4(I_4+1)}, \end{aligned} \tag{3.11}$$

$$\begin{aligned} &B(I_4|I_1I_2I_3)B(I_3|I_4I_1I_2)B(I_2|I_3I_4I_1)B(I_1|I_2I_3I_4) \\ &= \bar{q}^{2\{I_1(I_1+1)+I_2(I_2+1)+I_3(I_3+1)+I_4(I_4+1)\}}. \end{aligned} \tag{3.12}$$

Eq. (3.11) reflects the fact that the monodromy of a CVO of dimension Δ around the point at infinity is given by $e^{-4\pi i \Delta}$. (A more general statement is proven in Appendix A to [48].)
Define, in general, the *monodromy operators*

$$m_i = B_i^{I_1\ldots I_{i+1}I_i\ldots I_n} B_i^{I_1\ldots I_iI_{i+1}\ldots I_n} : \mathcal{J}^{I_1\ldots I_iI_{i+1}\ldots I_n} \to \mathcal{J}^{I_1\ldots I_iI_{i+1}\ldots I_n}. \tag{3.13}$$

Eqs. (3.4), (3.11) and (3.12) imply

$$m_1 = (B_1^{I_2I_1I_3I_4} B_1^{I_1I_2I_3I_4}) = q^{2\{I_3(I_3+1)+I_4(I_4+1)-I_1(I_1+1)-I_2(I_2+1)\}} m_3. \tag{3.14}$$

In order to relate B_1 and B_3 we need the following observation. The Jost version of the *TCP theorem* holds for chiral conformal blocks:

$$w(z_1, \zeta_1, I_1; z_2, \zeta_2, I_2; \ldots; z_n, \zeta_n, I_n)$$
$$= w(-z_n, -\zeta_n, I_n; \ldots; -z_2, -\zeta_2, I_2; -z_1, -\zeta_1, I_1). \quad (3.15)$$

Hence there is an involution operator Θ in the direct sum $\mathcal{L}(I_1, I_2, \ldots, I_n) \oplus \mathcal{L}(I_n, \ldots, I_2, I_1)$ which interchanges the two terms. For $n = 4$ it intertwines B_1 and B_3:

$$B_1^{I_4 I_3 I_2 I_1} \Theta = \Theta B_3^{I_1 I_2 I_3 I_4}. \quad (3.16)$$

The subgroup $\mathcal{B}(I, I', I', I)$ of exchange operators, that leave the order (I, I', I', I) unchanged, is generated by m_1 and B_2. Note that for this order Eq. (3.14) becomes simply $m_1 = m_3$.

The algebra of (4-point) exchange operators so constructed has two remarkable properties. Most importantly, B_i are well defined for the commonly used tree-like bases (see, e.g. [36]) which pretend to diagonalize non-diagonalizable matrices. Our second observation is a number theoretical one: there is a family of normalizations N_λ - including ours (3.8) (as well as the choice N_λ made in [40]) - such that the entries of the exchange matrices are *cyclotomic integers*:

$$B_i^{I_1 I_2 I_3 I_4} \in GL(2I + 1, Z(q^{1/2})), \quad (3.17)$$

where $Z(q^{1/2})$ is the ring of polynomials of integer coefficients in the cyclotomic unit $q^{1/2}$ (such that $(q^{1/2})^{2h} = -1$). It follows that the determinant $\det B_i$ is a power of $q^{1/2}$. This is also seen directly from our triangular matrix realization; for B_1 (3.2a) we have

$$\det B_1^{II'I'I} = (-1)^{I(2I+1)} \bar{q}^{\frac{2}{3} I(I+1)(2I+1)}. \quad (3.18)$$

3.2 Violation of Unitarity

There is a price for the regularity and the nice arithmetic properties of the above representation of the exchange algebra: it is, in general, not unitarizable. Whenever $I_1 + I_2 + I_3 + I_4 \geq h - 1$, the representation is actually indecomposable (hence, a fortiori, nonunitarizable). Postponing the general proof of this statement to Sec. 4, we shall verify it on the simplest example of four isospin 1/2 fields in a level 1 theory. For equal isospins the upper indices of B_i are superfluous, and we have

$$B_1 = \bar{q}^{1/2} - \bar{q} 1 0 q, \quad B_2 = \bar{q}^{1/2} q 0 1 - \bar{q} \quad (I_i = I = 1/2, q + \bar{q} = 1). \quad (3.19)$$

We leave it to the reader to verify that the vectors proportional to 11 form a 1-dimensional invariant subspace og $\mathcal{B}_4$ for $[2] = 1$ with no invariant complement.

3.3 Quantum Group Symmetry

The most natural way to come to the notion of quantum group symmetry starts with the Wess-Zumino-Novikov-Witten (WZNW) multivalued action [51] or, equivalently, with the closed 3-form [23]

$$\begin{aligned}\omega &= \frac{1}{4\pi} tr \left\{ \frac{1}{k} j\alpha dj^\alpha dx^0 dx^1 + i(dj^0 g^{-1} dg dx^1 + dj^1 dx^0 g^{-1} dg) \right. \\ & \left. -j^0 (g^{-1}dg)^2 dx^1 + j^1 dx^0 (g^{-1}dg)^2 - \frac{k}{3}(g^{-1}dg)^3 \right\}. \end{aligned} \tag{3.20}$$

The equations of motion (obtained by contracting ω with the vertical vector fields $\frac{\partial}{\partial j^\alpha}$ and $gX\frac{\partial}{\partial g}$, $X \in su_2$, and then *pulling back*, - i.e. setting $dj^\alpha = (\partial_\beta j^\alpha)dx^\beta$ etc.),

$$j_\alpha = -ikg^{-1}\partial_a\, g, \quad \partial_+\left((\partial_- g)g^{-1}\right) = 0 = \partial_-(g^{-1}\partial_+ g), \quad \partial_\pm = 1/2(\partial_1 \pm \partial_0), \tag{3.21}$$

yield a factorized solution

$$g(x) = g_-(x_-)g_+(x_+), \quad x_\pm = x^1 \pm x^0 \quad (g_\pm \in SU(2)). \tag{3.22}$$

Here g is a group valued field (in the case at hand $g(x) \in SU(2)$) defined on the cylindric space-time ($x^0 = t \in R$, $x^1 = x \in R/2\pi Z$), - i.e. periodic in the space coordinate x, which only implies a twisted periodicity for $g_\pm$:

$$\begin{aligned} g(t, x+2\pi) &= g(t,x) \Rightarrow g_-(x+2\pi) \\ &= g_-(x)M, g_+(x+2\pi) = M^{-1}g_+(x), \\ & \quad M \in SU(2). \end{aligned} \tag{3.23}$$

The symplectic form $\int_{x=-\pi}^{\pi} \omega(dt=0)$ splits into a sum of a left and a right invariant chiral symplectic form [23, 12]. Upon quantization [12, 17] the g-form yields the operator exchange relations

$$g_-^2(x_2)g_-^1(x_1) = g_-^1(x_1)g_-^2(x_2)R_{12}(x_{12}), \quad x_{12} = x_1 - x_2, \tag{3.24}$$

where R_{12} depends only on the sign of the coordinate difference ($R_{12}(x_{12}) = R_{12}^-\theta(x_{12}) + R_{12}^+\theta(x_{21})$). In order to make contact with our axiomatic results, we introduce the analytic picture chiral field $v(z)$ related to $g_-(x)$ by

$$g_-(x) = e^{-i\Delta x}v(e^{-ix}), \quad \Delta = \frac{3}{4h} \quad (h = k+2). \tag{3.25}$$

The exchange relation (3.24) can be recast as follows: for z_1 and z_2 exchanged along the path C_1^+(2.9) we have

$$v^2(z_2)v^1(z_1) = v^1(z_1)v^2(z_2)R_{12}^- \text{ or } Pv^1(z_2)v^2(z_1) = v^1(z_1)v^2(z_2)\check{R}_{12}, \tag{3.26}$$

where P is the permutation operator, $\check{R}_{12}$ is the *braid operator*:

$$(Pv^1(z_2)v^2(z_1))^{\alpha_1\alpha_2}\beta_1\beta_2 = v^{\alpha_2}_{\beta 1}(z_2)v^{\alpha_1}_{\beta 2}(z_1), \quad \check{R}_{12} = R^-_{12}P. \tag{3.27}$$

The exchange relation (3.26) admits a quantum group symmetry $v \to vT^{-1}$, where $T \in SL_q(2)$ - i.e., it is a 2×2 matrix with non-commuting entries satisfying

$$R^{\pm}_{12}T^1T^2 = T^2T^1R^{\pm}_{12} \text{ or } [\check{R}_{12}, T^1T^2] = 0 (\Rightarrow T^{\alpha}_1 T^{\alpha}_2 = qT^{\alpha}_2 T^{\alpha}_1, \text{ etc.}), \tag{3.28}$$

$$\det_q T := T^1_1 T^2_2 - qT^1_2 T^2_1 = T^2_2 T^1_1 - \bar{q} T^1_2 T^2_1 = 1. \tag{3.29}$$

The 4-point function of $v(z)$ can be expanded in terms of the basis w^λ ((2.24), (2.33), (2.34)), with coefficients $\mathcal{J}_\lambda$ that are quantum group invariant rank 4 tensors:

$$< 0|v(z_1,\zeta_1)v(z_2,\zeta_2)v(z_3,\zeta_3)v(z_4,\zeta_4)|0 >= w^0 \mathcal{J}_0 + w^1 \mathcal{J}_1. \tag{3.30}$$

The relation between B_1 (3.9) and $\check{R}_{12}$ (3.27) is then given by

$$\mathcal{J}_\lambda B^\lambda_{1\mu} = \mathcal{J}_\mu \check{R}_{12}, \quad B_1 = \bar{q}^{1/2} - \bar{q}10q \quad (I_i = 1/2, i = 1, \ldots, 4). \tag{3.31}$$

4. A Positive-Semidefinite Braid Invariant Hermitean Form

The construction of an (SU_2 and) quantum group covariant chiral field can be extended as follows. There exists for any I (not just for $I = 1/2$) a *chiral field* V_I, which transforms as an isospin I SU_2 tensor under left shifts, and as a q-spin I $U_q(sl_2)$ tensor under right shifts. It thus combines in a single entity all CVO V_I of a given I. Its correlation functions can be written as a finite sum of type (3.30) of solutions of the KZ equation multiplied by appropriate (n point) quantum group invariants, so that the exchange relations (3.26) are valid (for vacuum expectation values). There is a polynomial (coherent state) realization of $U_q(sl_2)$ representations [20] used in [40, 41], in which the quantum group invariants appear as a q-deformation of the SU_2 invariant monomials

$$\prod_{i<j} \zeta_{ij}^{\mu_{ij}}, \quad \zeta_{ij} = \zeta_i - \zeta_j, \quad \sum_{j \neq i} \mu_{ij} = 2I_i, \quad \mu_{ij}\mu_{kl} = 0 \text{ for } i < k < j < l. \tag{4.1}$$

For the general isospin configuration (2.23a) we have the following basis of 4-point $U_q(sl_2)$ invariants ([19, 48]):

$$\mathcal{J}_\lambda(u_1 I_1, u_2 I_2, u_3 I_3, u_4 I_4) = a_\lambda p_{12} p_{23} p_{34} p_{14} p_{13}, \quad a_\lambda = 2I, \tag{4.2a}$$

$$p_{ij} \equiv p(u_i, u_j; \mu_{ij}, \rho_{ij}) := \prod_{n=0}^{\mu_{ij}-1} (\bar{q}^{\rho_{ij}} u_i - q^{\mu_{ij}+\rho_{ij}-2n} u_j), \tag{4.2b}$$

where $2\rho_{ij}$ (and μ_{ij}) are (non-negative) integers given by

$$\begin{aligned}
0 \le \mu_{14} &\equiv \lambda \le 2I, \quad \mu_{23} = I_{34} - I_{12} + \lambda, \quad \mu_{12} = I_{14} + I_{23} + 2I - \lambda, \\
\mu_{34} &= 2I - \lambda, \mu_{13} = I_{12} + I_{34}, \quad 2\rho_{12} = \mu_{23}, \\
2\rho_{13} &= -I_{14} - I_{23}, 2\rho_{34} = \lambda, \\
2\rho_{23} &= I_{14} - I_{23} + 2I - \lambda, \quad 2\rho_{14} = \lambda - 2I;
\end{aligned} \tag{4.3a}$$

$$\begin{aligned}
\mu_{13} &= 0, \quad \mu_{12} = \mu_{34} = 2\rho_{23} = 2\rho_{14} = 2I - \lambda \\
\text{for } I_{34} &= I_{21} = I_{-} (\equiv I' - I), \quad I_{23} = I_{14} = 0.
\end{aligned} \tag{4.3b}$$

This basis is conjugate to the basis w^λ of solutions of the KZ equation constructed in Sec. 2 in the sense, that the 4-point correlation function has the form

$$\begin{aligned}
&< 0| \prod_{i=1}^{4} V_{I_i}(z_i, \zeta_i, u_i)|0 > \\
&= w^\lambda(z_1, \zeta_1, I_1; \ldots; z_4, \zeta_4, I_4)\mathcal{J}_\lambda(u_1 I_1, \ldots, u_4 I_4).
\end{aligned} \tag{4.4}$$

(Note that the product of the normalization constants N_λ (3.8) and a_λ (4.2a) is independent of λ.) The 4-point function (4.4) is characterized by the following property: The monodromy representation of the exchange generators B_i constructed in Sec. 3 reduces to the action of the quantum group braid operators $\check{R}_{i,i+1}$ on the $U_q(sl_2)$ invariants: an extension of Eq. (3.31) is valid for arbitrary isospins.

The space of all solutions of the KZ equation needed to define correlation functions with such a property is bigger than the space on n-point blocks obeying the *physical fusion rules*, if the sum of isospins exceeds $h-1$. According to [22, 24] the *physical n-point blocks* are singled out by the constraint

$$\frac{\partial^{l_1+\ldots+l_n}}{\partial\zeta_1^{l_1}\partial\zeta_n^{l_n}} w_n|_{\{\zeta_i = z_i\}} = 0 \text{ for } \sum_{i=1}^{n}(I_i - l_i) \ge h - 1. \tag{4.5}$$

The violation of this condition is the source of the breaking of full reducibility (and hence of unitarity) of the exchange algebra illustrated by (3.19). We shall now write down a braid invariant Hermitean form in the space of 4-point blocks which contains the unwanted pieces in its kernel. It corresponds to the following recipe for the inner product of monomials belonging to the same irreducible representation of q-spin I [48, 41]:

$$2In(\bar{u}^{\bar{n}}, u^n) = (-1)^n q^{(I-n)(2I-1)} \delta^{2I}_{n+\bar{n}}, \quad n = 0, 1, \ldots, 2I, \tag{4.6}$$

and is continued in a factorizable way to tensor products of representations. The overall normalization in (4.6) is chosen in such a way, that the norm square of the 2-point function

$$p(u_1, u_2; 2I, 0) = \sum_{m=0}^{2I} (-q)^m 2Imu_1^{2I-m}u_2^m, \tag{4.7}$$

obtained from (4.2) for $I_1 = I_2 = I$, $I_3 = I_4 = 0$, is given by the *quantum dimension*

$$(p(\bar{u}_1, \bar{u}_2; 2I, 0), p(u_1, u_2; 2I, 0)) = \sum_{n=0}^{2I} q^{2n-2I} = [2I+1]. \qquad (4.8)$$

The Hermitean form $Q = (Q_{\lambda\mu}^{I_1 \ldots I_4})$ is now given by the inner products of the quantum group invariants (4.2)

$$Q_{\lambda\mu}^{I_1 \ldots I_4} = (\mathcal{J}_\lambda(\bar{u}_1 I_1, \ldots, \bar{u}_4 I_4), \mathcal{J}_\mu(u_1 I_1, \ldots, u_4 I_4)) = S_{\nu\lambda} D_\nu S_{\nu\mu}. \qquad (4.9)$$

Here S, the matrix that diagonalizes Q, can be written as

$$S_{\lambda\mu} = (-1)^{\mu-\lambda} \mu \frac{[\mu - I_{12} + I_{34}]![2I_{34} + 2\lambda + 1]![2I - \lambda]![\lambda]!}{[\lambda - I_{12} + I_{34}]![2I_{34} + \lambda + \mu + 1]![\mu]![2I - \mu]!}, 0 \le \lambda \le m, \qquad (4.10a)$$

$$S_{\lambda\mu} = \delta_{\lambda\mu} \text{ for } m+1 \le \lambda \le 2I \quad (\text{if } 2I + 2I' > k), \qquad (4.10b)$$

$$m = m(k; I, I') = \min(2I, k - 2I') = 1/2(k - 2I_- - |2I + 2I' - k|). \qquad (4.11)$$

The diagonal matrix $D = (D_\lambda \delta_{\lambda\mu})$ is degenerate, if $2I + 2I' > k$, and has $2I + 2I' - k$ zero eigenvalues. First of all, if $4I > k$, then $D_\lambda = 0$ for $\lambda > m(k; I, I) = \min(2I, k - 2I)$. In the complementary region we find the expression

$$D_\lambda^{II'I'I} = \left(\frac{[2I' + \lambda + 1]![2I_- + \lambda]!}{[2I']![2I_- + 2\lambda + 1]!} \right)^2 [2I_- + 2\lambda + 1], \; 0 \le \lambda \le \min(2I, k - 2I), \qquad (4.12)$$

which is verified to vanish for $2I_-$ more values of λ (assuming $I_- > 0$):

$$D_\lambda^{II'I'I} = 0 \text{ for } k + 1 - 2I' \le \lambda \le \min(2I, k - 2I). \qquad (4.13)$$

The kernel $kerQ$ of $Q^{II'I'I}$ defines a braid invariant subspace of $\mathcal{L}_4 = \mathcal{L}(II'I'I)$ with no invariant complement. The factor space $\mathcal{L}_4/kerQ$ carries a unitarizable representation of the stabilizer $\mathcal{B}(I, I', I', I)$ of the configuration (2.23b) in the exchange algebra, provided the intermediate quantum dimensions are positive:

$$[2I_- + 2\lambda + 1] > 0 \text{ for } 0 \le \lambda \le m. \qquad (4.14)$$

That is indeed true for q satisfying

$$[2] = q + \bar{q} = 2\cos\frac{\pi}{h}. \qquad (4.15)$$

Note that the positivity requirement (4.14) is the first case in which the value of the complex number q among the primitive roots of the equation $q^h = 1$ does matter (cf. [47]). The braid invariance of the form (4.9-12) is verified by a direct computation ([48, 41]).

The factor representation of $\mathcal{B}(I, I', I,' I)$ in the 'physical' quotient space

$$\mathcal{P}(I, I', I', I) = \mathcal{L}(I, I', I', I)/\text{ker } Q \tag{4.16}$$

may still be reducible (then the braid invariant Hermitean form constructed here is not unique). Classifying its irreducible components should yield the ADE classification of $A^{(1)}$ current algebra CFTs [4]. (For the time being only the opposite path has been followed: using the classification of modular invariant $A^{(1)}$ partition functions to detect the cases of reducibility of the unitary monodromy representations of the mapping class group - see [42, 47].)

5. Concluding Remarks: Open Problems

We have reviewed in these lectures one of the best understood facets of RCFT. Even within this limited scope (leaving aside current efforts to classify higher rank current algebra theories [21, 38, 39, 6, 16]) the subject is far from being closed. We shall just mention two open problems directly related to the content of this paper.

Quantum group symmetry is only understood for diagonal (A-series) models. Its precise identification for the two infinite D series (even and odd) is still a challenge. Simple algebraic (even arithmetic) properties of ratios of structure constants of non-diagonal versus diagonal theories uncovered recently ([38, 37]) also await a better understanding. We should note in this connection, that the algebraic QFT approach [37] to the study of local extensions of conformal current algebras [34] (for an essay on the subject addressed to a wider audience - see [46]) has also remained out of the scope of the present paper.

The second problem concerns the existence of a consistent (continuum limit) operator formalism for chiral conformal fields satisfying exchange relations of the type (3.24). The difficulty stems from the fact that one mulitplies a (bilocal) operator valued distribution by a discontinuous step function. So far, it has only been treated in the framework of a lattice WZNW theory [12].

The authors thank Krzysztof Gawedzki for enlightening discussions. I. T. thanks Harald Grosse for his invitation to the 1995 Schladming Universitšatswochen and both him and Walter Thirring for their hospitality at the International Erwin Schrödinger Institute for Mathematical Physics during the course of this work. Y.S. thanks Università di Roma and INFN for hospitality and support. This work has been supported in part by the Bulgarian Foundation for Scientific Research under contract F-404

References

1. V. Bargmann and I.T. Todorov, *Spaces of analytic functions on a complex cone as carries for the symmetric tensor representations of $SO(n)$*, J. Math. Phys. **18** (1977) 1141-1148.
2. A.A. Belavin, A.M. Polyakov and A.B. Zamolodchikov, *Infinite conformal symmetry in two-dimensional quantum field theory*, Nucl. Phys. **B365** (1991) 98-120.
3. N.N. Bogolubov, A.A. Logunov, A.I. Oksak and I.T. Todorov, *General Principles of Quantum Field Theory*, Kluwer Academic Publ., Dordrecht (1990).
4. A. Cappelli, C. Itzykson and J.B. Zuber, *The A-D-E classification of minimal and $A_1^{(1)}$ conformal invariant theories*, Commun. Math. Phys. **113** (1987) 1-26.
5. P. Christie and R. Flume, *The four point correlations of all primary fields of the $d = 2$ conformally invariant $SU(2)$ σ-model with a Wess-Zumino term*, Nucl.Phys., **B282** (1987) 466-494
6. A. Coste and T. Gannon, *Galois symmetry in RCFT*, Phys. Lett. **B323** (1994) 316-321.
7. Vl.S. Dotsenko, V.A. Fateev, *Conformal algebra and multipoint correlation functions in 2D statistical models*, Nucl. Phys. **B240** (1984) 312-348.
8. S. Doplicher, J.E. Roberts, *Why there is a field algebra with a compact gauge group describing the superselection structure in particle physics*, Commun. Math. Phys. **131** (1991) 51-107.
9. V.G. Drinfeld, *Hopf algebras and the quantum Yang-Baxter equation*, Soviet Math. Dokl. **32** (1985) 254-258.
10. V.G. Drinfeld, *Quantum groups*, in Proc. 1986 I.C.M., Amer. Math. Soc. Berkeley, CA **1** (1987) 798-820.
11. V.G. Drinfield, *Quasihopf algebras and Knizhnik-Zamolodchikov equations*, Alg. i Anal. **1:6** (1989) 114-148; English transl.: Leningrad Math. J. **1** (1990), 1419–1457.
12. F. Falceto, K. Gawedzki, *Lattice Wess-Zumino-Witten model and quantum groups*, J. Geom. Phys. **11** (1993) 251-279.
13. L.D. Faddeev, N.Yu. Reshetikhin and L.A. Takhtajan, *Quantization of Lie groups and Lie algebras*, (English transl.: Leningrad Math. J. **1** (1990) 193-225) Alg. i Anal **1:1** (1989) 178-206.
14. D. Friedan and S. Shenker, *The analytic geometry of two-dimensional conformal field theory*, Nucl. Phys. **B281** (1987) 509-545.
15. J. Fröhlich and T. Kerler, *Quantum Groups, Quantum Categories and Quantum Field Theory*, Lecture Notes in Math. **1542** Springer Berlin et al., (1993)
16. J. Fuchs, A.N. Schellekens and C. Schweigert, *Galois modular invariants of WZW models*, Nucl. Phys. **B437** (1995) 667-694.
17. P. Furlan, L.K. Hadjiivanov and I.T. Todorov, *The price for quantum group symmetry: chiral versus 2D WZNW model*, Trieste-Vienna preprint (1995).
18. P. Furlan, G.M. Sotkov and I.T. Todorov, *Two-dimensional conformal quantum field theory*, Riv. Nuovo Cim. **12:6** (1989) 1-202.
19. P. Furlan, Ya.S. Stanev and I.T. Todorov, *Coherent state operators and n-point invariants for $U_q(sl(2))$*, Lett. Math. Phys. **22** (1991) 307-319.
20. A. Ch. Ganchev and V. B. Petkova, *$U_q(sl_{(2)})$ invariant operators and minimal theory fusion matrices*, Phys. Lett. **B233** (1989 374-382.
21. T. Gannon, *The classification of affine $SU(3)$ modular invariant partition functions*, Commun. Math. Phys. **161** (1994) 233-264.
22. K. Gawedzki, *Geometry of Wess-Zumino-Witten model of conformal field theory*, Nucl. Phys. B (Proc.Suppl.) **18B** (1990) 78-91.

23. K. Gawedzki, *Classical origin of quantum group symmetries in Wess-Zumino-Witten conformal field theory*, Commun. Math. Phys. **139** (1991) 210-213.
24. K. Gawedzki, A. Kupiainen, *$SU(2)$ Chern Simons theory of genus zero*, Comm. Math. Phys. **135** (1991) 531-546.
25. P. Goddard, D.Olive (eds.), *Kac-Moody and Virasoro Algebras, A reprint volume for Physicists*, World Scientific, Singapore (1988).
26. L.K. Hadjiivanov, R.R. Paunov and I.T. Todorov, *Quantum group extended chiral p-models*, Nucl. Phys. **B356** (1991) 387-438.
27. M. Jimbo, *A q-difference analogue of $U(g)$ and the Yang-Baxter equation*, Lett. Math. Phys.**10** (1985) 63-69.
28. M. Jimbo, *A q-analogue of $U(gl(N+1))$, Hecke algebras and the Yang-Baxter equation*, Lett. Math. Phys. **11** (1986) 247-252.
29. V.G. Kac, *Infinite Dimensional Lie Algebras*, Third edition, Cambridge Univ. Press, Cambridge, (1990).
30. V.G. Knizhnik and A.B. Zamolodchikov, *Current algebra and Wess-Zumino model in two dimension*, Nucl. Phys. **B247** (1984) 83-103.
31. R. de L. Kronig, *Zur Neutrino Theorie des Lichtes III*, Physica **2** (1935) 968-980.
32. G. Lusztig, *Introduction to Quantum Groups*, Progress in Mathematics v. 110 series eds. J. Oesterlé, A. Weinstein, Birkhäuser, Boston (1993).
33. G. Mack and V. Schomerus, *Quasi Hopf quantum symmetry in quantum theory*, Nucl. Phys. **B370** (1992) 185-230.
34. L. Michel, Ya.S. Stanev and I.T. Todorov, *D-E classification of the local extensions of the su_2 current algebras* (Theor. Math. Phys. **92** (1993) 1063), Teor. Mat. Fiz. **92** (1992) 507-521.
35. G. Moore and N. Seiberg, *Classical and quantum conformal field theory*, Commun. Math. Phys. **123** (1989) 177-254.
36. G. Moore and N. Seiberg, *Lectures on RCFT*, Superstrings 89, Proceedings of the Trieste 1989 Spring School, World Scientific Singapore (1990) 1-129.
37. K.-H. Rehren, Y.S. Stanev and I.T. Todorov, *Characterizing invariants for local extensions of currents algebras*, Hamburg-Vienna preprint DESY 94-164, ESI 132 (1994), hep-th/9409165, Commun. Math. Phys. (1995), to be published.
38. Ya.S. Stanev, *Classification of the local extensions of the $SU(3)$ chiral, current algebra*, J. Math Phys. **36** (1995) 2053-2069.
39. Ya.S. Stanev, *Classification of the local extensions of the $SU(2) \times SU(2)$ chiral current algebra*, J. Math Phys. **36** (1995) 2070-2084.
40. Ya. S. Stanev, I.T. Todorov and L.K. Hadjiivanov, *Braid invariant rational conformal models with a quantum group symmetry*, Phys. Lett. **B276** (1992) 87-94.
41. Ya.S. Stanev, I.T. Todorov and L.K. Hadjiivanov, *Braid invariant chiral conformal models with a quantum group symmetry*, *Quantum Symmetries*, ed. by H.-D. Doebner, V.K. Dobrev, World Scientific Singapore (1993), p. 24-40.
42. Ya. S. Stanev and I.T. Todorov, *On Schwarz problem for the $\widehat{su}_2$ Knizhnik-Zamolodchikov equation*, Lett. Math. Phys., ESI preprint 121 (1994)
43. W. Thirring, *A soluble relativistic field theory*, Ann. Phys. (N.Y.) **3** (1958) 91-112.
44. I.T. Todorov, *Infinite Lie algebras in 2-dimensional conformal field theory*, Talk in: International Conference on Differential Geomtric methods in Theoretical Physics (Shumen, Bulgaria, August 1984), in Differential Geometric Methods in Theoretical Physics, H.-D. Doebner and T.D. Palev, Eds., World Scientific Singapore (1986) pp. 297-347.
45. I.T. Todorov, *Current algebra approach to conformal invariant 2-dimensional models*, Phys. Lett. B153 (1985) 77-81.

46. I.T. Todorov, *What are we learning from 2-dimensional conformal models?* in: *Mathematical Physics Toward the 21st Century*, R.N. Sen, A. Gersten, Eds., Ben Gurion University of the Negev Press, Beer Sheva (1994), pp. 160-176.
47. I.T. Todorov, *Arithmetic features of rational conformal field theory*, Ann. Inst. Poincaré , Bures-sur-Yvette preprint IHES/P/95/10.
48. I.T. Todorov and Ya. S. Stanev, *Chiral Current Algebras and 2-Dimensional Conformal Models*, Troisième Cycle de la Physique en Suisse Romande, Lausanne (1992).
49. S. Tomonaga, *Remarks on Bloch's method of sound waves applied to many-fermion problems*, Prog. Theor. Phys. **5** (1950) 544-569.
50. A. Tsuchiya and Y. Kanie, *Vertex operators in the conformal field theory on P^1 and monodromy representations of the braid group*, Lett. Math. Phys. **13** (1987) 303-312, *Conformal field theory and solvable lattice modells*, Advanced Studies in Pure Mathematics **16** (1988) 297-372.
51. E. Witten, *Non-Abelian bosonization in two dimensions*, Commun. Math. Phys. **92** (1984) 455-472.
52. A.B. Zamolodchikov and V.A. Fateev, *Operator algebra and correlation functions in two-dimensional $SU(2) \times SU(2)$ chiral Wess-Zumino model*, Yad.Fiz. **43** (1986) 1031-1044; transl.: Sov. J. Nucl. Phys. **43** (1986), 657–664.

What are the Quantum Mechanical Lyapunov Exponents?

W. Thirring

Institut für Theoretische Physik Universität Wien

Dedicated to Paul Urban, teacher and friend

1. Introduction

In the past years it has been possible to widen the scope of ergodic theory so as to contain also quantum systems. It was rather easy to generalize the notions of mixing [1] and topological mixing [2,3]. The dynamical entropy was a long time ago defined for type II_1 systems [4], but now there are three different proposals for the general case [5,6,7]. They agree for commutative systems, but show different features in the quantum domain. The topological entropy could also be generalized to noncommutative systems [8]. Recently there have been proposals for a generalization of the Lyapunov exponents [9,10,11] for quantum systems. I shall follow here a definition first given in [12] for the quantum cat and elaborated in a general setting in [13]. It seems that the heart of the matter is a certain group. Its action immediately dictates sensitive dependence on initial conditions and exponential decay of correlations, provided the action is sufficiently continuous. We shall find many examples of the action of this group, starting from elementary mechanics to wave mechanics and to quantum field theory. They give feeling, when the continuity requirements are met, so that our general theorems become applicable.

The reader might be suspicious, since physical properties depend crucially on continuity properties. It is a sad fact that the key notions in ergodic theory are highly discontinuous objects, and this does by no means become better in the quantum domain. As example may serve the cat map on the noncommutative torus, which depends on a rotation parameter Θ ($= \hbar$ in appropriate units). This system is random in the sense of algorithmic complexity theory only for $\Theta = 0$ ([14] where this was called failure of the correspondence principle). The Connes–Størmer entropy is zero for almost all Θ's [15], but for rational Θ's it has the classical value, and for countably many Θ's it is positive [12]. Also the topological structure of this system considered is quite different for rational and irrational Θ's [16], in the first case it is a fibre bundle, in the latter it is not. On the other hand, the Alicki–Fannes entropy has the classical value irrespective of Θ, and the Voiculescu entropy has always at least half the classical value. The Lyapunov exponent does not show this nervous behaviour, and all definitions [9,10,11,12] give it its classical value for all Θ's.

2. The Anosov Groups

The intuition behind the sensitive dependence on initial conditions is, that next to any point on any trajectory there is a point at a distance s such that in the course of time their distance grows exponentially. In formulas it reads

$$\tau^t \circ \sigma^s = \sigma^{se^{-\lambda t}} \circ \tau^t , \tag{2.1}$$

which is the multiplication law of the semidirect product $\mathbf{R} \otimes \mathbf{R}$. From this we abstract the

Definition (2.2). The multiplication laws

(i) $(t,s) \circ (t',s') = (t+t', s+s'e^{-\lambda t})$
(ii) $e = (0,0)$
(iii) $(t,s)^{-1} = (-t, -e^{\lambda t}s)$

define for

(a) $(t,s) \in \mathbf{Z} \times \mathbf{R}^+$ the Anosov semigroup
(b) $(t,s) \in \mathbf{Z} \times \mathbf{R}$ the Anosov group
(c) $(t,s) \in \mathbf{R} \times \mathbf{R}$ the continuous Anosov group.

An Anosov dynamical system is a realization of these groups by endomorphisms τ^t, σ^s of a C*-algebra $\mathcal{A}$.

Remarks (2.3).

1. We shall consider only the biggest Lyapunov exponent, and in the continuous case the groups with different λ's are isomorphic. Then we shall scale λ to 1.
2. Rescaling s does not change λ. In fact σ^s should shift by the distance s, which has to be gauged by some metric. In [13] it is shown that λ is independent of the metric.
3. Since we are interested in the long time behaviour, $t \in \mathbf{Z}$ or $t \in \mathbf{R}$ are equally good. However we need τ to be invertible, so $t \in \mathbf{Z}^+$ would not do.
4. s has to be continuous, since on a compact manifold the final distance cannot be arbitrarily big, only the initial distance arbitrarily small. We do not only consider the case $s \in \mathbf{R}$, because we shall see that then an invariant state cannot be τ–KMS.
5. The systems we shall consider will be time–reversible, i.e. there is an automorphism (or antiautomorphisms) T with $T^{-1} \circ \tau^t \circ T = \tau^{-t}$. In this case $\sigma_-^s = T^{-1} \circ \sigma^s \circ T$ satisfies (2.1) with $\lambda \to -\lambda$.
6. In the physics we have in mind $\mathcal{A}$ is the algebra of observables and τ^t its true evolution. In spite of the nonlinearity of the dynamics τ^t is represented by linear transformations, namely isometries of the Banach space underlying $\mathcal{A}$. With the aid of invariant states we shall later construct unitary representations in a Hilbert space. There are finite–dimensional

non–unitary representations. Since σ^s generates an invariant subgroup, $(t,s) \to e^t$ is a representation which represents the factor group τ^t faithfully. $(t,s) \to \begin{pmatrix} 1 & 0 \\ s & e^{-t} \end{pmatrix}$ represents the whole group faithfully.

3. Examples

We shall first show that the Anosov groups appear quite frequently in physics. We begin with the simplest situation, namely classical linear systems. Contrary to the folklore sensitive dependence on initial conditions is not a characteristic of nonlinear systems.

3.1 Hamiltonian Flows

Abelian C*-algebras $\mathcal{A}$ are spaces of continuous functions, and any automorphism τ is induced by a homeomorphism τ_* of the underlying space: $\tau(f) = f \circ \tau_*^{-1}\ \forall\ f \in \mathcal{A}$. For the flows corresponding to τ_*^t and σ_*^s we shall take the Hamiltonian flows generated by $H, K \in C(T^*(\mathbf{R}))$, and because of $(\tau^t \circ \sigma^s)_* = \tau_*^t \circ \sigma_*^s$ the differential version of (2.1) implies

$$\{H, K\} = K \quad \text{and} \quad \tau^t K = e^{-t} K, \quad \sigma^s H = H + sK. \tag{3.1}$$

3.1.1 Repulsive Harmonic Forces. These are the prototype of sensitive dependence on initial conditions, since the particle runs with exponentially increasing velocity to infinity.

$$H = \tfrac{1}{2}(p^2 - x^2),\ K = p - x$$
$$\tau^t(x,p) = x\cosh t + p\sinh t, p\cosh t + x\sinh t)$$
$$\sigma^s(x,p) = (x + s, p + s)$$

which satisfy (3.1).

3.1.2 The Dilation. It represents the flow near a hyperbolic fixed point and corresponds to 3.1.1 written in the coordinates

$$X = \frac{x+p}{2},\ P = p - x$$
$$H = XP,\ K = P$$
$$\tau^t(X,P) = (Xe^t, Pe^{-t})$$
$$\sigma^s(X,P) = (X + s, P).$$

We see here that restriction to the half plane $(x,p) \in \mathbf{R} \times \mathbf{R}^+$ still gives an action of the continuous Anosov group, whereas on $(x,p) \in \mathbf{R}^+ \times (\mathbf{R} \text{ or } \mathbf{R}^+)$ we have an action of the semigroup $(s,t) \in \mathbf{R}^+ \times (\mathbf{R} \text{ or } \mathbf{Z})$. Thus Anosov $\not\Rightarrow$ ergodicity.

3.1.3 The Exponential of the Age Operator. If one wants to use H as a canonical variable, its conjugate, the "age operator", is the logarithm of K:

$$H = -p,\ K = e^x$$
$$\tau^t(x,p) = (x-t,p)$$
$$\sigma^s(x,p) = (x,p-se^x).$$

Note that now K is a positive function, in (3.1.1,2) it was not. However , in (3.1.2) we could have used $K = |p|$, which generates the flow $\sigma^s(x,p) = (x + sp/|p|, p)$, which also satisfies the Anosov relation.

3.1.4 Discretized Arnold Map on the Torus. So far our phase space was $\mathbf{R}^2$, and there was no invariant probability measure. We can also take the compact torus $\mathbf{R}^2/\mathbf{Z}^2$, but then τ^t will no longer be a diffeomorphism. However, for discrete time steps $\tau(x,p) \to (ax+bp, cx+dp)$, $a,b,c,d \in \mathbf{Z}$, $ad - bc = 1$, we obtain a symplectic transformation $T \in SL(2,\mathbf{Z})$ that stretches in one direction and contracts in another. If σ^s shifts in the expanding direction, we have again a realization of the Anosov group. It is not continuous, and

$$\lambda = \frac{t}{2} + \sqrt{\frac{t^2}{4} - 1}, \qquad t = \mathrm{Tr}\, T = a + d > 2.$$

3.2 Quantum Mechanics of one Particle

The examples (3.1.1,2,3) can be readily translated into quantum mechanics, where x and $p = \frac{1}{i}\frac{\partial}{\partial x}$ are operators on $L^2(\mathbf{R})$. In all three cases H and K are essentially selfadjoint on $C_0^\infty(\mathbf{R})$ and therefore define two families of unitaries e^{iHt}, e^{iKs}, which satisfy

$$e^{iKse^{-t}} e^{iHt} = e^{iHt} e^{iKs}.$$

Therefore they generate automorphisms of $\mathcal{B}(L^2(\mathbf{R}))$ by

$$\tau^t(A) = e^{iHt} A e^{-iHt}, \qquad \sigma^s(A) = e^{iKs} A e^{-iKs},$$

which satisfy (2.1). The quantum analogue of (3.1.4) is the Weyl algebra W generated by

$$e^{i(\alpha x + \beta p)} = W(\alpha,\beta), \qquad \alpha,\beta \in \Theta\mathbf{Z} \times \mathbf{Z},\ \Theta \in (0,2\pi).$$

The "cat map" $\tau(W(\alpha,\beta)) = W(T^t(\alpha,\beta))$ gives an automorphism of W, and together with a transversal shift in direction $\mathbf{e}$,

$$\sigma^s(W(\alpha,\beta)) = e^{is(\alpha e_1 + \beta e_2)} W(\alpha,\beta)\ ,$$

we have again a realization of the Anosov group with the same Lyapunov exponent λ as in the classical case. Note that λ does not depend on Θ which plays the rôle of Planck's constant.

3.3 Remarks

1. In the version (3.1.3) $\ln|K|$ and H appear as canonically conjugate variables, and if $t \to e^{iHt}$ is strongly continuous, the von Neumann uniqueness theorem of the representations of the CCR applies. This means that the spectrum of H and $\ln|K|$ has to be zero or absolutely continuous and fill all of $\mathbf{R}$ (resp. $\mathbf{R}^+$), only the multiplicity is free. Hence there is no chance to find the Anosov properties in finite quantum systems, where H has a discrete spectrum. If the representation is not faithful or not unitary, $\ln|K|$ may not exist as operator. In the first case we have the one–dimensional representations $(t,s) \to e^{i\gamma t}$, $\gamma \in \mathbf{R}$, so $K = 0$ and $\ln|K| = -\infty$. In the faithful two–dimensional representation given in (2.3.6) we have
$$H = \begin{pmatrix} 0 & 0 \\ 0 & -1 \end{pmatrix} \quad \text{and} \quad K = \begin{pmatrix} 0 & 0 \\ 1 & 0 \end{pmatrix}.$$
They do satisfy $[K,H] = K$, but $[\ln|K|, H] = -i$ cannot be represented by matrices, and indeed
$$|K| = |K^*K|^{1/2} = \begin{pmatrix} 0 & 0 \\ 0 & 1 \end{pmatrix}, \quad \text{so} \quad \ln|K| = \begin{pmatrix} -\infty & 0 \\ 0 & 0 \end{pmatrix}.$$

2. In these cases the dynamics is reversible, and there exists also a contracting direction. In the form (3.1.1) time reversal is simply $T(x,p) = (x,-p)$, and σ_-^s is generated by $-p + x$ and gives $(x,p) \to (x-s, p-s)$.

3.4 One–Dimensional Quantum Fields

The CCR and CAR algebras $\mathcal{A}$ are characterized by
$$[\psi_f, \psi_g^*]_\mp = \langle f|g\rangle, \qquad [\psi_f, \psi_g]_\mp = 0, \qquad f,g \in L^2(\mathbf{R}). \tag{3.4.1}$$
For fermions $\|\psi_f\| = \|f\|_2$, but for bosons ψ_f is not bounded and $\mathcal{A}$ is generated by the Weyl operators $e^{i(z\psi_f + z^*\psi_f^*)}$, $z \in C$, $f \in L^2(\mathbf{R})$. Since the unitary actions $f \to f_t = e^{-iHt}f$, $f \to f_s = e^{-iKs}f$ leave (3.4.1) invariant, they generate automorphisms $\tau^t(\psi_f) = \psi_{f_t}$, $\sigma^s(\psi_f) = \psi_{f_s}$, which again satisfy (2.1).

3.5 Classical Motion in extended Phase Space

Here the time becomes a dynamical variable x_0, and the energy p_0 is conjugate to it. The Poisson brackets are
$$\{x_i, p_k\} = \eta_{ik}, \qquad \eta_{ik} = \begin{pmatrix} -1 & 0 \\ 0 & 1 \end{pmatrix},$$
and physics requires the constraint $p^i p^k \eta_{ik} + m^2 = 0$.

3.5.1 Free Motion. The Lorentz transformation ("boost") generated by $L = x_1 p_0 - x_0 p_1$ and the shift along the light cone $K = (p_0 + p_1)/2$ satisfy $\{L, K\} = K$. Therefore their flows realize the Anosov group. This was to be expected, because the flow lines of the boost in configuration space are hyperbolas as the time evolution in phase space in (3.1.1). A physical realization of such a dynamics is given by a constant acceleration to be considered next.

3.5.2 The Constant Electric Field. A constant field is generated by a vector potential $A = \frac{1}{2}(-Ex_1, Ex_0)$, and the dynamics (with respect to proper time) are generated by the Hamiltonian

$$H = (p_i + eA_i)(p_k + eA_k)\frac{\eta^{ik}}{2m} = \frac{1}{2m}\left(\left(p_1 + \frac{eE}{2}x_0\right)^2 - \left(p_0 - \frac{eE}{2}x_1\right)^2\right).$$

The mass shell is given by $H = -m/2$. The two velocities

$$\dot{x}_0 = -\frac{1}{m}\left(p_0 - \frac{eE}{2}x_1\right) =: \frac{P}{m^{1/2}}, \qquad \dot{x}_1 = \left(p_1 + \frac{eE}{2}x_0\right)\frac{1}{m} =: \frac{eE}{m^{3/2}}Q$$

are up to constants conjugate, $\{Q, P\} = 1$, and H obtains the form (3.1.1):

$$H = -\frac{1}{2}(P^2 - \lambda^2 Q^2), \qquad \lambda = \frac{eE}{m}.$$

Therefore H and the shift generated by $K = Q + P$ satify $\{H, K\} = K$. Here is another pair of canonical coordinates [17] which do not appear in H. They describe the foci of the hyperbolas and are constant. Thus the 4 Lyapunov exponents are $\pm\lambda$, ± 0.

3.5.3 The de Sitter Universe. The inflationary universe is the prototype of an exponentially expanding space, and we shall now see how the Anosov group is at work. In two space–time dimensions the de Sitter space can be identified with the hyperboloid

$$-x_0^2 + x_1^2 + x_2^2 = 1 =: x_i x_k \eta^{ik} = (x|x) \tag{3.5.1}$$

in the three-dimensional Minkowski space, where

$$\eta = \begin{pmatrix} -1 & & \\ & 1 & \\ & & 1 \end{pmatrix}.$$

The geodesic flow follows from an Lagrangian

$$L = \frac{1}{2}(\dot{x}|\dot{x}), \tag{3.5.2}$$

and taking (3.5.1) into account by a Lagrangian multiplier c we get as Euler equations

$$\ddot{x} = cx. \tag{3.5.3}$$

(3.5.1) requires

$$0 = (x|\dot{x}) = (\ddot{x}|\dot{x}). \tag{3.5.4}$$

Thus $(\dot{x}|\dot{x})$ is constant, and we shall take the proper time as flow parameter, which normalizes this to

$$(\dot{x}|\dot{x}) = -1. \tag{3.5.5}$$

By (3.5.1, 4 and 5) the originally six–dimensional phase space $(x, \dot{x})$ is reduced to a three–dimensional mass shell $\mathcal{M}$. In configuration space the geodesics are simply the intersections of the hyperboloid (3.5.1) with planes through the origin and steeper than 45^0. This follows from the constancy of the generators of $SO(2,1)$

$$\ell_i = \varepsilon_{ik\ell} x^k \dot{x}^\ell \ . \tag{3.5.6}$$

Since $(\ell|x) = 0$, any point $x(t)$ on the trajectory remains confined to the plane $\perp \ell$. Any point on $\mathcal{M}$ can be reached by a Lorentz transformation $L \in SO(2,1)$ from the point $x(0) = (0,1,0)$, $\dot{x}(0) = (1,0,0)$. This induces a diffeomorphism of $\mathcal{M}$ and $SO(2,1)$. The geodesic through this point is according to the above $x(t) = M(t)x(0)$, $\dot{x}(t) = \dot{M}(t)x(0)$, with

$$M(t) = \begin{pmatrix} \cosh t & \sinh t & 0 \\ \sinh t & \cosh t & 0 \\ 0 & 0 & 1 \end{pmatrix} \in SO(2,1). \tag{3.5.7}$$

Since Lorentz transformations L are isometries, they transform geodesics into geodesics. Thus the geodesic through the point $(Lx(0), L\dot{x}(0))$ will be $(LM(t)x(0), L\dot{M}(t)x(0))$, or with the identification of $\mathcal{M}$ and $SO(2,1)$ we have

$$\tau^t(L) = LM(t). \tag{3.5.8}$$

The transversal shift is easily found. With the Lie algebra of $SO(2,1)$

$$(\ell_0, \ell_1, \ell_2) = \left(\begin{pmatrix} 0 & 0 & 0 \\ 0 & 0 & -1 \\ 0 & 1 & 0 \end{pmatrix}, \begin{pmatrix} 0 & 0 & 1 \\ 0 & 0 & 0 \\ 1 & 0 & 0 \end{pmatrix}, \begin{pmatrix} 0 & 1 & 0 \\ 1 & 0 & 0 \\ 0 & 0 & 0 \end{pmatrix} \right), \tag{3.5.9}$$

$M(t)$ can be written as $e^{t\ell_2}$, and since one calculates $[\ell_0 - \ell_1, \ell_2] = \ell_1 - \ell_0$, we see that τ^t and $\sigma^s(L) = Le^{s(\ell_1 - \ell_0)}$ represent the continuous Anosov group.

To get the flow explicitly it is convenient to use the isomorphism $SO(2,1) = SL(2,\mathbf{R})/\{1,-1\}$, which gave the two–dimensional representation of section 2. A point on (3.5.1) is given by a 2×2 matrix

$$X = \begin{pmatrix} x_2 & x_1 + x_0 \\ x_1 - x_0 & -x_2 \end{pmatrix},$$

which is characterized by Tr $X = 0$, $\det X = -1$. $SO(2,1)$ is represented by the action $X \to L^{-1}XL$, $L \in S(2,\mathbf{R})$, which leaves Tr and det invariant. The geodesic through $X(0) = \begin{pmatrix} 0 & 1 \\ 1 & 0 \end{pmatrix}$ with $\dot{X}(0) = \begin{pmatrix} 0 & 1 \\ -1 & 0 \end{pmatrix}$ is given by $X(t) = m^{-1}(t)X(0)m(t)$, $m(t) = \begin{pmatrix} e^{-t/2} & 0 \\ 0 & e^{t/2} \end{pmatrix}$. Thus the geodesic through $\widetilde{X} = L^{-1}X(0)L$ is given by $\widetilde{X}(t) = L^{-1}m^{-1}(t)X(0)m(t)L$. A family of geodesics with $\dot{X}(0) = \begin{pmatrix} 0 & 1 \\ -1 & 0 \end{pmatrix}$, which covers every point on (3.5.1), is obtained with $L = \begin{pmatrix} \cos\varphi & \sin\varphi \\ -\sin\varphi & \cos\varphi \end{pmatrix}$, and one calculates $(x_0, x_1, x_2) = (\sinh t, \cos 2\varphi \cosh t, \sin 2\varphi \sinh t)$. In this representation the transversal shift has the action

$$\sigma^s(L^{-1}X(0)L) = L^{-1}\gamma_s^{-1}X(0)\gamma_s L \qquad \text{with} \qquad \gamma_s = \begin{pmatrix} 1 & s \\ 0 & 1 \end{pmatrix}.$$

Because of $\gamma_s m_t = m_t \gamma_{se^t}$ we also have

$$\sigma^s(\tau^t(L^{-1}X(0)L)) = \tau^t(\sigma^{se^t}(L^{-1}X(0)L)).$$

3.6 Remarks

1. The spacelike hyperboloid $-x_0^2 + x_1^2 + x_2^2 = -1$ is locally isometric to $\mathbf{R} \times \mathbf{R}^+$ with $g = (dx^2 + dy^2)/y^2$ ("Poincaré half plane"), and exactly the same arguments show that there the geodesic flow also furnishes an Anosov dynamics. This still holds classically, if one compactifies $\mathcal{M}$ by going to the coset space $SL(2,\mathbf{R})/SL(2,\mathbf{Z})$. However quantum mechanically this is no longer the case, since the Laplacian on this space of finite volume has a discrete spectrum [13].
2. In section 3.5.1 the wedge $x_1 > |x_0|$ is invariant under τ^t, but σ^s maps the wedge into the wedge only for $s > 0$. Thus the algebra of functions supported on the wedge give an example of an Anosov semigroup.
3. We shall not go into the quantum mechanical generalizations of the examples 3.5. Since we have identified the Anosov group as a subgroup of both the Poincaré and the de Sitter groups, any quantum field theory with an action of these groups gives a realization of an Anosov dynamics.

4. Sensitive Dependence on Initial Conditions

The unlimited expansion, as it happened in some of our examples, can certainly not happen on a compact space. So it is better to look at the Anosov property, as saying that something arbitrarily small can become finite rather than that something finite can become arbitrarily big. So we will consider the limit and require the continuity

$$\lim_{s\to 0} D_s(A) = 0 \quad \forall\, A \in \mathcal{A} \quad \text{with} \quad D_s(A) = \|\sigma^s(A) - A\|. \tag{4.1}$$

$D_s(A)$ will not always be differentiable in s, but one can construct a norm–dense set $\mathrm{d} \subset \mathcal{A}$ where it is. In this case we get the

Sensitive dependence on initial conditions (4.2).

$$\lim_{s\to 0}\frac{D_s(\tau^t A)}{D_s(A)} = e^{\lambda t} \qquad \forall\, A \in \mathrm{d}.$$

Proof:. Since τ^t is norm preserving, we see $\forall\, A \in \mathrm{d}$:

$$\begin{aligned}\lim_{s\to 0}\frac{\|\sigma^s\tau^t(A)-\tau^t(A)\|}{D_s(A)} &= \lim_{s\to 0}\frac{\|\sigma^{se^{\lambda t}}(A)-A\|}{D_s(A)}\\ &= \lim_{s\to 0}\frac{D_{se^{\lambda t}}(A)}{D_s(A)} = e^{\lambda t}\,.\end{aligned}$$

Remarks (4.3).

1. We do not need continuity in t, $t \in \mathbf{Z}$ is just as good. s has to be continuous, but $s \in \mathbf{R}^+$ is also sufficient.
2. Differentiability means, that there is a derivation $\delta : \mathrm{d} \to \mathcal{A}$ such that

$$\lim_{s\to 0}\left\|\frac{\sigma^s(A)-A}{s} - \delta A\right\| = 0.$$

This implies weak differentiability, and therefore we have for all τ–invariant states ω

$$\lim_{s\to 0}\frac{\omega(\sigma^s\tau^t(A)-\tau^t(A))}{\omega(\sigma^s(A)-A)} = e^{\lambda t} \qquad \forall\, A \in \mathrm{d}.$$

Example (4.4). Going back to section 3 we see that the continuity (4.1) is not realized in the examples (3.1.1,2,3), only in (3.1.4). The reason is that for continuous functions f the translates $f_s(x) = f(x+s)$ are an equicontinuous family on compact sets, but not necessarily in an infinite space. Similarly in the quantum version $s \to \sigma^s(A) = e^{iKs}Ae^{-iKs}$ will be strongly continuous, but not necessarily norm continuous. Only in the second quantized version for fermions τ^t and σ^s will be norm continuous. The reason is that

$$\|\sigma^s(\psi_f)-\psi_f\| = \|\psi_{e^{-iKs}f}-\psi_f\| = \|e^{-iKs}f - f\| \to 0$$

for $s \to 0$. In this case d is norm–dense in $\mathcal{A}$, since K has a dense set of analytic vectors and (4.2) is fully applicable.

Continuity can be gained by restriction to the subalgebra of quasiperiodic functions generated

$$W(\alpha,\beta) = e^{i(\alpha x+\beta p)}, \qquad (\alpha,\beta) \in \mathbf{R}^2. \tag{4.5}$$

The shifts of (3.1.1,2,4) are continuous, since

$$\|\sigma^s W(\alpha,\beta) - W(\alpha,\beta)\| = \|(e^{i(\alpha+\beta)s}-1)W(\alpha,\beta)\| = |e^{i(\alpha+\beta)s}-1| \to 0$$

for $s \to 0$. Exactly the same holds in quantum mechanics, (4.5) generates the Weyl algebra W, and the noncommutativity

$$W(\alpha,\beta)W(\alpha',\beta') = e^{i\hbar(\alpha\beta'-\beta\alpha')}W(\alpha+\alpha',\beta+\beta')$$

does not change the Anosov properties. In fact, λ is independent of $\hbar$.

It is instructive to verify that (4.2) holds on W, and not on all of $\mathcal{B}(L^2(\mathbf{R}))$. Consider $e^{i\alpha x} \in W$ and $e^{i\alpha x^2} \in \mathcal{B}(L^2(\mathbf{R}))$. We have

$$\begin{aligned}\lim_{s\to 0}\frac{\|\sigma^s\tau^t(e^{i\alpha x}) - \tau^t(e^{i\alpha x})\|}{\|\sigma^s(e^{i\alpha x}) - e^{i\alpha x}\|} &= \lim_{s\to 0}\frac{\sup_{x\in\mathbf{R}}|e^{i\alpha e^t(x+s)} - e^{i\alpha e^t x}|}{\sup_{x\in\mathbf{R}}|e^{i\alpha(x+s)} - e^{i\alpha x}|}\\ &= \lim_{s\to 0}\frac{|e^{i\alpha e^t s}-1|}{|e^{i\alpha s}-1|} = e^t\,,\end{aligned}$$

but

$$\begin{aligned}\lim_{s\to 0}\frac{\|\sigma^s\tau^t(e^{i\alpha x^2}) - \tau^t(e^{i\alpha x^2})\|}{\|\sigma^s(e^{i\alpha x^2}) - e^{i\alpha x^2}\|} &= \lim_{s\to 0}\frac{\sup_{x\in\mathbf{R}}|e^{i\alpha e^{2t}(x+s)^2} - e^{i\alpha(e^t x)^2}|}{\sup_{x\in\mathbf{R}}|e^{i\alpha(x+s)^2} - e^{i\alpha x^2}|} = \frac{2}{2}\\ &= 1 \quad \forall\, t\in\mathbf{R}.\end{aligned}$$

To see how (4.2) works in relativistic quantum field theory, we consider finally a massless Dirac field in two space–time dimensions. It has two components, the right and left movers $\psi_j(x_1 + (-)^j x_0)$, $j = 1, 2$, which satisfy the anti-commutation relations

$$\{\psi_j(x), \psi_k^*(x')\} = \delta(x-x')\delta_{jk}. \tag{4.6}$$

The actions of τ^t, σ^s from section 3.5.1 (boost and shift on the light cone) are

$$\tau^t(\psi_j(x) = e^{(-)^j t/2}\psi_j(xe^{-t}), \qquad \sigma^s(\psi_j(x)) = \begin{matrix}\psi_j(x+s) & \text{for } j=1\\ \psi_j(x) & \text{for } j=2.\end{matrix}$$

The $\psi_j(x)$ do not form an algebra, but the smeared $\psi_{j,f} = \int dx f(x)\psi_j(x)$, $f \in L^2(\mathbf{R})$, do. For the $\psi_{j,f}$, $f' \in L^2(\mathbf{R})$ (4.2) holds, and they generate a norm dense set in the CAR algebra:

$$\begin{aligned}\lim_{r\to 0}\frac{\|\sigma^s\tau^t(\psi_f) - \tau^t(\psi_f)\|^2}{\|\sigma^s(\psi_f) - \psi_f\|^2} &= \lim_{s\to 0}\frac{\int dx e^t|f(e^t(x+s)) - f(e^t x)|^2}{\int dx|f(x+s) - f(x)|^2}\\ &= e^{2t}\frac{\int dx e^t|f'(xe^t)|^2}{\int dx|f'(x)|^2} = e^{2t}.\end{aligned}$$

5. Exponential Decay of Time Correlations

It is sometimes considered as characteristic for chaos, that time correlations decay exponentially. To see whether this happens in Anosov systems, we have first to verify that there are invariant states $\omega = \omega\circ\tau^t = \omega\circ\sigma^s$. A priori this is assured, because the Anosov group is amenable. Rather then proving this assertion we make the following observation. Abelian groups are always amenable, so there exists an invariant state $\omega = \omega \circ \tau^t$, and therefore τ^t is unitarily implemented, $\tau^t(a) = V_{-t} a V_t$. V_t can be chosen such that $V_t|\Omega\rangle = |\Omega\rangle$, where $|\Omega\rangle$ is the cyclic vector in the GNS construction with ω. Now we have the simple

Lemma (5.1). If σ^s is unitarily implemented, $\sigma^s(a) = U_{-s} a U_s$, and $s \to U_s$ is strongly continuous, then $V_t|\Omega\rangle = |\Omega\rangle$ implies $U_s|\Omega\rangle = |\Omega\rangle$.

Proof:. $\forall\, t \in \mathbf{R}$ we have

$$U_s|\Omega\rangle = U_s V_t|\Omega\rangle = V_t U_{e^t s}|\Omega\rangle \overset{t \to -\infty}{\longrightarrow} V_t|\Omega\rangle = |\Omega\rangle.$$

This has the sad consequence that ω cannot be an equilibrium (= KMS) state for the time evolution τ^t. This follows from the following well-known

Lemma (5.3). If ω is τ^t–KMS and σ^s–invariant for an automorphism group σ^s, then $[\tau^t, \sigma^s] = 0$.

Proof:. For any s ω is KMS for the automorphism group $\bar{\tau}^t = \sigma^{-s}\tau^t\sigma^s$, since

$$\omega(b\bar{\tau}^{i\beta}(a)) = \omega(b\sigma^{-s}\tau^{i\beta}\sigma^s(a)) = \omega(\sigma^s(b)\tau^{i\beta}\sigma^s(a)) = \omega(\sigma^s(a)\sigma^s(b)) = \omega(ab).$$

However the modular automorphism of a state is unique, thus $\bar{\tau}^t = \tau^t$ or $\tau^t\sigma^s = \sigma^s\tau^t$. We have given the standard proof, because it shows the way out of this dilemma. If σ^s, $s \geq 0$, is only a semigroup of endomorphisms, then it is possible that ω is τ^t–KMS.

Example (5.3). (The poor man's Bisognano–Wichmann theorem.) We take the example of the end of section 4, namely a Dirac field in $1+1$ dimensions,

$$\psi = \begin{pmatrix} \psi_1(x_1 - x_0) \\ \psi_2(x_1 + x_0) \end{pmatrix},$$

where τ^t is the boost and σ^s the shift in a light–like direction $u = x_0 + x_1$. We consider only one component:

$$\psi(u) = \int_{-\infty}^{\infty} \frac{dp}{2\pi} a(p) e^{ipu},$$

$$[\psi(u), \psi^*(u')]_+ = \delta(u - u')[a(p), a^*(p')]_+ = \frac{1}{2\pi}\delta(p - p').$$

The C* algebra $\mathcal{A}$ is generated by the $\psi_f = \int_{-\infty}^{\infty} du \psi(u) f(u)$, $f \in L^2(\mathbf{R})$. The automorphisms act as $\tau^t(\psi_f) = \psi_{f_t}$, $f_t(u) = e^{-t/2} f(e^{-t}u)$, and $\sigma^s(\psi_f) = \psi_{f_s}$, $f_s(u) = f(u+s)$. We know that the spectrum of the generator H of τ^t is all of $\mathbf{R}$, whereas the generator K of σ^s can have positive (or negative) spectrum. Formally

$$H = \int_{-\infty}^{\infty} \frac{dp}{2\pi} : a^*(p) \frac{1}{2i}(p\frac{\partial}{\partial p} + \frac{\partial}{\partial p} p) a(p) :, \qquad K = \int_{-\infty}^{\infty} \frac{dp}{2\pi} : a^*(p) p a(p) :,$$

and to realize the first possibility we let the : : refer to a vacuum, where the negative p–states are filled:

$$\omega(a^*(p)a(p')) = \frac{1}{2\pi}\theta(-p)\delta(p-p') ,$$

and therefore

$$\omega(a(p')a^*(p)) = \frac{1}{2\pi}\theta(p)\delta(p-p') ,$$

which means

$$: a^*(p)a(p) : = -a(p)a^*(p) \qquad \text{for } p < 0.$$

Next we consider the subalgebra $\mathcal{A}^\pm$ generated by ψ_f with supp $f \in \mathbf{R}^\pm$. ω is faithful over $\mathcal{A}^+$, since

$$\|\psi_f|\Omega\rangle\|^2 = \langle\Omega|\psi_f^*\psi_f|\Omega\rangle = \int_{-\infty}^0 \frac{dp}{2\pi}|\tilde{f}(p)|^2 ,$$

and for supp $f \in \mathbf{R}^+$, $\tilde{f}(p) = \int_0^\infty du e^{ipu} f(u)$ is analytic in the upper half plane. $\tilde{f}(p) = 0 \ \forall \ p < 0$ implies then $f = 0$ or $\psi_f = 0$. On $\mathcal{A}^+$ τ^t acts as automorphism, and σ^s for $s > 0$ as endomorphism. ω is invariant under both, since these actions imply $\tilde{f}(p) \to e^{t/2}\tilde{f}(e^t p)$ or $\tilde{f}(p) \to e^{ips}\tilde{f}(p)$. Furthermore $\mathcal{A}^+|\Omega\rangle$ is dense in $\mathcal{H}_\omega$, the representation space of $\Pi_\omega(\mathcal{A})$, since

$$\langle\Omega|\psi_g^*\psi_f|\Omega\rangle = \int_{-\infty}^0 \frac{dp}{2\pi}\tilde{g}^*(p)\tilde{f}(p) = 0 \qquad \forall\, \psi_f \in \mathcal{A}^+$$

implies supp $g \subset \mathbf{R}^+$, and thus $\psi_g|\Omega\rangle = 0$. Thus $\Pi_\omega|_{\mathcal{A}^+}(\mathcal{A}^+) = \Pi_\omega(\mathcal{A}^+)$, and we claim that $\omega|_{\mathcal{A}^+}$ is $\tau^{2\pi i}$–KMS and $\Pi_\omega(\mathcal{A}^+)' = \Pi_\omega(\mathcal{A}^-)''$, but ω is not a product state over $\mathcal{A} = \mathcal{A}^+ \otimes \mathcal{A}^-$. To show this note that $\tau^t\psi_f|\Omega\rangle = \psi_{f_t}|\Omega\rangle$ can with $\tilde{f}_t(p) = e^{t/2}\tilde{f}(e^t p)$ be continued in the upper half plane to $t = i\pi$, and similarly $\psi_f^*|\Omega\rangle$ in the lower half plane to $-i\pi$, since $\tilde{f}^*(p) = \int_0^\infty du e^{-ipu} f^*(u)$ is analytic for Im $p < 0$. $\tau^{i\pi}\psi_f = \psi_{f_-}$, $\tilde{f}_-(p) = i\tilde{f}(-p)$, and the KMS condition says

$$\langle\Omega|\psi_g^*\psi_f|\Omega\rangle = \int_{-\infty}^0 \frac{dp}{2\pi}\tilde{g}^*(p)\tilde{f}(p)$$

$$= \langle\Omega|\tau^{-i\pi}\psi_f\tau^{i\pi}\psi_g^*|\Omega\rangle = \int_0^\infty \frac{dp}{2\pi}\tilde{g}^*(-p)\tilde{f}(-p)$$

and is therefore satisfied for the two point function. Since the two point function determines here the n–point function, these considerations can be extended to all of $\mathcal{A}^+$.

Remark:. Note that $\Pi_\omega(\mathcal{A}^+)''$ is of type III$_1$, whereas $\Pi_\omega(\mathcal{A})''$ is type I$_\infty$, so this feature appears already in this simple example.

Theorem (5.4). Let ω be τ^t invariant, τ^t and σ^s represented by strongly continuous unitaries $V_t = e^{iHt}$, $U_s = e^{iKs}$, resp. Let $|\Omega\rangle$ be the cyclic vector in Π_ω such that $H|\Omega\rangle = 0$ and therefore $K|\Omega\rangle = 0$. Let χ_Δ be the characteristic function of $\Delta = (-\infty, -a) \cup (a, \infty)$, $a > 0$ and $A, B \in \mathcal{A}$ such that

(i) $A|\Omega\rangle \in D(K^r)$ the domain of K^r, $r > 0$, and
(ii) $\chi_{\Delta^c}(K)B|\Omega\rangle = 0$, then

$$|\omega(A\tau^t B)| \leq e^{-tr} a^{-r} \| B|\Omega\rangle \| \cdot \| K^r A|\Omega\rangle \|.$$

Remarks.

1. Without loss of generality we may assume that $K \geq 0$, otherwise K^r may be read as $|K|^r$.
2. Since $K|\Omega\rangle$ is zero (ii) implies $\langle\Omega|B|\Omega\rangle = 0$ which can be achieved by $B \to B - \langle\Omega|B|\Omega\rangle$. Thus the left hand side can be replaced by $|\omega(A\tau^t B) - \omega(A)\omega(B)|$ which shows that the theorem proves decay of correlations.
3. (i) can be achieved by replacing A by $A_f = \int ds f(s)\sigma^s(A)$. In fact by our continuity assumptions A can be approximated in norm by A_f's which qualify for the theorem.
4. Because of $\omega(A\tau^t B) = \omega(\tau^{-t}(A)B) = \omega(B^*\tau^{-t}A^*)^*$ the decay is in both time directions.

Proof:. We use $\chi_\Delta(x) \leq \left(\frac{x}{a}\right)^r \; \forall\, r > 0$ to conclude

$$\begin{aligned} |\langle\Omega|A\tau^{-t}B|\Omega\rangle| &= |\langle\Omega|AV_{-t}\chi_\Delta B|\Omega\rangle| = |\langle\Omega|A\chi_{e^t\Delta}V_{-t}B|\Omega\rangle| \\ &\leq \langle\Omega|A\chi_{e^t\Delta}A^*|\omega\rangle^{1/2}\langle\Omega|B^*B|\Omega\rangle^{1/2} \\ &\leq (ae^t)^{-r}\|B|\Omega\rangle\| \cdot \|K^r A|\Omega\rangle\|. \end{aligned}$$

Examples.

1. The Weyl algebra
   We consider the algebra $\mathcal{W}$ which consists of polynomials of $W(z) = e^{i(\alpha x + \beta p)}$, $z = (\alpha, \beta) \in \mathbf{R}^2$ with the multiplication law

$$W(z)W(z') = e^{i\sigma(z,z')}W(z+z'),$$

   with the symplctic form $\sigma(z, z') = \frac{1}{2}(\alpha\beta' - \beta\alpha')$. The unitaries $V(t) = \exp[i\frac{1}{2}(px + xp)t] \notin \mathcal{A}$ and $U(s) = e^{ips} \in \mathcal{A}$ generate

$$\tau^t(W(\alpha,\beta)) = W(e^t\alpha, e^{-t}\beta), \qquad \sigma^s(W(\alpha,\beta)) = e^{i\alpha s}W(\alpha,\beta).$$

   The action τ^t is not norm–continuous. A state ω is characterized by $E(z) = \omega(W(z))$ and has to satisfy $E(0) = 1$, and $e^{-i(\sigma(z,z')}f(z-z')$ has to be a positive integral kernel. (This implies $|E(z)| \leq 1$.) If ω is to be invariant under σ^s we need $E(\alpha, \beta) = 0 \; \forall\, \alpha \neq 0$ and then invariance and τ^t requires that $E(0, \beta)$ is constant on $\mathbf{R}^+$ or $\mathbf{R}^-$ resp. The positivity requirement leaves two possibilities

a) The tracial state $E_0 = 0 \; \forall \; (\alpha, \beta) \neq (0,0)$. This state is well-known from the Hilbert space of quasiperiodic functions which is not separable. Here the actions of τ^t and σ^s on $W(z)$ are not weakly continuous since the typical matrix elements of $W(z)$ are

$$\begin{aligned}\langle z'|W(z)|z''\rangle : &= \langle \Omega|W^*(z')W(z)W(z'')|\Omega\rangle \\ &= e^{i\sigma(z,z''-z')-i\sigma(z',z'')}E(z+z''-z').\end{aligned}$$

Thus neither H nor K exists. The representation is type II_1 but not hyperfinite.

b) The other state is $E(\alpha,\beta) = 0 \; \forall \; \alpha \neq 0, = 1$ for $\alpha = 0$, and again τ is not weakly continuously represented, so Theorem (5.4) does not apply. Nevertheless in both cases there is some extreme clustering, since

$$\langle \Omega|W^*(z)\tau^t W(z')|\Omega\rangle = 0$$

for all t except if $\alpha = e^t \alpha'$.

2. Here we consider the subalgebra $\mathcal{W}_c = \{W(z), z \in \Theta\mathbf{Z} \times \mathbf{Z}\}$. As in the classical case the τ^t of (2.1) is not an automorphism of $\mathcal{W}_c$ but if we choose other directions for expansion and contraction it can happen that for a discrete set of times we get the following automorphism of $\mathcal{W}_c$, $\tau W(z) = W(Tz)$, $T \in SL(2,\mathbf{Z})$. Of course, for $\mathcal{W}$ there exists a τ^t such that $\tau^1 = \tau$ but $\tau^t(\mathcal{W}_c) \subset \mathcal{W}_c$ only for $t \in \mathbf{Z}$. In this case the Lyapunov exponent cannot be scaled to 1 but turns out to have the classical value [11,12,13]

$$\lambda = \frac{c}{2} + \sqrt{\frac{c^2}{4} - 1}, \qquad c = \operatorname{Tr} T > 2.$$

E_0 gives the only invariant state and a regular representation Π_0. It acts in a separable Hilbert space and $\Pi_0(\mathcal{W}_c)''$ is type II_1 hyperfinite. If σ is the shift in transversal direction X such that $\sigma^s(W(z)) = e^{is\langle x|z\rangle}W(z)$, then σ^s acts even norm continuously and the theorem applies.

References

1. H. Narnhofer, W. Thirring, J. Stat. Phys. **57**, 811 (1989)
2. R. Longo, C. Peligrad, J. Funct. Anal. **58**, 157 (1984)
3. A. Kishimoto, J. Op. Th. **13**, 237 (1985)
4. A. Connes, E. Størmer, Acta Math. **134**, 289 (1975)
5. A. Connes, H. Narnhofer, W. Thirring, Commun. Math. Phys. **111**, 691 (1987)
6. R. Alicki, M. Fannes, Lett. Math. Phys. **32(1)**, 75 (1994)
7. D. Voiculescu, Berkeley preprint (1994)
8. T. Hudetz, J. Math. Phys. **35(8)**, 4303 (1994)
9. W.A. Majewski and M. Kuna, J. Math. Phys. **34** 5007 (1993)
10. M. Kuna and A.W. Majewski, Rep. Math. Phys. **33**, 111 (1993)
11. R. Vilela Mendes, Phys. Lett. **A 171**, 253 (1992)

12. H. Narnhofer, J. Math. Phys. **33**, 1502 (1992)
13. G. Emch, H. Narnhofer, G. Sewell, W. Thirring, J. Math. Phys. **35**, 5582 (1994)
14. J. Ford, private communication
15. H. Narnhofer, W. Thirring, ESI preprint, Vienna (1994)
16. P. Michor, private communication
17. W. Thirring, *A Course in Mathematical Physics* (I.5.2.16), Springer, New York (1992)

Area-Preserving Structure of Massless Matter-Gravity Fields in 1+1 Dimensions

G. Amelino-Camelia, D. Bak, and D. Seminara

Center for Theoretical Physics, Laboratory for Nuclear Science and Department of Physics, Building 6, Massachusetts Institute of Technology, Cambridge, Massachusetts 02139, U.S.A.

1. Introduction

1+1 dimensional ("lineal") quantum gravity is one of the areas of low-dimensional quantum field theory, which has attracted more attention in recent years. The role played by symmetries in these theories is obviously very important, and here we shall be concerned with the role of symmetries in the quantization of the 1+1-dimensional matter-gravity field theory with action[1]

$$\mathcal{I}(X,g) = \frac{1}{2}\int d^2\xi \; \sqrt{-g}\; g^{\mu\nu}\; \partial_\mu X_A \, \partial_\nu X^A \; , \tag{1.1}$$

where $g_{\mu\nu}$ is a metric tensor with signature (1,-1), $A=1,2,...,d$, and X is a d-component massless scalar field.

In the quantization of this theory one necessarily encounters anomalies that break part of the symmetry of the classical theory, which, as seen from $\mathcal{I}(X,g)$, has Weyl and diffeomorphism invariance[1, 2].

In the conventional quantization approach[1-4] diffeomorphism invariance is preserved, while renouncing Weyl invariance. Integrating out the matter degrees of freedom, using a measure[1, 2] with the appropriate symmetries, one obtains an effective pure gravity theory with action[1]

$$\begin{aligned} \Gamma^D(g) \;\; = \;\; & \frac{d}{96\pi}\int d^2\xi_1\, d^2\xi_2\, \sqrt{-g(\xi_1)}\, R(g(\xi_1))\, \Box^{-1}(\xi_1,\xi_2)\, \sqrt{-g(\xi_2)} \\ & R(g(\xi_2))\,, \end{aligned} \tag{1.2}$$

where $\Box^{-1}$ is the inverse of the Laplace-Beltrami operator.

Since $\Gamma^D(g)$ is diffeomorphism-invariant, but not Weyl-invariant, the energy-momentum tensor

$$T^D_{\mu\nu} \equiv (2/\sqrt{-g})(\delta\Gamma^D(g)/\delta g^{\mu\nu})$$

is covariantly conserved, but possesses non-vanishing trace:

[1] Note that, for simplicity, we set the cosmological constant to zero.

$$\nabla_\mu (g^{\mu\nu} T^D_{\nu\alpha}) = 0 \ , \qquad g^{\mu\nu} T^D_{\mu\nu} = \frac{d}{24\pi} R(g) \ . \tag{1.3}$$

Recently, an alternative approach to the quantization of the classical theory (1.1) has been considered[5-7], in which the functional measure for the integration over the matter fields is Weyl-invariant and invariant under area-preserving diffeomorphisms (i.e. diffeomorphisms of unit Jacobian), but is not invariant under non-area-preserving diffeomorphisms. This leads to the following effective action:[2]

$$\Gamma^W(\gamma) = \frac{d}{96\pi} \int d^2\xi_1\, d^2\xi_2 \; R(\gamma(\xi_1))\, \Box^{-1}(\xi_1, \xi_2)\, R(\gamma(\xi_2)) \ , \tag{1.4}$$

where $\gamma^{\mu\nu} \equiv \sqrt{-g}\, g^{\mu\nu}$. The fact that $\Gamma^W(\gamma)$ is Weyl-invariant, but is not invariant under general diffeomorphisms, leads to the anomaly relations

$$\hat{\nabla}_\mu (\gamma^{\mu\nu} T^W_{\nu\alpha}) = -\frac{d}{48\pi} \partial_\alpha R(\gamma) \ , \tag{1.5}$$

$$\gamma^{\mu\nu} T^W_{\mu\nu} = 0 \ , \tag{1.6}$$

where $\hat{\nabla}$ is the covariant derivative computed with the metric $\gamma_{\mu\nu}$, and

$$T^W_{\mu\nu} \equiv \frac{2}{\sqrt{-g}} \frac{\delta \Gamma^W(\gamma)}{\delta g^{\mu\nu}} = 2 \frac{\delta \Gamma^W(\gamma)}{\delta \gamma^{\mu\nu}} - \gamma_{\mu\nu} \gamma^{\alpha\beta} \frac{\delta \Gamma^W(\gamma)}{\delta \gamma^{\alpha\beta}} \ , \tag{1.7}$$

while the invariance of $\Gamma^W(\gamma)$ under area-preserving diffeomorphisms is encoded in the relation

$$\hat{\nabla}_\mu \hat{\nabla}_\nu (\gamma^{\beta\nu} \epsilon^{\mu\alpha} T^W_{\alpha\beta}) = 0 \ , \tag{1.8}$$

which is consistent with (1.5).

In the following we shall derive the anomalous Ward identities both for the Weyl-invariant approach and the conventional diffeomorphism invariant approach, and observe that, although the difference in the symmetries leads to several differences at intermediate steps of the derivation, the final results are equivalent.

[2] Note that, by appropriate choice of measure, one can obtain more general effective actions, that are invariant when $g_{\mu\nu}$ is transformed as $\delta g_{\mu\nu} = \xi^\alpha \partial_\alpha g_{\mu\nu} + g_{\alpha\nu} \partial_\mu \xi^\alpha + g_{\alpha\mu} \partial_\nu \xi^\alpha + a g_{\mu\nu} \partial_\alpha \xi^\alpha$, where a is a fixed real parameter. It is seen that this combination of diffeomorphisms and Weyl transformations is equivalent to the statement that $g_{\mu\nu}$ is a tensor density of weight a. Such a modification of the standard formula ($a = 0$) leaves the classical action invariant, because the combination $\sqrt{-g} g^{\mu\nu}$ is insensitive to the weight of $g^{\mu\nu}$. The Weyl-invariant approach considered in the present paper corresponds to the limit $a \to \infty$, with the prescription that $\partial_\mu \xi^\mu \to w/a$ for $a \to \infty$, where w is an arbitrary function.

2. Anomalous Ward Identities

We start by considering the functional integrals

$$Z^D[J] = \int \frac{\mathcal{D}g}{\Omega_{diff}} \exp(i\Gamma^D(g) + i\int \sqrt{-g}g^{\mu\nu}J^D_{\mu\nu}) , \tag{2.1}$$

$$Z^W[J] = \int \frac{\mathcal{D}\gamma}{\Omega_{Sdiff}} \exp(i\Gamma^W(\gamma) + i\int \gamma^{\mu\nu}J^W_{\mu\nu}) , \tag{2.2}$$

where $J^D_{\mu\nu}$ and $J^W_{\mu\nu}$ are sources, Ω_{diff} is the volume of the diffeomorphism group, and Ω_{Sdiff} is the volume of the group of the area-preserving diffeomorphisms. The volume of the Weyl group does not appear in (2.2), because the functional integral is already written in terms of the Weyl-invariant field γ. $Z^D[J]$ and $Z^W[J]$ are the generating functionals for the Green's functions of the diffeomorphism-invariant approach and the Weyl-invariant approach, respectively.

In order to factorize out the gauge volume, one can fix the gauge and introduce the corresponding action for the ghost fields. We choose to work in the light-cone gauge, and, after integrating out the ghost fields, Z^D and Z^W take the following form:

$$Z^D[J] = \int \mathcal{D}g_{++} \exp\left(i\Gamma^D(g) + i\Gamma^D_{gh}(g) + i\int g^{++}J^D_{++}\right) , \tag{2.3}$$

$$Z^W[J] = \int \mathcal{D}\gamma_{++} \exp\left(i\Gamma^W(\gamma) + i\Gamma^W_{gh}(\gamma) + i\int \gamma^{++}J^W_{++}\right) , \tag{2.4}$$

where $\Gamma^D+\Gamma^D_{gh}$ and $\Gamma^W+\Gamma^W_{gh}$ are gauge-fixed actions for gravity.

Our choice of gauge is motivated by the fact, that[5-7] in the light-cone gauge $\Gamma^W+\Gamma^W_{gh}$ takes the same form as $\Gamma^D+\Gamma^D_{gh}$, and we intend to exploit this correspondence in the investigation of the anomalous Ward identities. Still, in the analysis we shall need to take into account the fact that the measure $\mathcal{D}\gamma$, which is Weyl-invariant but not diffeomorphism-invariant, is different from the diffeomorphism-invariant but not Weyl-invariant measure $\mathcal{D}g$. Moreover, since γ and g have different transformation properties, $\Gamma^W+\Gamma^W_{gh}$ and $\Gamma^D+\Gamma^D_{gh}$ satisfy different anomaly relations; specifically[7]

$$\nabla_\mu(g^{\mu\nu}\Theta^D_{\nu\alpha}) = 0 , \qquad g^{\mu\nu}\Theta^D_{\mu\nu} = \frac{d-28}{24\pi}R(g) , \tag{2.5}$$

$$\hat{\nabla}_\mu(\gamma^{\mu\nu}\Theta^W_{\nu\alpha}) = -\frac{d-28}{48\pi}\partial_\alpha R(\gamma) , \qquad \gamma^{\mu\nu}\Theta^W_{\mu\nu} = 0 , \tag{2.6}$$

where

$$\Theta^{D,W}_{\mu\nu} \equiv \frac{2}{\sqrt{-g}}\frac{\delta(\Gamma^{D,W} + \Gamma^{D,W}_{gh})}{\delta g^{\mu\nu}} . \tag{2.7}$$

We now consider the following infinitesimal shifts in the functional variables of integration:

$$\delta_f g_{++} = (2\partial_+ - g_{++}\partial_-)\,\delta f + \delta f\,\partial_- g_{++}\ , \tag{2.8}$$
$$\delta_f \gamma_{++} = (2\partial_+ - \gamma_{++}\partial_-)\,\delta f + \delta f\,\partial_- \gamma_{++}\ , \tag{2.9}$$

and observe that

$$\begin{aligned}\int \frac{\delta[\Gamma^D(g)+\Gamma^D_{gh}(g)]}{\delta g_{++}}\delta_f g_{++} &= \int [\nabla_\mu(g^{\mu\nu}\Theta^D_{\nu-})-\frac{1}{2}\nabla_-(g^{\mu\nu}\Theta^D_{\mu\nu})]\delta f \\ &= \int \frac{28-d}{48\pi}\partial^3_- g_{++}\delta f,\end{aligned} \tag{2.10}$$

$$\begin{aligned}\int \frac{\delta[\Gamma^W(\gamma)+\Gamma^W_{gh}(\gamma)]}{\delta \gamma_{++}}\delta_f \gamma_{++} &= \int [\nabla_\mu(g^{\mu\nu}\Theta^W_{\nu-})-\frac{1}{2}\nabla_-(g^{\mu\nu}\Theta^W_{\mu\nu})]\delta f \\ &= \int \frac{28-d}{48\pi}\partial^3_- \gamma_{++}\delta f,\end{aligned} \tag{2.11}$$

where we used the anomaly relations (2.5) and (2.6).

Following a standard procedure[4], the relations (2.10) and (2.11) lead to the following anomalous Ward identities:

$$\sum_i^n \langle g_{++}(\xi_1)\ldots\delta_f g_{++}(\xi_i)\ldots g_{++}(\xi_n)\rangle + \frac{d-28+\lambda^D}{i48\pi} \int d\xi^2 \delta f(\xi)\,\langle \partial^3_- g_{++}(\xi) g_{++}(\xi_1)\ldots g_{++}(\xi_n)\rangle = 0; \tag{2.12}$$

$$\sum_i^n \langle \gamma_{++}(\xi_1)\ldots\delta_f \gamma_{++}(\xi_i)\ldots \gamma_{++}(\xi_n)\rangle + \frac{d-28+\lambda^W}{i48\pi} \int d\xi^2 \delta f(\xi)\,\langle \partial^3_- \gamma_{++}(\xi) \gamma_{++}(\xi_1)\ldots \gamma_{++}(\xi_n)\rangle = 0. \tag{2.13}$$

Here λ^D is the additional contribution to the anomaly, which is due to the fact that $\delta_f g_{++}$ is a composition of a diffeomorphism and a Weyl transformation on g_{++}, and therefore the diffeomorphism-invariant but not Weyl-invariant measure $\mathcal{D}g_{++}$ is not invariant under $g_{++} \to g_{++} + \delta_f g_{++}$. Analogously, the presence of λ^W is due to the fact, that $\delta_f \gamma_{++}$ is an infinitesimal (not area-preserving) diffeomorphism transformation on γ_{++}, and therefore the measure $\mathcal{D}\gamma_{++}$ is not invariant under $\gamma_{++} \to \gamma_{++} + \delta_f \gamma_{++}$. The values of λ^D and λ^W can be fixed by requiring, that the theory be independent of the choice of gauge. In Ref.[2] the class of gauges $g_{--} = g^B_{--}$, $g_{+-} = 1$ is considered, and it is found that the independence of the partition function on the choice of g^B_{--} requires, that

$$d - 28 + \lambda^D = \frac{d - 13 - \sqrt{(d-1)(d-25)}}{2}\ . \tag{2.14}$$

Following the corresponding procedure for the Weyl invariant approach, one finds that also λ^W must satisfy Eq.(2.14), *i.e.* $\lambda^W = \lambda^D$. This observation, together with the results (2.12) and (2.13), indicates that the anomalous Ward identities satisfied by γ_{++} in the Weyl-invariant approach are identical to the ones satisfied by g_{++} in the diffeomorphism-invariant approach. Since these Ward identities completely determine[2] the Green's functions, also the Green's functions are identical.

3. Conclusion

The investigation of the anomalous Ward identities indicates, that the two approaches are equivalent, and this is consistent with the results[7] of the (classical) Dirac Hamiltonian analysis. It appears that the physics described by the model is independent of the local term[6], which one needs to add to the action in order to convert the Weyl anomaly into a diffeomorphism anomaly. This does not always happen in anomalous quantum field theories. For example in the chiral Schwinger model[8], the mass emergent at the quantum level as a result of the anomaly does depend on the coefficient of one such local term.

It is also interesting to notice, that in deriving the equivalence of the two approaches at the level of the anomalous Ward identities a key role is played by the minus component of the combination $\nabla_\mu(g^{\mu\nu}\Theta_{\nu\alpha}) - \nabla_\alpha(g^{\mu\nu}\Theta_{\mu\nu})/2$ (see Eqs.(2.10) and (2.11)), which (in the chosen gauges) takes the same form in both approaches. Clearly this combination of the anomaly relations encodes some essential feature of the model, but its physical interpretation is not yet clear to us.

Finally, we want to point out, that 1+1-dimensional quantum gravities of the type here considered and their supersymmetric extensions are related[2, 3, 9] to some low-dimensional models in statistical physics, such as the Ising model, random surfaces, percolation, tree-like polymers, and self-avoiding polymers. In several occasions results first obtained in the study of the quantum field theories have been useful also in the context of the statistical models and vice versa. Since only recently there has been increased interest in the Weyl invariant approach, the possibility of application of this new viewpoint to the study of statistical models has not yet been investigated.

Acknowledgements
We are very grateful to R. Jackiw for several stimulating conversations and useful comments. This work is supported in part by funds provided by the U.S. Department of Energy (D.O.E.) under cooperative agreement #DE-FC02-94ER40818, as well as in part by the National Science Foundation under contracts #INT-910559 and #INT-910653, and by Istituto Nazionale di Fisica Nucleare (INFN, Frascati, Italy).

References

1. A.M. Polyakov, Phys. Lett. **103B** (1981) 207.
2. A.M. Polyakov, Mod. Phys. Lett. **A2** (1987) 893; V. Knizhnik, A.M. Polyakov and A. Zamolodchikov, Mod. Phys. Lett. **A3** (1988) 819; A. Polyakov, in *Fields, Strings and Critical Phenomena*, Les Houches Lectures (1988), eds. E. Brezin and J. Zinn-Justin (Elsevier Science Publishers, 1989).
3. J. Distler and H. Kawai, Nucl. Phys. **B321** (1988) 171;F. David, Mod. Phys. Lett. **A3** (1988) 1651.
4. A. Bilal and I.I. Kogan, Report No. LPTENS-95-14/OUTP-95-11P (1995).
5. D.R. Karakhanian, R.P. Manvelian, R.L. Mkrtchian, Phys. Lett. **B329** (1994) 185.
6. R. Jackiw, Report No. MIT-CTP-2377/hep-th-9501016 (1995).
7. G. Amelino-Camelia, D. Bak, and D. Seminara, to appear in Physics Letters **B**.
8. R. Jackiw and R. Rajaraman, Phys. Rev. Lett. **54** (1985) 1219; for a review see R. Jackiw in *Quantum Mechanics of Fundamental Systems 1*, C. Teitelboim ed. (Plenum, New York, 1988).
9. C. Itzykson and J.-M. Drouffe, *Statistical Field Theory*, (Cambridge University Press, 1989).

Solitons in the Calogero-Sutherland Collective Field Model

I. Andrić, V. Bardek, and L. Jonke

Department of Theoretical Physics, Rudjer Bošković Institute, P.O. Box 1016, 41001 Zagreb, Croatia

1. Introduction

The Calogero-Sutherland model [1, 2, 3] (CSM) formally describes a gas of spinless non-relativistic particles in one dimension, interacting with the scale-invariant inverse-square potential:

$$H = \frac{1}{2m}\sum_{i=1}^{N} p_i^2 + \frac{\hbar^2\lambda(\lambda-1)}{2m}\sum_{i\neq j}^{N}\frac{1}{(x_i - x_j)^2}. \tag{1.1}$$

The increasing interest in the CSM is due to its relation to various problems. Let us mention a few recent topics. The CSM is related to Brownian motion in a random matrix ensemble [4]. Two-dimensional QCD, compactified to a circle, can be reduced to a unitary matrix model, which is, in a singlet sector, equivalent to the CSM (see a review by Douglas [5]).

There is a interesting physical interpretation of the scale invariant, long-range potential in (1.1) as a statistical interaction. This interaction is not introducing any intrinsic length scale. The model can be exactly solved [3, 6], and it is shown that particles obey the generalized Pauli principle. In fact, the CSM represents the simplest model for Haldane generalized exclusion statistics [7]. Therefore, the effective field theory of the CSM must represent a non-Fermi liquid [8].

Let us present results on the large-N limit [9] of the CSM, where N is the number of particles in the system described by (1.1). The system becomes dense, and it is known that the large-N limit is semiclassical [10]. There is a general procedure [10] to express the N-particle Hamiltonian in terms of density and its conjugate field, and to obtain the $1/N$ expansion for the first few terms. For the Hamiltonian (1.1), the result is [11, 12]

$$H_{coll} = \frac{1}{2}\int dx\rho(x)\left(\partial_x\pi\right)^2 + V_{eff}(\rho), \tag{1.2}$$

where $\partial_x\pi(x)$ is the conjugate field,

$$[\partial_x\pi(x), \rho(y)] = -i\partial_x\delta(x-y).$$

The effective potential is built out of several terms:

$$\begin{aligned} V_{eff} &= \frac{\lambda(\lambda-1)}{2} \fint dx dy \frac{\partial_x \rho(x)\rho(y)}{x-y} \\ &+ \frac{\lambda^2\pi^2}{6}\int dx \rho^3(x) + \frac{(\lambda-1)^2}{8}\int dx \frac{(\partial_x\rho(x))^2}{\rho(x)}. \end{aligned} \tag{1.3}$$

The first term, with the principal part integral, is the collective-field generalization of the statistical interaction. The term cubic in $\rho(x)$ represents the one-dimensional analogue of the Thomas-Fermi energy and vanishes for bosons ($\lambda = 1$). The last term (the entropy term) arises due to the change of the variables, when the density field is introduced. There are some additional terms in (1.3), but they will not contribute to the leading terms in the $1/N$ expansion. If the particles are restricted to the circle and the interaction is trigonometrical, we obtain the V_{eff} with the integration kernel $\cot(2\pi(x_i - x_j)/L)$.

2. Soliton

What is the ground-state density configuration of the Hamiltonian (1.2)? In the large-N limit (continum limit), which is semiclassical, H is a functional in the density $\rho(x)$ and the velocity field $\partial_x\pi(x)$. Before minimizing the collective Hamiltonian, we must introduce the Lagrange multiplier μ in order to take care of the normalization condition for density. The equations to be solved are

$$\dot{\rho} = -\partial_x(\rho(x)\partial_x\pi), \tag{2.1}$$

$$\dot{\pi} = -\frac{1}{2}(\partial_x\pi)^2 - \frac{\delta V_{eff}}{\delta\rho}, \tag{2.2}$$

where V_{eff} is given by (1.3). We are looking for constant-profile solutions, propagating at the speed v, depending only on $\xi = x - vt$. After some algebra, we obtain the equation for a moving-soliton solution:

$$\begin{aligned} \frac{v^2}{2}\left(\frac{\rho_0^2}{\rho^2} - 1\right) &+ \frac{\pi^2\lambda^2}{2}(\rho^2 - \rho_0^2) - \frac{(\lambda-1)^2}{4}\partial_\xi\left(\frac{\partial_\xi\rho}{\rho}\right) \\ &- \frac{(\lambda-1)^2}{8}\left(\frac{\partial_\xi\rho}{\rho}\right)^2 - \lambda(\lambda-1)t, \end{aligned} \tag{2.3}$$

where ρ_0 is a constant solution defined by

$$\frac{\pi^2\lambda^2}{2}\rho_0^2 = \mu.$$

The solution of equation (2.3) is

$$\rho(\xi) = \rho_0 + \frac{a^2 - \rho_0 b^2}{\xi^2 + b^2}, \tag{2.4}$$

where the parameters a and b are

$$b = \frac{\lambda(\lambda-1)\pi\rho_0}{v^2 - \rho_0^2\pi^2\lambda^2}, \tag{2.5}$$

$$a^2 = b^2\rho_0 + \frac{\lambda-1}{\lambda\pi}b = \frac{(\lambda-1)^2 v^2 \rho_0}{[v^2 - \rho_0^2\pi^2\lambda^2]^2}. \tag{2.6}$$

Notice that the above formulae are valid only for $b > 0$. It can be easily seen from equation (2.4) that two basically different soliton profiles emerge, depending on the sign of the numerator $a^2 - \rho_0 b^2$. For $a^2 - \rho_0 b^2 < 0$ $(\lambda < 1)$, we have a hole excitation propagating at the speed $|v| < \pi\rho_0\lambda$. For $a^2 - \rho_0 b^2 > 0$ $(\lambda > 1)$, we have a lump solution propagating at the speed $|v| > \pi\rho_0\lambda$. The static $(v = 0)$ soliton exists only for $\lambda < 1$. The charge of the soliton $\rho(\xi)$ is defined by

$$\Delta N = \int d\xi(\rho(\xi) - \rho_0) = (a^2 - \rho_0 b^2)\int \frac{d\xi}{\xi^2 + b^2} = \frac{\lambda - 1}{\lambda}. \tag{2.7}$$

A duality relation between the soliton solutions with the statistical parametars λ and $1/\lambda$ can be established [13]. Namely, it follows from the relations (2.4) and (2.6) that, upon the exchange $\lambda \to 1/\lambda$, hole-like excitations (with velocity v) transform into particle-like excitations (with velocity v/λ) and vice versa.

The energy of the moving soliton is defined with respect to the stationary background given by ρ_0:

$$E = H(\rho(\xi)) - H(\rho_0). \tag{2.8}$$

Using the equation of motion (2.13) we obtain

$$E = \frac{\lambda-1}{\lambda}\left(\frac{v^2}{2} - \mu\right). \tag{2.9}$$

In the static case $(v = 0)$ the energy reduces to the expression $E = -\mu(\lambda-1)/\lambda$. Let us now find the momentum of the moving soliton. The collective-field theory gives

$$P = \sum_i p_i = \int dx \partial_x \rho(x)\pi(x). \tag{2.10}$$

After partial integration, the momentum (2.10) takes the form

$$P = -\int dx\rho(x)\partial_x\pi + \rho_0(\pi(\infty) - \pi(-\infty)), \tag{2.11}$$

and we easily obtain

$$P = \frac{\lambda-1}{\lambda}(\sqrt{2\mu} - v). \tag{2.12}$$

If we want to interpret our soliton in terms of quasi-particle and quasi-hole excitations of the exact solution [6], we are led to conjecture that our soliton is, effectively, made of one quasi-particle (with unit mass) and one quasi-hole (with mass equal to $-1/\lambda$) excitation in the Fermi sea, defined by the Fermi momentum $k_f = \sqrt{2\mu}$. From the relations (2.9) and (2.12) we have the following dispersion law for the moving soliton:

$$E(P) = \frac{\lambda}{\lambda - 1}\frac{P^2}{2} + \sqrt{2\mu}|P|. \tag{2.13}$$

The above dispersion relation has a natural interpretation in terms of new effective Fermi-sea excitations (with mass equal to $1-1/\lambda$). A single particle-like excitation is obtained by moving a state from the lower Fermi surface $(-k_f^*)$ to a level below. A single hole-like excitation is obtained by moving a state from the Fermi sea to the Fermi surface $(-k_f^*)$, where $k_f^* = (1-1/\lambda)k_f$.

Finally, let us comment on the classical soliton solution of references [14, 15]. In the classical limit $\hbar \to 0$, $g = 2\hbar^2\lambda(\lambda-1) = \text{const.}$, and therefore $\lambda \to \infty$, our moving-soliton solution (2.4) is in agreement with the classical soliton found in reference [14]. Furthermore, it has qualitatively the same behaviour as the solution found in [15], but only in the $v \to \lambda\pi\rho_0$ regime. We conclude by noting that our moving-soliton solutions exist only for λ different from zero or one, i.e. for generic intermediate statistics.

References

1. F. Calogero, *J. Math. Phys.* **10** (1969) 2191,2197 ; *J. Math. Phys.* **12** (1971) 419.
2. B. Sutherland, *Phys. Rev.* **A4** (1971) 2019; *Phys. Rev.* **A5** (1972) 1372; *Phys. Rev. Lett.* **34** (1975) 1083.
3. V. Pasquier, *On Models of the Calogero-Sutherland Type and their relevance to Condesed Matter Physics*, these Proceedings.
4. E. Brézin, C. Itzykson, G. Parisi, and J. B. Zuber, *Comm. Math. Phys.* **59** (1978) 35.
5. M. R. Douglas, preprint RU-93-13, INSF-ITP-93-39, hep-th/9303159.
6. Z. N. C. Ha, *Nucl. Phys.* **B435** (1995) 604.
7. F. M. D. Haldane, *Phys. Rev. Lett.* **67** (1991) 937.
8. R. Caracciolo, A. Lerda and G. R. Zemba, preprint DFTT 18/95, hep-th/9503229.
9. I. Andrić, V. Bardek and L. Jonke, preprint RBI-TH/13-94, hep-th/9411136.
10. A. Jevicki and B. Sakita, *Nucl. Phys.* **B165** (1981) 511.
11. I. Andrić, A. Jevicki and H. Levine, *Nucl. Phys.* **B215** (1983) 307.
12. I. Andrić and V. Bardek, *J. Phys.* **A21** (1988) 2847; *J. Phys.* **A24** (1991) 353.
13. J. A. Minahan and A. P. Polychronakos, preprint CERN-TH 7243/94, USC-94/008, hep-th/9404192.
14. A. P. Polychronakos, preprint CERN-TH 7496/94, hep-th/9411054.
15. B. Sutherland and J. Campbell, *Phys. Rev.* **B50** (1994) 888.

$W_{1+\infty}$ Minimal Models and the Hierarchy of the Quantum Hall Effect

A. Cappelli[1], C. A. Trugenberger[2], and G. R. Zemba[3]

[1] I.N.F.N. and Dipartimento di Fisica, Largo E. Fermi 2, I-50125 Firenze, Italy
[2] Département de Physique Théorique, Univ. de Genève, 24 quai E. Ansermet, CH-1211, Genève 4, Switzerland; supported by a Profil 2 fellowship of the Swiss NSF
[3] I.N.F.N. and Dipartimento di Fisica Teorica, Via P. Giuria 1, I-10125 Torino, Italy

1. Introduction

We would like to briefly report on a research program on the quantum Hall effect, which we have been developing over the last three years. The main points of our program are:

- The high degree of *universality* and *precision* in the measures of the fractional Hall conductivity [1] strongly motivates an effective field theory approach, to describe the low-energy, long-distance physics of the quantum Hall effect.
- The Laughlin theory of the *incompressible quantum fluid* of electrons at the Hall plateaus [2] is assumed here. Classical incompressible fluids are characterized by the *dynamical* symmetry of *area-preserving diffeomorphisms*, obeying the w_∞ algebra [3]. Correspondingly, the effective field theory for the quantum incompressible fluid possesses the quantum analogue of this symmetry, obeying the $W_{1+\infty}$ algebra.
- The low-energy excitations of a *droplet* of incompressible fluid reside at its boundary. Thus, the effective field theory is $(1+1)$-dimensional [4] - actually, it is a conformal field theory [5].
- The conformal field theories with $W_{1+\infty}$ symmetry can be exactly solved by *algebraic* methods and classified completely [6]. This gives an algebraic characterization of the quantum Hall fluids. The necessary mathematical tools were provided by the works [7].
- The special class of *minimal* $W_{1+\infty}$ field theories (minimal models) was recently constructed [8] and shown to give a new hierarchy of the fractional Hall fluids. These models reproduce the Jain hierarchy [9] and predict new, *non-Abelian*, properties for the low-energy edge excitations.
- The detailed experimental predictions of the $W_{1+\infty}$ minimal models are being developed. They must be compared with the experimental data and with the predictions of the standard, Abelian theory of edge excitations [10].

2. The $W_{1+\infty}$ Dynamical Symmetry of the Laughlin Incompressible Fluid

In a seminal work, Laughlin proposed [2] that the ground-state density for the electrons at the plateaus of the quantum Hall effect is uniform *(fluid character)* and has an energy gap *(incompressibility)*. The latter property accounts for the lack of low-energy modes for the longitudinal conduction ($\sigma_{xx} = 0$), while the Hall conductivity is realised as a rigid motion of the uniform droplet. Most of the characteristics of the incompressible fluid can be understood at the (semi)-classical level [2][3]. A *classical* incompressible fluid is defined by its distribution function,

$$\rho(z, \bar{z}, t) = \rho_0 \, \chi_{S_A(t)} \ , \qquad \rho_0 \equiv \frac{N}{A} \ , \tag{2.1}$$

where $\chi_{S_A(t)}$ is the characteristic function for a surface $S_A(t)$ of area A, and $z = x + iy$, $\bar{z} = x - iy$ are complex coordinates on the plane. Since the particle number N and the average density ρ_0 are constant, the area A is also *constant.* The only possible change in response to external forces is in the shape of the surface. The shape changes at constant area can be generated by *area-preserving diffeomorphisms* of the two-dimensional plane. Thus, the configuration space of a *droplet* of incompressible fluid can be generated by applying these transformations to a reference droplet. This can be the ground state in a rotation-invariant potential,

$$\rho_{GS}(z, \bar{z}) = \rho_0 \, \Theta\left(R^2 - z\bar{z}\right) \ . \tag{2.2}$$

Next we recall the Liouville theorem, which states that canonical transformations preserve the phase-space volume. Area-preserving diffeomorphisms are, therefore, canonical transformations of a two-dimensional *phase space.* If we identify the $(z, \bar{z})$ coordinate plane with a phase space, we can describe the area-preserving diffeomorphisms in terms of familiar formulas of canonical transformations. We define the dimensionless Poisson brackets,

$$\{f, g\} \equiv \frac{i}{\rho_0} \ \left(\partial f \bar{\partial} g - \bar{\partial} f \partial g\right) \ , \tag{2.3}$$

where $\partial \equiv \partial / \partial z$ and $\bar{\partial} \equiv \partial / \partial \bar{z}$. Thus, area-preserving diffeomorphisms are defined by $\delta z = \{\mathcal{L}, z\}$ and $\delta \bar{z} = \{\mathcal{L}, \bar{z}\}$, in terms of generating functions $\mathcal{L}(z, \bar{z})$ of both "coordinate" and "momentum". The basis of generators $\mathcal{L}^{(cl)}_{n,m} \equiv \rho_0^{n+m/2} \ z^n \bar{z}^m$ satisfy the w_∞ algebra [3],

$$\left\{\mathcal{L}^{(cl)}_{n,m}, \mathcal{L}^{(cl)}_{k,l}\right\} = -i \ (mk - nl) \ \mathcal{L}^{(cl)}_{n+k-1,m+l-1} \ . \tag{2.4}$$

The *small* excitations above the ground state are given by the infinitesimal w_∞ transformations of ρ_{GS} in (2.1), $\delta\rho_{n,m} \equiv \left\{\mathcal{L}^{(cl)}_{n,m}, \rho_{GS}\right\}$. Using the Poisson brackets (2.3) , we obtain

$$\delta\rho_{n,m} = i\left(\rho_0 R^2\right)^{\frac{n+m}{2}} (m-n)\ \mathrm{e}^{i(n-m)\theta}\ \delta\left(R^2 - z\bar{z}\right) . \tag{2.5}$$

These correspond to density fluctuations localized on the sharp boundary (which is parametrized by the angle θ) of the classical droplet. Due to the dynamics provided by the energy function, they will propagate on the boundary with a frequency ω_k dependent on the angular momentum $k \equiv (n-m)$, thereby turning into *edge waves*.

Another type of excitations are classical *vortices* in the bulk of the droplet, which correspond to localized holes or dips in the density. The absence of density waves, due to incompressibility, implies that any localized density excess or defect is transmitted completely to the boundary, where it is seen as a further edge deformation. For each given vorticity in the bulk, we can then construct the corresponding basis of edge waves as in (2.5). Thus, the configuration space of the excitations of a classical incompressible fluid (of a given vorticity) is spanned by infinitesimal w_∞ transformations. This is the *dynamical symmetry* of classical incompressible fluids.

The classical edge waves and vortices have quantum analogues in the Laughlin theory [2]. These are easily understood in the simplest case of incompressible quantum fluid, corresponding to the fully filled Landau level (filling fraction $\nu = 1$) [3]. The edge waves are particle-hole excitations across the Fermi surface represented by the edge of the droplet of radius R. The vortices correspond to localized quasi-particle and quasi-hole excitations in the bulk of the fluid, the so-called *anyons* with fractional charge, spin and statistics. As in the classical case, they manifest themselves as charged excitations at the edge, owing to incompressibility.

In the quantum theory, there is also a relation between edge excitations and the generators of the quantum version of w_∞, called $W_{1+\infty}$ [3]. This algebra is obtained by replacing the Poisson brackets with quantum commutators: $i\{\ ,\ \} \rightarrow [\ ,\ ]$, and by taking the thermodynamic limit of large droplets [4]. In this limit, the radius of the droplet grows as $R \propto \ell\sqrt{N} \rightarrow \infty$, where $\ell = \sqrt{2/(eB)}$ is the magnetic length and B the magnetic field. Quantum edge excitations, instead, are confined to a boundary annulus of finite size $O(\ell)$. Therefore, edge excitations become the particle hole excitations of the $(1+1)$-dimensional relativistic theory of the Weyl (chiral) fermion [4]. The quantum incompressible fluid becomes the Dirac sea for this long-distance, "scaling" theory. Moreover, the charged states in the relativistic theory represent the edge effect of quasi-particle excitations.

It is not difficult to understand the $W_{1+\infty}$ symmetry of the Weyl fermion. The generators of the $W_{1+\infty}$ algebra are built by bilinears of the Weyl fermion field $F(\theta)$:

$$V_n^j = \int_0^{2\pi} \frac{d\theta}{2\pi}\ :\ F^\dagger(\theta)\ \mathrm{e}^{-in\theta}\ g_n^j\left(i\partial_\theta\right)\ F(\theta)\ :\ , \tag{2.6}$$

where $g_n^j\left(i\partial_\theta\right)$ are j-th order polynomials in $i\partial_\theta$, whose form specifies the basis of V_n^i generators. The $W_{1+\infty}$ algebra reads [4]

$$[\, V_n^i, V_m^j \,] = (jn - im)\, V_{n+m}^{i+j-1} + q(i,j,m,n)\, V_{n+m}^{i+j-3} + \ldots + c\, d^i(n)\, \delta^{i,j} \delta_{n+m,0} \,. \tag{2.7}$$

Here, $i+1 = h \geq 1$ represents the "conformal spin" of the generator V_n^i, while $-\infty < n < +\infty$ is the angular momentum (the Fourier mode on the circle). The first term on the right-hand-side of (2.7) reproduces the classical w_∞ algebra (2.4) by the correspondence $\mathcal{L}_{i-n,i}^{(cl)} \to V_n^i$, and identifies $W_{1+\infty}$ as the algebra of "quantum area-preserving diffeomorphisms". There are also a finite number of quantum operator corrections with polynomial coefficients $q(i,j,n,m)$. Finally, the c-number term $c\ d^i(n)$ is the quantum *anomaly*, a relativistic effect [5]. All terms in the commutation relations (2.7) are known and uniquely determined by the closure of the algebra; only the *central charge* c is a free parameter, which takes the value one in the Weyl fermion theory.

The generators V_n^0 are Fourier modes of the fermion density evaluated at the edge $|z| = R$, and V_0^0 measures the edge charge. Instead, the V_n^1 generate angular momentum transformations on the edge, such that V_0^1 measures the angular momentum of edge excitations. The generators V_n^0 and V_n^1 obey the Abelian current algebra $\widehat{U(1)}$ and the Virasoro algebra, respectively [5].

In conformal field theory [5], the Weyl fermion theory can be defined *algebraically* as the Hilbert space made by a set of irreducible, highest-weight representations of the $W_{1+\infty}$ algebra, which is closed under the *fusion rules* for making composite excitations. Any representation contains an infinite number of states, corresponding to all the particle-hole excitations above a bottom state, the so-called *highest weight* state. This can be, for example, the ground-state $|\Omega\rangle$ corresponding to the incompressible quantum fluid. The particle-hole excitations can be written as

$$V_{-n_1}^{i_1}\, V_{-n_2}^{i_2} \cdots V_{-n_s}^{i_s} |\, \Omega \,\rangle \,, \quad n_1 \geq n_2 \geq \cdots \geq n_s > 0 \,, \quad i_1, \ldots, i_s \geq 0 \,, \tag{2.8}$$

while the positive modes ($n_i < 0$) annihilate $|\ \Omega\ \rangle$. Here $k = \sum_j n_j$ is the total angular momentum of the edge excitation [3].

Furthermore, any charged edge excitation, together with its tower of particle-hole excitations, also forms an irreducible, highest-weight representation of $W_{1+\infty}$. The states in this representation have the same form of (2.8), but the bottom state $|Q\rangle$ now represents a quasi-particle inside the droplet. The charge Q and the spin J of the quasi-particle are given by the eigenvalues of the operators V_0^0 and V_0^1, respectively:

$$V_0^0 \,|\, Q\,\rangle \;=\; Q \,|\, Q\,\rangle \,,\; V_0^1 \,|\, Q\,\rangle \;=\; J \,|\, Q\,\rangle \,. \tag{2.9}$$

Moreover, the statistics θ/π of quasi-particles is equal to twice the spin J. For $W_{1+\infty}$ representations , all the operators V_0^i are simultaneously diagonal and assign other quantum numbers to the quasi-particle,

$$V_0^i |\, Q\,\rangle = m_i(Q) |\, Q\,\rangle, \; i \geq 2 \,,$$

which are known polynomials in the charge Q [7]. These quantum numbers measure the radial moments of the charge distribution of the quasi-particles [4].

3. Classification of QHE Universality Classes

Besides the previous explicit construction for $\nu = 1$, we have found a general method for constructing the effective edge theories corresponding to fractional fillings [6][8]. A known mathematical result is that the $W_{1+\infty}$ algebra (2.7) is the unique quantization of the w_∞ algebra in the $(1+1)$-dimensional field theory on the circle - the droplet boundary [7]. This implies that the effective quantum theories of the incompressible fluids are completely specified by the $W_{1+\infty}$ symmetry, for any value of the density, i.e., of the filling fraction. Although these general theories cannot be derived explicitly in terms of electrons, they can be completely constructed by using the algebraic methods of conformal field theory [5] and the $W_{1+\infty}$ representation theory [7].

We shall therefore *postulate* that *all* quantum Hall incompressible fluids are in one-to-one correspondence with $W_{1+\infty}$ theories [6]. This characterization provides a powerful classification scheme for quantum Hall universality classes.

These classes are specified by the *kinematical data* of the incompressible fluid, which can be obtained from the quantum numbers (weights) of the $W_{1+\infty}$ representations, as follows:

i) the charge Q and fractional statistics θ/π of the quasi-particle excitations are given by the eigenvalues of V_0^0 and V_0^1, respectively (see (2.9)) (the other $V_0^i, i > 1$ eigenvalues are of secondary interest);

ii) the Hall conductivity $\sigma_H = (e^2/h)\nu$, proportional to the filling fraction ν of the ground state, is obtained from the chiral anomaly of the $W_{1+\infty}$ $(1+1)$-dimensional theory [4] [6];

ii) the number of particle-hole excitations of given angular momentum, *i.e.*, the *degeneracies* of states above the ground state (2.8), are obtained from the characters of the $W_{1+\infty}$ representations [8].

4. The Standard Theory of Edge Excitations and the Hierarchy

There is nowadays a "standard model" for the description of edge excitations, which is given by the $(1+1)$-dimensional theory of the chiral boson field $\phi(\theta)$ [10]. An equivalent description is given by Abelian Chern-Simons theories on $(2+1)$-dimensional open domains. While one field describes the edge excitations of the Laughlin fluid $\nu = 1, 1/3, 1/5, \ldots$, many components have been introduced for describing more general fractional fillings. Each boson gives rise to an independent current $J^i \propto \partial_\theta \phi^i$, which satisfy the algebra,

$$[J^i(\theta_1), J^k(\theta_2)] = \delta^{ik}\, \delta'(\theta_1 - \theta_2)\ , \qquad i,k = 1,\ldots,m\ . \tag{4.1}$$

This multi-component Abelian current algebra is denoted by $\widehat{U(1)}^{\otimes m}$ and implies the Virasoro algebra with integer central charge equal to the number of current components $c = m$ [5].

Besides the central charge, each Abelian edge theory is specified by a $(m \times m)$ symmetric, integer-valued matrix K_{ij} (with K_{ii} odd); K is pseudo-Euclidean, if the theory has excitations with both types of chiralities. The observables quantities of edge excitations are obtained as follows [10]: the spectrum of charges and fractional statistics are given by

$$Q = \sum_{i,j=1}^{m} K_{ij}^{-1} n_j , \qquad \frac{\theta}{\pi} = \sum_{i,j=1}^{m} n_i K_{ij}^{-1} n_j , \qquad n_i \in \mathbf{Z} , \tag{4.2}$$

respectively, where the integers n_i label the edge excitation. The Hall conductivity, given by the chiral anomaly [4], is similarly parametrized by

$$\sigma_H = \frac{e^2}{h} \nu , \qquad \nu = \sum_{i,j=1}^{m} K_{ij}^{-1} . \tag{4.3}$$

The one-component theory has one free parameter (K_{11} = odd), which spans the simplest Laughlin fluids, and its predictions have been confirmed by recent experiments[1]. Instead, the multi-component theory has many free parameters (K_{ij}) and is thus *too generic.* Actually, the experimental values of the fractional Hall conductivity show a characteristic *hierarchical pattern* (see figure 1). This pattern is not captured by the standard edge theory, and new physical inputs must be introduced for making this theory predictive [10].

On the other hand, this pattern has been understood in the completely different language and methods of trial wave functions and microscopic dynamics, which has culminated in the Jain hierarchical theory [9]. For filling fractions,

$$\nu = \frac{m}{mp \pm 1} , \qquad m = 1, 2, \ldots , \qquad p = 2, 4, \ldots , \tag{4.4}$$

Jain devised a generalization of the Laughlin wave functions based on the physical picture of the *composite fermion,* a local bound state of the electron, and an even number (p) of flux quanta. The strongly-interacting electrons can be mapped into weakly-interacting composite fermions at effective integer filling (m). The stability of the fluids at the fractional fillings (4.4) can be related with the stability of m completely filled Landau levels. Therefore, at fixed p, equation (4.4) gives one-parameter families of approximately equally stable Hall fluids, which accumulate at $\nu \to (1/p)$. This is the hierarchical

[1] See the references quoted in [8].

pattern shown by the experimental points in figure 1 (**bold** fractions); the other points (*italic* fractions) are not presently well understood [10].

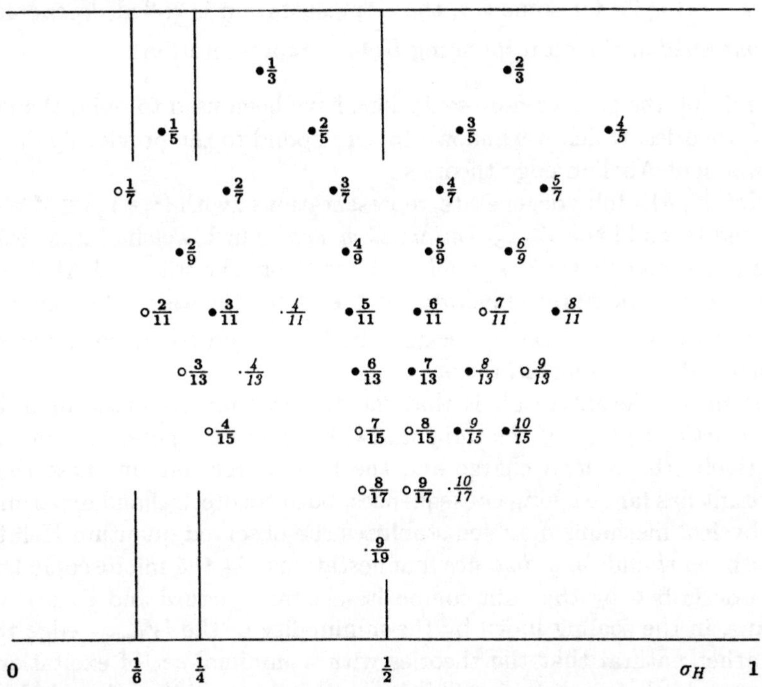

Fig.1.: Experimentally observed plateaus in the range $0 < \nu < 1$: their Hall conductivity $\sigma_H = (e^2/h)\nu$ is displayed in units of (e^2/h). The points denote stability: ($\bullet$) very stable, ($\circ$) stable, and ($\cdot$) less stable plateaus. Theoretically understood plateaus are in **bold**, unexplained ones are in *italic*. Observed cases of coexisting fluids are displayed as $\nu = 2/3, 6/9, 10/15$, $\nu = 3/5, 9/15$, and $\nu = 5/7, 15/21$ (but 15/21 is not displayed).

5. $W_{1+\infty}$ Minimal Models

The $W_{1+\infty}$ classification of incompressible quantum Hall fluids outlined in section two can be carried through, because all unitary, irreducible, highest-weight representations of the $W_{1+\infty}$ algebra (2.7) were recently obtained in the mathematical work [7]. These representations exist for positive integer central charge $c = m = 1, 2, \ldots$, and are labeled by a m-component weight vector $\mathbf{r} = \{r_1, \ldots, r_m\}$, which is the multi-component generalization of the charge Q in (2.9). There exist two kinds of representations, depending on the value of the weight vector:

- if $(r_i - r_j) \notin \mathbf{Z}\ \forall\, i \neq j$, the representation is called *generic* and is actually *equivalent* to a representation of the $\widehat{U(1)}^m$ algebra (4.1) having the same weight;
- if $(r_i - r_j) \in \mathbf{Z}$ for some i, j, the representation is called *degenerate* and is *contained* in the corresponding $\widehat{U(1)}^m$ representation.

In ref. [6], the generic representations have been used to build the *generic* $W_{1+\infty}$ theories, which were shown to correspond to the previously discussed m-component Abelian edge theories.

In ref. [8], the fully degenerate representations, with $(r_i - r_j) \in \mathbf{Z}\ \forall\, i \neq j$, were used to build the $W_{1+\infty}$ *minimal models*, which we shall now describe. The minimal models are maximally different from the standard Abelian edge theories, being *quantum reductions* of the latter theories. This means that they have less excited states, because the $\widehat{U(1)}^m$ representations decompose into many $W_{1+\infty}$ representations.

The most relevant result is that *the minimal models exists only for the filling fractions (4.4) of the Jain fluids* [8]. The parameters m and p are, respectively, the central charge and the free charge unit in these theories. This result has far-reaching consequences, both theoretical and experimental. The physical mechanism, which stabilizes the observed quantum Hall fluids, has both *short* and *long distance* manifestations. At the microscopic level, it can be described by the Jain composite-electron picture and by the size of the gaps; in the scaling limit, by the minimality of the $W_{1+\infty}$ edge theory. It is rather natural that the theories with a minimal set of excitations are also dynamically more stable. This long-distance stability principle leads to a logically self-contained edge theory of the fractional Hall effect: a thorough derivation of experimental results is obtained from the principle of $W_{1+\infty}$ symmetry, which is the basic property of the Laughlin incompressible fluid. This independent hierarchical construction is the main result of our approach.

The spectrum of charge and fractional statistics of edge excitations of the $W_{1+\infty}$ minimal models is given by [8]

$$\begin{aligned} Q &= \frac{1}{pm \pm 1} \sum_{i=1}^{m} n_i \,, \qquad p = 2,4,\dots, \qquad m = 1,2,\dots, \\ \frac{\theta}{\pi} &= \pm \left(\sum_{i=1}^{m} n_i^2 - \frac{p}{mp \pm 1} \left(\sum_{i=1}^{m} n_i \right)^2 \right), \\ & \qquad n_1 \geq n_2 \geq \cdots \geq n_m \,, \quad n_i \in \mathbf{Z} \,. \end{aligned} \tag{5.1}$$

Let us remark that the spectra (5.1) of quantum numbers have also been obtained in the standard Abelian $\widehat{U(1)}^m$ theory [10], by choosing a specific form of the K_{ij} matrix in (4.2-4.3), inspired by the wave-function construction [9]. However, the properties of the excitations in the two theories are rather different, due to the degeneracy of $W_{1+\infty}$ representations .

Degenerate representations are rather common in conformal field theory: if the central charge and the weight of a given representation satisfy certain algebraic relations, some of its states decouple, and should be projected out from the Hilbert space. The minimal models are made of sets of degenerate representations , which are closed under the fusion rules [5]. There are specific minimal models for any symmetry algebra: the well-known ones are the $c < 1$ Virasoro minimal models; larger symmetry algebras, like $W_{1+\infty}$, have $c > 1$ minimal models.

The $W_{1+\infty}$ minimal models are *not* realised by the multi-component chiral boson theories with $\widehat{U(1)}^{\otimes m}$ symmetry, because the latter do not incorporate the projection that makes irreducible the $W_{1+\infty}$ degenerate representations. They are instead realised by the $\widehat{U(1)} \otimes \mathcal{W}_m(p = \infty)$ conformal theories [7], where the $\mathcal{W}_m(p)$ are the Zamolodchikov-Fateev-Lykyanov models with $c = (m-1)\left[1 - m(m+1)/p(p+1)\right]$. The weight $\mathbf{r}$ of these representations belongs to the weight lattice of the Lie algebra $U(1) \times SU(m)$. As a consequence, the $W_{1+\infty}$ minimal models give a natural explanation of the $SU(m)$ "symmetry" of the Jain fluids often observed before [10]. Moreover, this non-Abelian character implies the following properties of the edge excitations:

i) There is a *single* Abelian current, instead of m independent ones, and therefore a single elementary (fractionally) charged excitation; there are neutral excitations, but they cannot be associated to $(m-1)$ independent edges.

ii) The dynamics of these neutral excitations is new: they have associated an $SU(m)$ (not $\widehat{SU(m)}_1$ [10]) *"isospin"* quantum number, given by the highest weight [8]

$$\Lambda = \sum_{a=1}^{m-1} \Lambda^{(a)} \; (n_a - n_{a+1}) \ , \tag{5.2}$$

where $\Lambda^{(a)}$ are the fundamental weights of $SU(m)$, and $\{n_i\}$ are the integer labels of eq. (5.1). Therefore, the neutral excitations are quark-like, and their statistics is non-Abelian. For example, the edge excitation corresponding to the electron is a composite carrying both the additive electric charge and the fundamental "quark" $SU(m)$ representation . It is possible that a direct experimental test of the non-Abelian properties of the neutral excitations could be made in the future [8].

iii) The degeneracy of particle-hole excitations at fixed angular momentum is modified by the projection of the minimal models. This counting of states is provided by the characters of degenerate $W_{1+\infty}$ representations, which are known [7]. If the neutral $SU(m)$ excitations have a bulk gap, the particle-hole degeneracy of the ground state (the Wen topological order on the disk) is different from the corresponding one of $\widehat{U(1)}^{\otimes m}$ excitations [10]. This can be tested in numerical diagonalizations of few electron systems [8].

6. Conclusions

These results show that the hierarchical pattern of fractional Hall fluids emerges naturally from the $W_{1+\infty}$ symmetry. Moreover, they suggest that the $W_{1+\infty}$ minimal models are the correct edge theories corresponding to the Jain wave functions. This hypothesis can be confirmed by specific experimental and numerical tests of the properties of edge excitations in the minimal models [8]. Finally, there is a hierarchical structure of intermediate $W_{1+\infty}$ theories, in between the generic and minimal theories, which has yet to be explored and shows some similarities with the higher orders in the Jain hierarchy.

References

1. For a review see: R. A. Prange, S. M. Girvin, *The Quantum Hall Effect*, Springer Verlag, New York (1990).
2. R. B. Laughlin, Phys. Rev. Lett. **50** (1983) 1395; for a review see: R. B. Laughlin, *Elementary Theory: the Incompressible Quantum Fluid*, in [1].
3. A. Cappelli, C. A. Trugenberger and G. R. Zemba, Nucl. Phys. **396 B** (1993) 465; Phys. Lett. **306 B** (1993) 100.
4. A. Cappelli, G. V. Dunne, C. A. Trugenberger and G. R. Zemba, Nucl. Phys. **398 B** (1993) 531; A. Cappelli, C. A. Trugenberger and G. R. Zemba, *$W_{1+\infty}$ Dynamics of Edge Excitations in the Quantum Hall Effect*, cond-mat/9407095.
5. A. A. Belavin, A. M. Polyakov and A. B. Zamolodchikov, Nucl. Phys. **B 241** (1984) 333; for a review see: P. Ginsparg, *Applied Conformal Field Theory*, in *Fields, Strings and Critical Phenomena*, Les Houches School 1988, E. Brezin and J. Zinn-Justin eds., North-Holland, Amsterdam (1990).
6. A. Cappelli, C. A. Trugenberger and G. R. Zemba, Phys. Rev. Lett. **72** (1994) 1902.
7. V. Kac and A. Radul, Comm. Math. Phys. **157** (1993) 429; E. Frenkel, V. Kac, A. Radul and W. Wang, preprint hep-th/9405121.
8. A. Cappelli, C. A. Trugenberger and G. R. Zemba, *Stable hierarchical Quantum Hall Fluids as $W_{1+\infty}$ minimal models*, hep-th/9502021, to appear in Nucl. Phys. **B**.
9. J. K. Jain, Adv. in Phys. **41** (1992) 105.
10. X. G. Wen, Int. J. Mod. Phys. **6 B** (1992) 1711; J. Fröhlich, U. M. Studer and E. Thiran, *A Classification of Quantum Hall Fluids*, preprint KUL-TF-94/35; see also Fröhlich's lectures in Schladming 1995.

Creation and Annihilation Operators for $b - c$ Systems on General Algebraic Curves

F. Ferrari[1] and J. Sobczyk[2]

[1] Dipartimento di Fisica, Universitá di Trento, 38050 Povo, (TN), Italy and INFN, Gruppo Collegato di Trento
[2] Institute for Theoretical Physics, Wroclaw University, pl. Maxa Borna 9, 50204 Wroclaw, Poland.

1. Introduction

We are going to review the basic characteristics of a new operator formalism on higher genus Riemann surfaces developed in [1]. The language used here to describe the Riemann surfaces (RS's) is that of algebraic curves. It is well known that every RS can be represented by means of a so-called Weierstrass polynomial [2]:

$$F(z,y) = P_n(z)y^n + P_{n-1}(z)y^{n-1} + ... + P_1(z)y + P_0(z) = 0, \qquad (1.1)$$

where the $P_j(z)$ are polynomials in z of degree at most $n-j$, $j = 0, \ldots, n$. Solving (1.1) with respect to y, one obtains a very useful and intuitive picture of a RS as a branched covering of the complex sphere $\mathbf{C}P(1)$. The RS is parametrized by the coordinates z and $y(z)$, where $y(z)$ is a multivalued function with n branches. We will develop our operator formalism for the simple $b - c$ systems with integer spin λ and action:

$$S_{\mathrm{bc}} = \int_{\Sigma_g} d^2\xi \left(b\bar{\partial}c + \mathrm{c.c.}\right), \qquad (1.2)$$

where ξ and $\bar{\xi}$ are complex coordinates on Σ_g. For convenience, we will assume here that $\lambda > 1$. The case of $\lambda = 1$ can be treated in a similar way [3]. There is already an extensive literature about the $b-c$ systems [4], but, up to our knowledge, this is the first time that an explicit operator formalism has been constructed on algebraic curves. The main advantages of working on an algebraic curve are: *a*) a simple expression of the amplitudes of the $b - c$ systems as rational functions of z and y; *b*) the possibility of treating the fields as multivalued tensors on $\mathbf{C}P(1)$. Point *a* is relevant for string theory applications. Thanks to point *b*), we are also able to establish a connection between the $b - c$ systems on RS's and conformal field theories on the complex plane. The remarkable features of the latter, like braid group statistics, relations with the Riemann monodromy problem etc. have been discussed in [5]. Another important element of our construction is related to the explicit representation obtained for the Hilbert space of the $b - c$ systems, which is splitted into n "independent" Fock spaces. In each Fock space only modes with the same monodromy characteristics are allowed to propagate. The notion of a normal ordering naturally arises, and with the help of the Wick theorem it is possible to reproduce all the correlation functions of the theory.

2. The Operator Formalism

It can be shown that the most general solutions of the classical equations coming from the action (1.2) admit the following generalized Laurent expansion:

$$b(z)dz^{\lambda} = \sum_{k=0}^{n-1} b_k(z)dz^{\lambda} = \sum_{k=0}^{n-1} \sum_{i=-\infty}^{\infty} b_{k,i} z^{-i-\lambda} f_k(z), \tag{2.1}$$

$$c(z)dz^{1-\lambda} = \sum_{k=0}^{n-1} c_k(z)dz^{1-\lambda} = \sum_{k=0}^{n-1} \sum_{i=-\infty}^{\infty} c_{k,i} z^{-i+\lambda-1} \phi_k(z), \tag{2.2}$$

with f_k and ϕ_l, $(k, l = 0, ..., n-1)$, chosen as follows:

$$f_k(z) = \frac{y^{n-1-k}(z)dz^{\lambda}}{(F_y(z, y(z)))^{\lambda}}, \tag{2.3}$$

$$\phi_l(w)dw^{1-\lambda} = \frac{dw^{1-\lambda}}{(F_y(w, y(w)))^{1-\lambda}}$$
$$(y^l(w) + y^{l-1}(w)P_{n-1}(w) + y^{l-2}(w)P_{n-2}(w) + ... + P_{n-l}(w)) . \tag{2.4}$$

We have used the notation $F_y = \frac{\partial F}{\partial y}$, where F is defined in eq. (1.1). The zero modes (holomorphic differentials) can be identified as:

$$\Omega_{k,i} dz^{\lambda} = f_k(z) z^{-i-\lambda}, \tag{2.5}$$

under the assumption that

$$i \leq -\lambda, \qquad k + \lambda(n-2) - (n-1) + i \geq 0. \tag{2.6}$$

To quantize the $b-c$ systems, we impose the commutation relations:

$$\{b_{k,i}, c_{k',i'}\} = \delta_{kk'} \delta_{i+i',0}. \tag{2.7}$$

The identification of creation, annihilation and "zero modes" is done in the following way:

$$\begin{aligned} b_{k,j} \;=\; & \Big\{ b^{-}_{k,j};\; j \geq 1-\lambda b^{0}_{k,j}\,; \\ & j = -\lambda - N_{b_k} + 1, \ldots, -\lambda b^{+}_{k,j}; \\ & j \leq -\lambda - N_{b_k}\; c_{k,j} \;= c^{-}_{k,j};\; j \geq \lambda c^{+}_{k,j};\; j \leq \lambda - 1 N_{b_k} \Big\} \end{aligned} \tag{2.8}$$

is the number of zero modes in a given sector k. The total number of zero modes:

$$N_b = \sum_{k=0}^{n-1} N_{b_k} = \frac{(2\lambda-1)(n-3)n}{2} = (2\lambda-1)(g-1) \tag{2.9}$$

is in agreement with the Riemann-Roch theorem for RS of genus g. Creation and annihilation operators labeled by different indices k act on the bra and ket vacua ${}_k\langle 0|$ and $|0\rangle_k$ satisfying the relations:

$${}_k\langle 0|0\rangle_k = 0 \qquad \text{if } N_{b_k} \neq 0; \qquad\qquad {}_k\langle 0| \prod_{i=1}^{N_{b_k}} b_{k,i}|0\rangle_k = 1. \tag{2.10}$$

The total vacuum is given by:

$$|0\rangle = \bigotimes_{k=0}^{n-1} |0\rangle_k. \tag{2.11}$$

The above prescriptions yield the following normal ordering between the fields $b_k(z)$ and $c_k(z)$ of eqs. (2.1) and (2.2):

$$b_k(z)c_k(w) =: b_k(z)c_k(w) : + \frac{1}{z-w} f_k(z)\phi_k(w). \tag{2.12}$$

Summing over k, we obtain:

$$b(z)c(w) =: b(z)c(w) : + K_\lambda(z,w)dz^\lambda dw^{1-\lambda}. \tag{2.13}$$

The tensor $K_\lambda(z,w)$ appearing in eq. (2.13) is the so-called Weierstrass kernel. It is the basic "building block" of the correlation functions thanks to these properties: (i) it is a λ differential in the z variable and $1-\lambda$-differential in the w variable; (ii) it has a unique pole with residue $+1$ in the limit $z \to w$. For a given Weiertrass polynomial $F(z,y)$, the Weierstrass kernel can be explicitly expressed as

$$K_\lambda(z,w) = \frac{[F_y(w,y(w))]^{\lambda-1}}{[F_y(z,y(z))]^\lambda} \frac{1}{z-w} \frac{F(w,y(z))}{y(z)-y(w)} dz^\lambda dw^{1-\lambda}. \tag{2.14}$$

Exploiting the recipes given above, it is now only a matter of algebraic manipulations to obtain all the correlation functions of the $b-c$ systems:
a) The zero point function is

$$\langle 0|b(z_1)\ldots b(z_{N_b})|0\rangle = \det|\Omega_I(z_J)|\,. \tag{2.15}$$

b) The most general $M-$ point functions ($M - N = N_b$, $M \neq 0$) are

$$\frac{< 0| \prod_{s=1}^{M} b(z_p) \prod_{t=1}^{N} c(w_t)|0 >}{\langle 0| \prod_{I=1}^{N_b} b(z_I)|0\rangle} =$$

$$\det \begin{vmatrix} \Omega_1(z_1) & \dots & \Omega_{N_b}(z_1) & K_\lambda(z_1, w_1) & \dots & K_\lambda(z_1, w_N) \\ \vdots & \ddots & \vdots & \vdots & \ddots & \vdots \\ \Omega_1(z_M) & \dots & \Omega_{N_b}(z_M) & K_\lambda(z_M, w_1) & \dots & K_\lambda(z_M, w_N) \end{vmatrix}. \qquad (2.16)$$

3. Final Remarks

The successful construction of the Fock space for fermionic $b - c$ systems on general Riemann surfaces rises the question, if it is possible to perform similar computations in the case of the bosonic fields as well. There are strong indications that this can be done. The key point of our approach sketched in this report is in fact the following factorization property of the Weierstrass kernel:

$$(z - w)K_\lambda(z, w) = \sum_{j=0}^{n-1} f_j(z)\phi_j(w). \qquad (3.1)$$

On the other hand, it is known that an analogous decomposition of the current-current correlation function holds [6]:

$$G(z, w) = \langle \partial_z X(z, \bar{z})\partial_w X(w, \bar{w})\rangle = \sum_{nm} \frac{P_{nm}(z, w)}{(z - w)^2} I_n(z)I_m(w), \qquad (3.2)$$

where P is a polynomial in z and w, and the I_n are meromorphic differentials with possible poles at infinity of degree ≤ 2.

References

1. F. Ferrari, J. Sobczyk and W. Urbanik, *Operator Formalism on the Z_n Symmetric Algebraic Curves*, Preprint ITF UWr 856/93, to appear in Jour. Math. Phys.; F.Ferrari, J.Sobczyk, *Operator formalism on general algebraic curves*, Preprint UTF 333-94, hep-th 9409127, to be published in J. Geom. Phys..
2. P. Griffiths and J. Harris, *Principles of Algebraic Geometry*, Johm Wiley & Sons, New York 1978.
3. F. Ferrari and J. Sobczyk, *Operator Formalism for $b - c$ systems with $\lambda = 1$ on General Algebraic Curves*, Preprint UTF 333-95, IFT UWr 879/95, hep-th 9501125.

4. D. Friedan, E. Martinec and S. Shenker, Nucl. Phys. **B271** (1986) 93; E. Verlinde and H. Verlinde, Nucl. Phys. **B288** (1987) 357; L. Bonora, A. Lugo, M. Matone and J. Russo, Comm. Math. Phys. **123** (1989) 329; P. Di Vecchia, F. Pezzella, M. Frau, K. Hornfleck, A. Lerda and A. Sciuto, Nucl. Phys. **B332** (1989) 317; ibid. **B333** (1990) 635; A. K. Raina, Comm. Math. Phys. **122** (1989) 625; ibid. **140** (1991) 373; A. M. Semikhatov, Phys. Lett. **B212** (1988) 357; V. G. Knizhnik, Comm. Math. Phys. **112** (1987) 587; Sov. Phys. Usp. **32** (11) (1989) 945.
5. J. Sobczyk and W. Urbanik, Mod. Phys. Lett. **21** (1991) 1; J. Sobczyk, Int. Journ. Mod. Phys. **A6** (1991) 1103; ibid. **A8** (1993) 1153; F. Ferrari, Comm. Math. Phys. **156** (1993) 179; Phys. Lett. **277** (1992) 423; Int. Jour. Mod. Phys. **A9** (1994) 313; Int. Jour. Mod. Phys. **A5** (1990) 2799.
6. M. Bershadsky, A. Radul, Phys. Lett. B **193** (1987) 213.

Interacting Electrons and Localized Spins: Exact Results from Conformal Field Theory

P. Fröjdh[1] and H. Johannesson[2]

[1] Department of Physics, University of Washington, P.O. Box 351560, Seattle, WA 98195, USA

[2] Institute of Theoretical Physics, Chalmers University of Technology and Göteborg University, S-412 96 Göteborg, Sweden

1. Introduction

Much of traditional condensed matter physics falls under the Fermi liquid paradigm. The notion goes back to Landau and implies, that a liquid of fermions (such as ^{3}He or conduction electrons in a metal) can be treated as a system of essentially free particles [1]. Certain conditions apply: symmetries must remain unbroken, and the energies probed should be "small." But with these provisos the only price to pay for removing the interactions is to keep track on how certain parameters (quasiparticle masses, lifetimes, etc.) renormalize. The Fermi-liquid picture has proven enormously successful and explains, how we can get away with single-particle quantum mechanics when doing elementary solid state physics. Fortunately — to keep us busy! — experimentalists have found a number of lab systems that violate the standard Fermi liquid picture. (An outstanding example is the metallic phase of the high-T_c superconductors.) The design of concepts and methods to handle this intriguing and growing class of problems — often nicknamed "strongly correlated systems" — is a major challenge for the theorist. By the very nature of the problem one here confronts matter nonperturbative. Lacking a universal paradigm, the most sensible thing to do is to practice on simple models of non-Fermi liquids, work out the consequences, compare with experiments, build up intuition, and collect that "critical mass" of knowledge necessary for the emergence of an effective, unified approach to correlated systems.

Two particularly clean realizations of non-Fermi liquids are the Luttinger liquid and (certain generalized versions of) the Kondo effect. "Luttinger liquid" is a code name for the low-energy, long-wavelength physics of interacting electrons in one dimension (1D) [2]. The topology of the 1D Fermi surface for free electrons (it consists of two distinct points!) yields dramatic effects when interactions are included: The single-particle poles of the propagators get wiped out and are replaced by branch cuts. The resulting spectral density develops a two-peak structure, with a broad band in between: A single electron added to the system "decays" into two distinct collective excitations, one carrying the charge, the other the spin. Remarkably, the two excitations propagate with different velocities, leading to a spatial spin-charge separation.

The Kondo effect is a different story [3]. Here one considers the spin exchange interaction between a localized magnetic impurity and a band of free conduction electrons in a metal. By symmetry, the problem is identical to that of 1D free fermions coupled to a local spin. Surprisingly, this mundane setup leads to quite spectacular physics. As one scales to low temperatures and large distances, the electron-impurity coupling flows to a strong-coupling fixed point, where electrons are effectively interacting with each other *via the impurity.* In certain versions of the problem, where the impurity couples to several degenerate bands of conduction electrons (multi-channel Kondo effect), the induced interaction among electrons cannot be "renormalized away," and one finds a new kind of low-temperature critical behavior. Yet another breakdown of the simple Fermi-liquid picture!

The theories of Luttinger liquids and the multi-channel Kondo effect are important landmarks in the physics of correlated systems. Their construction did not come easy, though, and has required decades of shrewd work by many people, exploiting some of the most powerful tools of theoretical physics: Bosonization, the Renormalization Group, Bethe's *Ansatz*, and most recently Conformal Field Theory (CFT).

2. Boundary Conformal Field Theory

The use of CFT to study the Kondo effect was pioneered by Affleck and Ludwig in a series of remarkable papers [4]. The basic idea goes back to Nozières [5]. In the simplest case — with a spin-$\frac{1}{2}$ impurity coupled antiferromagnetically to a single channel of conduction electrons — Nozières showed that the impurity may be traded for a boundary condition: At the strong-coupling fixed point the impurity traps an electron into a bound singlet state, with the remaining electrons being excluded from entering the impurity site. In other words, the impurity disappears from the problem and is replaced by a boundary condition on the free electron wave functions. Affleck and Ludwig argued that all quantum impurity problems essentially work this way, and showed how to take advantage of CFT, once the problem is reformulated in terms of a new boundary condition.

By introducing a boundary in CFT, say at $x = 0$, one imposes constraints on any field $\varphi(t,x)$ [6]. The left and right moving pieces get identified via analytic continuation beyond the boundary, such that

$$\varphi(t,x) = \varphi_L(t,x)\varphi_R(t,x) \ \rightarrow \ \varphi_L(t,x)\varphi_L(t,-x). \tag{2.1}$$

Hence, the boundary has effectively turned the field nonlocal. As a consequence, the one-point function in the presence of a boundary develops a profile given by the corresponding two-point function in the bulk,

$$\langle\varphi(t,x)\rangle \ \rightarrow \ \langle\varphi_L(t+x)\varphi_L(t-x)\rangle \ \sim \ x^{-\Delta_{bulk}}, \tag{2.2}$$

with Δ_{bulk} the ordinary scaling dimension of φ. Correlation functions are also altered. In particular, the autocorrelation function for φ close to the boundary is governed by a boundary-condition dependent exponent Δ called the *boundary scaling dimension* of φ:

$$\langle \varphi(t,x)\varphi(0,x)\rangle - \langle \varphi(t,x)\rangle\langle \varphi(0,x)\rangle \sim t^{-2\Delta}\,, \qquad |t| >> |x|. \tag{2.3}$$

This sets the strategy for treating quantum impurity problems: Identify first the particular boundary condition that plays the role of the impurity interaction. (In most cases this is a highly nontrivial task. The CFT scheme gives some guidance, though, and it has turned out that *conformal fusion* often corresponds to such a change of boundary condition.) Given that the new boundary condition is indeed scale invariant (together with the original bulk theory), one can then use the machinery of CFT to extract the corresponding boundary scaling dimensions. As these determine the asymptotic autocorrelation functions, the finite-temperature properties due to the presence of the boundary (*alias* the impurity!) are easily accessed from standard finite-size scaling by treating (Euclidean) time as an inverse temperature.

3. Kondo Effect in a Luttinger Liquid

What happens, if one puts a magnetic impurity into a Luttinger liquid (LL)? As for the Kondo effect in a Fermi liquid, we expect that the impurity also here will induce effective interactions among the electrons. But in a LL, the electrons are already strongly correlated by the mutual Coulomb interactions. Then, what happens? Does the interplay between "induced" and "direct" correlations lead to novel effects? Or do we recover the same Kondo physics as in a Fermi liquid? This is not an easy problem, but again CFT can be put to productive use [7].

The "standard model" for low-energy electrons in 1D is the Tomonaga-Luttinger (TL) Hamiltonian [2]

$$\begin{aligned}\mathcal{H}_{TL} = \frac{1}{2\pi}\int dx\Big\{ & v_F\,\Big[:\psi^{\dagger}_{L,\sigma}(x)i\frac{d}{dx}\psi_{L,\sigma}(x): - :\psi^{\dagger}_{R,\sigma}(x)i\frac{d}{dx}\psi_{R,\sigma}(x):\Big] \\ & + \frac{g}{2}\sum_{k,l=L,R} :\psi^{\dagger}_{k,\sigma}(x)\psi_{k,\sigma}(x)::\psi^{\dagger}_{l,-\sigma}(x)\psi_{l,-\sigma}(x): \\ & + g\,:\psi^{\dagger}_{R,\sigma}(x)\psi_{L,\sigma}(x)\psi^{\dagger}_{L,-\sigma}(x)\psi_{R,-\sigma}(x):\Big\}, \\ & g > 0. \end{aligned} \tag{3.1}$$

Here $\psi_{L/R,\sigma}(x)$ are the left/right moving components of the electron field $\Psi_\sigma(x)$, expanded about the Fermi points $\pm k_F$, and we implicitly sum over repeated indices for spin $\sigma = \uparrow, \downarrow$. The first term in (3.1) is that of free relativistic fermions, while the second and third terms describe forward and backward electron-electron scattering, respectively. Normal ordering is carried out w.r.t. the filled Dirac sea.

The TL Hamiltonian is conveniently written on diagonal Sugawara form, using the charge and spin currents

$$\begin{aligned} j_{L/R}(x) &= \cosh\theta :\psi^{\dagger}_{L/R,\sigma}(x)\psi_{L/R,\sigma}(x): \\ &+ \sinh\theta :\psi^{\dagger}_{R/L,\sigma}(x)\psi_{R/L,\sigma}(x):, \end{aligned} \tag{3.2}$$

$$\boldsymbol{J}_{L/R}(x) = :\psi^{\dagger}_{L/R,\sigma}(x)\tfrac{1}{2}\psi_{L/R,\mu}(x):, \tag{3.3}$$

with $\tanh 2\theta = g/(v_F+g)$. Dropping a marginally irrelevant term $-(g/\pi)\boldsymbol{J}_L \cdot \boldsymbol{J}_R$, one obtains the critical bulk Hamiltonian

$$\mathcal{H}^*_{TL} = \frac{1}{2\pi}\int_{-\ell}^{\ell} dx \sum_{i=1,2} \left\{ \frac{v_c}{4} :j^i_L(x)j^i_L(x): + \frac{v_s}{3} :\boldsymbol{J}^i_L(x)\cdot \boldsymbol{J}^i_L(x): \right\}, \tag{3.4}$$

where we have confined the system to $x \in [-\ell, \ell]$, and replaced the right-moving currents with a second channel of left-moving currents. The last step involves folding the interval in half to $[0, \ell]$, and analytically continuing the left-handed currents back to the full interval. Note that the spin and charge separation in (3.4) yields two dynamically independent theories, each Lorentz invariant with a characteristic velocity, $v_c = v_F(1+2g/v_F)^{\frac{1}{2}}$ and $v_s = v_F - g$. The charge and spin currents satisfy the $U(1)$ and (level-1) $SU(2)_1$ Kac-Moody algebras, respectively, and it follows that $\mathcal{H}^*_{TL}$ is invariant under the chiral symmetry $U(1) \times U(1) \times SU(2)_1 \times SU(2)_1$.

We now couple a localized spin-$\frac{1}{2}$ impurity $\boldsymbol{S}$ to the electrons. As a warm-up let us first consider the simple case of only *forward scattering* off the impurity:

$$\mathcal{H}_F = \sum_{k=L,R} \lambda :\psi^{\dagger}_{k,\sigma}(0)\tfrac{1}{2}\boldsymbol{\sigma}_{\sigma\mu}\psi_{k,\mu}(0): \cdot \boldsymbol{S}. \tag{3.5}$$

Using the spin currents in (3.3), and replacing the right-handed one by a second left-handed current, Eq. (3.5) turns into

$$\mathcal{H}_F = \lambda_F[\boldsymbol{J}^1_L(0) + \boldsymbol{J}^2_L(0)] \cdot \boldsymbol{S}. \tag{3.6}$$

As the two currents are coupled via $\boldsymbol{S}$, $\mathcal{H}_F$ breaks the $SU(2)_1 \times SU(2)_1$ symmetry of $\mathcal{H}^*_{TL}$ down to the diagonal level-2 subalgebra $SU(2)_2$ spanned by $\boldsymbol{J}(x) = \boldsymbol{J}^1_L(x) + \boldsymbol{J}^2_L(x)$. To adopt to this fact, we use the coset construction [8] to write the spin part of the Hamiltonian as a sum of an $SU(2)_2$ Sugawara Hamiltonian and a free Majorana fermion ($\Leftrightarrow$ 2D Ising model). We can then absorb $\mathcal{H}_F$ into $\mathcal{H}^*_{TL}$ by the canonical transformation $\boldsymbol{J}(x) \to \boldsymbol{J}'(x) \equiv \boldsymbol{J}(x) + \boldsymbol{S}\delta(x)$, $\boldsymbol{J}'(x)$ being the spin current of electrons *and* impurity. The impurity thus disappears from the Hamiltonian and is replaced by a nontrivial boundary condition on $\mathcal{H}^*_{TL}$.

In CFT, a boundary condition is equivalent to a selection rule for quantum numbers of a conformal embedding. In our case, the conformal embedding is $U(1) \times U(1) \times SU(2)_2 \times$ *Ising*, with quantum numbers $Q, \Delta Q \in \mathbf{Z}$ (sum and difference of net charge in the two channels), $j = 0, \frac{1}{2}, 1$ (spin of primary states), and $\phi = \mathbf{1}, \sigma, \epsilon$ (*Ising* primary fields). Without the impurity, i.e. with a trivial boundary condition at $x = 0$, the eigenstates of $\mathcal{H}^*_{TL}$ organize into products of four conformal towers, labeled by these quantum numbers. The allowed combinations can be described by a selection rule that reproduces the spectrum of interacting fermions. *With* the impurity (nontrivial boundary condition), the spin current gets redefined as that of electrons *and* impurity: Effectively, a spin-$\frac{1}{2}$ degree of freedom is added to the $SU(2)_2$ towers, which get shifted according to the CFT fusion rule $j = 0 \to \frac{1}{2}, \frac{1}{2} \to 0$ or $1, 1 \to \frac{1}{2}$. A new spectrum emerges, with the $U(1) \times U(1)$, $SU(2)_2$, and Ising conformal towers combined differently according to a new selection rule. It describes *interacting fermions + impurity* and yields immediate information about the boundary scaling dimensions (which are really what we are after!). This follows from a well-known result by Cardy [9]: The eigenstates of a Hamiltonian $\mathcal{H}$ on a finite interval are in 1-1 correspondence to the boundary scaling dimensions of the same theory boosted to Euclidean space-time (with the time axis defining the boundary).

Having thus obtained the spectrum of boundary scaling dimensions, we identify the leading correction-to-scaling boundary operator (*LCBO*) $\mathcal{O}$, that respects all symmetries of the problem. This operator is added to $\mathcal{H}^*_{TL}$ to give an effective Hamiltonian in the scaling region of the new fixed point:

$$\mathcal{H}_{scaling} = \mathcal{H}^*_{TL} + \mu\mathcal{O}(0), \tag{3.7}$$

with μ the scaling field conjugate to $\mathcal{O}$. Unlike for ordinary critical phenomena, $\mathcal{O}$ produces the leading scaling in temperature, although it is in fact an irrelevant boundary operator, given by the first Kac-Moody descendant in the $j = 1$ conformal tower: $\boldsymbol{J}_{-1} \cdot \boldsymbol{\phi}$, of dimension $\frac{3}{2}$. This is the same operator that drives critical scaling in the two-channel Kondo effect for noninteracting electrons [10]. Specifically, the impurity contributions to the specific heat δC and spin susceptibility $\delta\chi$ are given to leading order by

$$\delta C = \frac{\mu^2 9\pi^2}{v_s^3} T \ln(\frac{1}{\tau_0 T}), \qquad \delta\chi = \frac{\mu^2 18}{v_s^3} \ln(\frac{1}{\tau_0 T}), \qquad T \to 0, \tag{3.8}$$

with τ_0 a short-time cutoff. With the known bulk response for the Tomonaga-Luttinger model, $C = \pi(v_c^{-1} + v_s^{-1})T/3$ and $\chi = 1/2\pi v_s$, we predict a Wilson ratio

$$R_W = \frac{\delta\chi/\chi}{\delta C/C} = \frac{4}{3}(1 + \frac{v_s}{v_c}). \tag{3.9}$$

For $g \to 0$ $(v_c,\ v_s \to v_F)$, this reduces to the universal number 8/3 characterizing the ordinary two-channel Kondo effect [10].

Let us now turn to the more realistic case with forward (F) *and* backward (B) scattering off the impurity. To keep things simple we assume the F and B amplitudes to be the same, and hence replace H_F by

$$\mathcal{H}_K \equiv \mathcal{H}_F + \mathcal{H}_B = \lambda \sum_{k,l=L,R} :\psi^\dagger_{k,\sigma}(0)\tfrac{1}{2}\boldsymbol{\sigma}_{\sigma\mu}\psi_{l,\mu}(0): \cdot \boldsymbol{S}. \tag{3.10}$$

The extra terms, mixing left and right, break the chiral $SU(2)$ *and* chiral $U(1)$ invariance of $\mathcal{H}^*_{TL}$, and this produces a boundary operator of dimension $\leq \frac{1}{2}$ at the forward scattering fixed point. Back scattering is thus a relevant perturbation and drives the system to a new fixed point describing *Kondo interaction* in a Luttinger liquid.

With no electron-electron interaction ($g = 0$ in (3.1)) we have a free bulk Hamiltonian $\mathcal{H}_0$ together with $\mathcal{H}_K$. Passing to a basis spanned by definite-parity fields $\psi_{\pm,\sigma}(x) = [\psi_{L,\sigma}(x) \pm \psi_{R,\sigma}(-x)]/\sqrt{2}$, $\mathcal{H}_0 + \mathcal{H}_K$ transforms into a two-channel theory, but with the impurity coupled to the electrons in only one of the channels. This renormalizes to a local Fermi liquid (like the ordinary 3D Kondo problem), with response functions scaling analytically with temperature. However, a different approach must be used for the interacting problem, since $\mathcal{H}^*_{TL}$ is non-local in this basis. Here we exploit the expectation that *any* local impurity interaction, including the Kondo interaction $\mathcal{H}_K$, can be substituted by a renormalized boundary condition on the critical bulk theory [4]. The equivalent selection rule defines a fixed point, and by demanding that any associated *LCBO* must respect the symmetries of the problem *and* correctly reproduce the non-interacting limit as $g \to 0$, the possible critical theories can be deduced. (Note that a selection rule here defines a *boundary* fixed point, and is valid for all values of the marginal bulk coupling g. Hence, given a selection rule, Fermi liquid scaling must emerge in the limit $g \to 0$.)

To have a generally applicable formalism, we replace Q and ΔQ by new quantum numbers, treating the two diagonalized $U(1)$ towers as *independent.* (In the previous case these two towers were tangled up via the labeling by Q and ΔQ, as an implicit part of the corresponding selection rule for combining conformal towers.) This allows us to formulate a general criterion for selecting operators that respect global, but not necessarily chiral, $U(1)$ invariance. Together with invariance under channel exchange ($1 \leftrightarrow 2$), this leaves only certain possibilities for the $U(1) \times U(1)$ part of the candidate *LCBOs*. The complete set of possible scaling dimensions is then obtained by also including the $SU(2)_2$ and Ising conformal towers, as allowed by symmetry.

By requiring Fermi-liquid scaling for the impurity specific heat δC and susceptibility $\delta\chi$ as $g \to 0$, we find that there are only *two* possible types of critical behavior. *Either* Fermi-liquid scaling persists for $g \neq 0$, *or* there is a non-Fermi liquid behavior given by

$$\delta C = c_1((1/K_\rho) - 1)^2 T^{(1/K_\rho)-1} + c_2 T, \tag{3.11}$$

$$\delta\chi = c_3 T^0, \tag{3.12}$$

as $T \to 0$. Here $K_\rho = (1 + 2g/v_F)^{-1/2}$ and $c_{1,2,3}$ are amplitudes depending on the scaling fields and velocities. The *LCBO* driving the anomalous scaling in (3.11) and (3.12) is given by the composite operator

$$\mathcal{O}_{LCBO} = [V^1_{2,0} \times V^2_{-2,0} + V^1_{-2,0} \times V^2_{2,0}] \times \epsilon,$$

where $V^i_{C,D}$ is a $U(1)$ primary (vertex) operator in channel i, and ϵ the Ising energy density.

The scaling in (3.11) and (3.12) agrees exactly with a conjecture by Furusaki and Nagaosa [11], in support of the non-Fermi liquid scenario. However, a simplified model (neglecting backward spin diagonal and forward spin off-diagonal Kondo scattering) suggests that in fact the other scenario (Fermi liquid) may be realized [12]! Whatever is the case, our exact CFT analysis shows that *no other fixed point theories are possible.* Note that in none of the two cases the electron-electron interaction influences $\delta\chi$: the impurity remains completely screened for $g \neq 0$.

This work is supported by NSF grants DMR-9205125 and DMR-91-120282 (P. F.), and a grant from the Swedish Natural Science Research Council (H. J.).

References

1. For a classic treatise, see A. A. Abrikosov, L. P. Gorkov and I. E. Dzyaloshinski, *Methods of Quantum Field Theory in Statistical Physics* (Dover, New York, 1963).
2. For a recent review, see H. J. Schulz, *Fermi liquids and non-Fermi liquids,* `cond-mat/9503150`, in 1994 Les Houches lecture notes, to appear.
3. For a review, see e.g. A. C. Hewson, *The Kondo Problem to Heavy Fermions* (Cambridge University Press, Cambridge, 1993).
4. I. Affleck, in *Correlation Effects in Low-Dimensional Electron Systems*, edited by A. Okiji and N. Kawakami (Springer-Verlag, Berlin Heidelberg, 1994), and references therein.
5. P. Nozières, in Proceedings of the 14th International Conference on Low-Temperature Physics, edited by M. Krusius and M. Vuorio (North-Holland, Amsterdam, 1974).
6. J. L. Cardy, Nucl. Phys. **B240**, 514 (1984); in *Infinite Lie Algebras and Conformal Invariance in Condensed Matter and Particle Physics*, edited by K. Dietz and V. Rittenberg (World Scientific, Singapore, 1986).
7. P. Fröjdh and H. Johannesson, `cond-mat/9505100`, Phys. Rev. Lett., in press (1995).
8. P. Goddard, A. Kent and D. Olive, Commun. Math. Phys. **103**, 105 (1986).
9. J. L. Cardy, J. Phys. A: Math. Gen. **17**, L385 (1984); Nucl. Phys. **B275**, 200 (1986).
10. I. Affleck and A. W. W. Ludwig, Nucl. Phys. **B360**, 641 (1991).
11. A. Furusaki and N. Nagaosa, Phys. Rev. Lett. **72**, 892 (1994).
12. A. Schiller and K. Ingersent, Phys. Rev. B **51**, 4676 (1995).

On the U(1)-Problem of QED_2

Ch. Gattringer

Max-Planck-Institut für Physik, Föhringer Ring 6, D-80805 Munich, Germany, and
Institut für theoretische Physik der Universität Graz, Universitätsplatz 5, A-8010 Graz, Austria

1. Introduction

QED_2 is a model which has many features in common with (4-dimensional) QCD. U(1) gauge fields in two dimensions allow for topologically nontrivial configurations, which are the counterparts of the instantons of 4-dimensional Yang-Mills theory. Thus the formal linear combination of such sectors to the θ-vacuum of QCD [1] can be repeated for QED_2. Furthermore, the U(1)-axial current acquires an anomaly and the corresponding pseudoscalar particle turns out to be massive as it is the case for QCD. The quoted features allow to take over the formal reasoning that leads to the formulation of the U(1)-problem [2].

The purpose of this work is to construct QED_2 with mass and flavor, and to study the announced topics (θ-vacua, U(1)-problem) *without* making use of poorly defined concepts, such as the formal superposition of topologically nontrivial sectors to the θ-vacuum, or the definition of non-gauge invariant currents (as it is done in the standard formulation of the U(1)-problem).

Clustering θ-vacua will be defined using a mathematically rigorous limit procedure, which replaces the formal instanton arguments. The properties of this vacuum state will be compared to the θ-vacuum of QCD. Using the method of bosonization, we investigate the axial U(1)-symmetry on the physical Hilbert space and discuss the status of the U(1)-problem. Finally Witten-Veneziano-type formulas [3], that relate the masses of the pseudoscalars to the topological susceptibility, will be tested.

2. The Model and Techniques Applied

The Euclidean action of the model we consider is given by
$S = S_G + S_h + S_F + S_M$. The gauge field action reads

$$S_G[A] = \int d^2x \left[\frac{1}{4}F_{\mu\nu}(x)F_{\mu\nu}(x) + \frac{1}{2}\lambda\Big(\partial_\mu A_\mu(x)\Big)^2\right]. \tag{2.1}$$

A gauge fixing term is included, that will be considered in the limit $\lambda \to \infty$, which ensures $\partial_\mu A_\mu = 0$ (transverse gauge). As usual, $F_{\mu\nu} = \partial_\mu A_\nu - \partial_\nu A_\mu$

denotes the field strength tensor. The fermion action is a sum over N flavor degrees of freedom,

$$S_F[\overline{\psi},\psi,A,h] = \sum_{f=1}^{N} \int d^2x\, \overline{\psi}^{(f)}(x)\gamma_\mu\Big(\partial_\mu - ieA_\mu(x) - i\sqrt{g}h_\mu(x)\Big)\psi^{(f)}(x)\,. \tag{2.2}$$

Since the mass term will be treated perturbatively, we denote it separately:

$$S_M[\overline{\psi},\psi] = -\sum_{f=1}^{N} m^{(f)} \int d^2x\, \chi_\Lambda(x)\, \overline{\psi}^{(f)}(x)\psi^{(f)}(x); \tag{2.3}$$

$m^{(f)}$ are the fermion masses for the various flavors. In order to prove the convergence of the mass perturbation series for more than one flavor, the fields in the mass term have to be cut off outside some finite rectangle Λ in space-time. χ_Λ denotes the characteristic function of Λ.

In addition to the gauge field, an auxiliary vector field h_μ is coupled to the fermions (compare Equation (2)). Its action is given by

$$S_h[h] = \frac{1}{2}\int d^2x h_\mu(x)\Big(\delta_{\mu\nu} - \lambda'\partial_\mu\partial_\nu\Big)h_\nu(x)\,. \tag{2.4}$$

$S_h[h]$ is simply a white noise term plus a term that makes h_μ transverse in the limit $\lambda' \to \infty$. When integrating out h_μ it generates a Thirring term for the U(N) flavor singlet current. The purpose of this Thirring term is to make the short distance singularity of $\overline{\psi}^{(f)}(x)\psi^{(f)}(x)\, \overline{\psi}^{(f)}(y)\psi^{(f)}(y)$ integrable. This expression is a typical term showing up in a power series expansion of the mass term (3). It has to be integrated over d^2xd^2y, which is possible only if an ultraviolet regulator such as the Thirring term is included.

The approach we adopted is the quantization via Euclidean functional integrals. After having expanded the exponential of the mass term of the action $\exp(-S_M[\overline{\psi},\psi])$, vacuum expectation values of operators $\mathcal{O}$ are given by

$$\langle \mathcal{O}[\overline{\psi},\psi,A,h]\rangle = \frac{1}{Z}\sum_{n=0}^{\infty}\frac{(-1)^n}{n!}\langle \mathcal{O}[\overline{\psi},\psi,A,h]\Big(S_M[\overline{\psi},\psi]\Big)^n\rangle_0\,, \tag{2.5}$$

where the expectation values of the massless model are formally defined as $\langle \mathcal{P}\rangle_0 = Z_0^{-1}\int \mathcal{D}h\mathcal{D}A\mathcal{D}\overline{\psi}\mathcal{D}\psi\; \mathcal{P}\; e^{-S_G-S_h-S_F}$. Of course also the normalization constant Z showing up in (5) has to be expanded in terms of the massless theory.

When giving a precise mathematical meaning to the so far poorly defined functional integral, one usually starts with integrating out the fermions. This gives rise to the fermion determinant, which can be constructed properly when the fermions are massless (this is the reason why the mass term is treated perturbatively). The renormalized determinant can be found in e.g.

[4] and reads (the potentials A_μ, h_μ are assumed to be transverse and to satisfy some mild regularity and falloff conditions at infinity)

$$\det_{ren}[A_\mu, h_\mu] = \exp\left((2\pi)^{-1}\|eA^T + \sqrt{g}h^T\|_2^2\right) ,$$

where the superscript T denotes the transverse part of the vector fields. Adding the exponent of this expression to the action, one ends up with the effective action for the gauge field and the auxiliary field. Both those terms are Gaussian, and the vacuum expectation values obtain a precise mathematical meaning through Gaussian functional integrals. Since the dependence of the propagators on the external fields is exponential, the functional integrals can be computed.

The vacuum expectation functional constructed for the massless model so far violates the cluster decomposition property. In particular it turns out (see e.g. [5]), that operators which are singlets under $\mathrm{U(1)}_V \times \mathrm{SU(N)}_L \times \mathrm{SU(N)}_R$, but transform nontrivially under $\mathrm{U(1)}_A$, do not cluster. This implies that the vacuum state is not unique. In order to obtain a proper vacuum, we use a limit process [5]. The original operator $\mathcal{P}$ is correlated with a test operator $\mathcal{U}_\tau(\mathcal{P})$ depending on the chiral charge $Q_5(\mathcal{P})$ of $\mathcal{P}$, and then the limit of shifting the test operator to timelike infinity is considered:

$$\langle \mathcal{P}(\{x\})\rangle_0^\theta := e^{i\theta\frac{Q_5(\mathcal{P})}{2N}} \lim_{\tau\to\infty} \langle \mathcal{U}_\tau(\mathcal{P})\, \mathcal{P}(\{x\})\rangle_0 \, . \tag{2.6}$$

The set of 'test operators' $\mathcal{U}_\tau(\mathcal{B})$ is defined by

$$\mathcal{U}_\tau(\mathcal{P}) := \begin{cases} \mathcal{N}^{(n)}(\{y\}) \prod_{i=1}^n \mathcal{O}_\mp(\{y+\hat{\tau}\}) \text{ for } Q_5(\mathcal{P}) = \pm 2nN,\ n \geq 1 , \\ 1 \text{ otherwise} , \end{cases} \tag{2.7}$$

where

$$\mathcal{O}_\pm(\{y\}) := \prod_{f=1}^N \overline{\psi}^{(f)}(y^{(f)})\, \frac{1}{2}[1 \pm \gamma_5]\, \psi^{(f)}(y^{(f)}).$$

The space-time arguments $y^{(f)}$ are arbitrary, and the left hand side of (7) does not depend on them, due to a normalizing factor $\mathcal{N}^{(n)}(\{y\})$. θ is a real parameter in the range $[0, 2\pi)$. For this decomposition prescription the following theorem was proved [5].

Theorem 1.
i) *The cluster decomposition property holds for* $\langle..\rangle_0^\theta$.
ii) *The state* $\langle..\rangle_0$ *constructed initially is recovered by averaging over* θ, *and thus it is a mixture of the pure states* $\langle..\rangle_0^\theta$.

The decomposition (6) replaces the mathematically poorly defined concept of superimposing topological sectors to the θ-vacuum [1].

In two dimensions one has at hand the powerful tool of bosonization. It will be used to construct the massive model. By evaluating a generating functional including vector currents and chiral densities, one can establish the following bosonization prescriptions:

$$J_\nu^{(I)}(x) \longleftrightarrow \begin{cases} -\frac{1}{\sqrt{\pi+gN}}\, \varepsilon_{\mu\nu}\partial_\mu \Phi^{(1)}(x) & I=1 \\ -\frac{1}{\sqrt{\pi}}\, \varepsilon_{\mu\nu}\partial_\mu \Phi^{(I)}(x) & I=2,\ \dots N \end{cases} \tag{2.8}$$

The currents are defined as

$$J_\mu^{(I)} := \sum_{f,f'=1}^{N} \overline{\psi}^{(f)}\gamma_\mu H_{ff'}^{(I)}\, \psi^{(f')}.$$

The mixing matrices $H^{(I)}$ in flavor space are chosen to be proportional to the unit matrix for $H^{(1)}$, and proportional to the generators of a Cartan subalgebra of SU(N) for $H^{(I)}$, $I = 2, 3, .., N$ (see [5] for the explicit definition). The chiral densities are bosonized by

$$\overline{\psi}^{(f)}(x)\, \frac{1}{2}[1 \pm \gamma_5]\psi^{(f)}(x) \longleftrightarrow$$

$$\frac{1}{2\pi} c^{(f)} : e^{\mp i 2\sqrt{\pi}\sqrt{\frac{\pi}{\pi+gN}}U_{1f}\Phi^{(1)}(x)} :_{M^{(1)}} \prod_{I=2}^{N} : e^{\mp i2\sqrt{\pi}U_{If}\Phi^{(I)}(x)} :_{M^{(I)}} e^{\pm i\frac{\theta}{N}} . \tag{2.9}$$

U_{If} are some real factors that are related to the entries of the matrices $H^{(I)}$. Normal ordering of the exponentials with respect to some reference mass M (see e.g. [6]) is denoted by $: .. :_M$.

The Lagrangian of the bosonic theory was obtained by bosonizing the terms of the mass expansion (5) and then summing up the perturbation series. One ends up with the following generalized Sine Gordon model:

$$\mathcal{L}_{GSG} = \frac{1}{2}\sum_{I=1}^{N} \partial_\mu \Phi^{(I)}\partial_\mu \Phi^{(I)} + \frac{1}{2}\left(\Phi^{(1)}\right)^2 \frac{e^2 N}{\pi+gN}$$
$$- \frac{1}{\pi}\sum_{f=1}^{N} m^{(f)}c^{(f)} : \cos\left(2\frac{\pi}{\sqrt{\pi+gN}}U_{1f}\Phi^{(1)} + 2\sqrt{\pi}\sum_{I=2}^{N} U_{If}\Phi^{(I)} - \frac{\theta}{N}\right) : . \tag{2.10}$$

Normal ordering of the cosine is understood in the sense of the mass-expansion and thus reduces to the normal ordering of exponentials. It has to be stressed, that this normal ordering is essential for the correct bosonization of the clustering states $\langle .. \rangle^\theta$. In particular, the normal ordered exponentials of the massless fields $\Phi^{(I)}, I \geq 2$, have to obey the neutrality condition [6], to give nonvanishing expectation values. This ensures the correct assignment of expectation values to operators with nonvanishing chiral charge. A model similar to (10), but without Thirring term, was discussed in [7], using an operator approach.

Using the bosonized version of the model we were able to prove the following theorem [5], that generalizes the N=1 proof by J. Fröhlich [6].

Theorem 2.

The mass perturbation series converges for finite Λ and $m^{(b)} < r(\Lambda)$.

Due to the massless degrees of freedom the applied techniques only allowed to establish a radius of convergence $r(\Lambda)$, which shrinks to zero when sending Λ to infinity. We believe that the proof can be made better (i.e. r becomes independent from Λ) by applying some generalized cluster expansion methods.

This concludes the construction of the model, and one can start to discuss its physical implications.

3. Lessons on the U(1)-Problem

The following three lessons formulate the physically interesting results that were obtained for QED_2, and shed light on the corresponding problems in *QCD*.

Lesson 1 : *The structure of the vacuum functional, that has been suggested within the instanton picture, is recovered from the mathematically rigorous construction given in formula* (6).

In particular only operators with chirality $2N\nu$, $\nu \in \mathbb{Z}$, have nonvanishing vacuum expectation values. This can be seen immediately from the prescription (6) for the θ-vacuum. The same has been claimed by 't Hooft for QCD [8]. Furthermore an analysis of the Lagrangian (10) shows, that as long as one of the fermion masses $m^{(f)}$ vanishes physics is independent of θ (see [1] for the case of QCD).

Lesson 2 : *The axial U(1)-symmetry is not a symmetry on the physical Hilbert space, and there is no U(1)-problem for QED_2.*

Under a U(1)-axial transformation all the chiral densities $\overline{\psi}^{(f)}(1 \pm \gamma_5)/2\psi^{(f)}$ obtain phases $\exp(\pm i2\varepsilon)$. This transformation property is not compatible with the structure of the coefficients U_{If} that show up in the bosonization prescription (9) (see [5] for the explicit proof). Thus even in the case of vanishing fermion masses $m^{(f)}$, the axial U(1)-symmetry is not realized on the physical Hilbert space, and thus the U(1)-problem is not there at all. The same could be true for QCD, since the current that is used to construct the axial symmetry is not gauge invariant.

Lesson 3 : *The masses of the particles that correspond to the vector currents obey a Witten-Veneziano type formula.*

This result was established using a semiclassical approximation of the Lagrangian (10). The approximation is necessary due to the problem with removing the cutoff Λ. Nevertheless this approximation is under good control, since for small fermion masses $m^{(f)}$ it reduces to the exact solution.

References

1. C.G. Callan, R.F. Dashen, D.J. Gross, Phys. Lett. **B63** (1976) 334; R. Jackiw, C. Rebbi, Phys. Rev. Lett. **37** (1976) 172
2. S. Weinberg, Phys. Rev. **D11** (1975) 3583
3. E. Witten, Nucl. Phys. **B160** (1979) 57; G. Veneziano, Nucl. Phys. **B159** (1979) 213; J. Smit, J.C. Vink, Nucl. Phys. **B284** (1987) 234; E.Seiler, I.O.Stamatescu, Preprint (unpublished) **MPI-PAE/PTh 10/87** (1987)
4. E. Seiler, *Gauge Theories as a Problem of Constructive Quantum Field Theory and Statistical Mechanics*; Springer Lecture notes in Physics 159, New York 1982
5. C. Gattringer, E. Seiler, Ann. Phys. **233** (1994) 97; C. Gattringer, *QED_2 and U(1)-problem* (thesis, 1995) **hep-th/9503137**; C. Gattringer, work in preparation
6. J. Fröhlich, *Quantum Sine-Gordon Equation Quantum Solitons in two Space-Time Dimensions*, in: *Renormalization Theory* (Erice 1975), G. Velo, A.S. Wightman (eds.); Reidel publishing company, Dordrecht 1976
7. L.V. Belvedere, J.A. Swieca, K.D. Rothe, B. Schroer, Nucl. Phys. **B153** (1979) 112
8. G. 't Hooft, Phys. Rev. Lett. **37** (1976) 8; G. 't Hooft, Phys. Rev. **D14** (1976) 3432

The Finite Gauge Model on the Truncated Sphere

H. Grosse[1], C. Klimčík[2], and P. Prešnajder[3]

[1] Institute for Theoretical Physics, University of Vienna, Boltzmanngasse 5, A-1090 Vienna, Austria
[2] Theory Division CERN, CH-1211 Geneva 23, Switzerland
[3] Department of Theoretical Physics, Comenius University, Mlynská dolina, SK-84215 Bratislava, Slovakia

1. Introduction

The basic notions of non-commutative geometry were developed in [1, 2], and in the form of matrix geometry in [3, 4]. The essence of this approach consists in reformulating first the geometry in terms of commutative algebras and modules of smooth functions, and then generalizing them to their non-commutative analogues.

In what follows we shall perform such a construction for the Schwinger model on S^2. Its commutative version was described in details in [5], first steps towards the non-commutative version were made in [6] and [7]. Recently, we have found a proper setting for investigating the issues connected with the chiral symmetry symmetry [8]. Here we sketch how to extend the model including the gauge degrees of freedom, thus arriving at the Schwinger model on the truncated S^2.

2. The Schwinger Model on the Standard S^2

The action of the Schwinger model on the standard sphere is given as (see [5])

$$S = \frac{1}{4\pi r^2}\int_{S^2} dx\ [\tilde{\Psi}^+ i(\tilde{D}_0 + A)\tilde{\Psi} + \frac{1}{4}\mathrm{tr}F^2]\ . \tag{2.1}$$

Here $dx = d\Omega\sqrt{g}$ is a Haar measure on S^2, and $\tilde{\Psi} = (\tilde{\Psi}_\alpha)$, $\alpha = 1, 2$, is a two-component spinor field on S^2 belonging to a bundle with a fixed topological winding number $k \in \mathbf{Z}$. The free Dirac operator $\tilde{D}_0$ corresponding to a given k is

$$\tilde{D}_0 = \frac{1}{i}\sigma^k e^\mu_k(\partial_\mu + \frac{1}{4}\omega_{\mu ij}[\sigma^i, \sigma^j] + ikA^0_\mu)\ , \tag{2.2}$$

where σ^i are Pauli matrices, e^μ_i are zweibeins, $\omega_{\mu ij}$ is spin-connection, and A^0_μ is a topological gauge field (the field on S^2 of the magnetic monopole sitting in the middle of S^2) defined separately on the upper and lower hemisphere. Furthermore, $A = \sigma^i e_i^\mu A_\mu(x)$ is the gauge field defined globally on S^2, which can be decomposed to exact and co-exact part:

$$A_\mu(x) = \partial_\mu \Phi_1(x) + \sqrt{g}\varepsilon_{\mu\nu} g^{\nu\rho}\partial_\rho \Phi_2(x) , \tag{2.3}$$

where $\Phi_1(x)$ and $\Phi_2(x)$ are real scalar fields defined globally on S^2. Finally, the field strenght is defined by the relation

$$F = \frac{1}{4} F_{\mu\nu}[\sigma^i, \sigma^j] e_i^\mu e_j^\nu = (\tilde{D}_0 + A)^2 - \mathrm{tr}(\tilde{D}_0 + A)^2 , \tag{2.4}$$

where tr denotes the trace in the space of 2×2 matrices.

It is useful to rewrite the model in terms of the more algebraic $SU(2)$-invariant formalism (see [5, 7]), where the basic variables are auxiliary two-component spinors:

$$\chi = \begin{pmatrix} \chi_1 \\ \chi_2 \end{pmatrix} = \begin{pmatrix} r^{1/2} \cos\frac{\theta}{2} e^{-\frac{i}{2}(\varphi+\psi)} \\ -r^{1/2} \cos\frac{\theta}{2} e^{\frac{i}{2}(\varphi-\psi)} \end{pmatrix} . \tag{2.5}$$

For fixed $r > 0$, $0 < \theta < \pi, -\pi < \varphi, \psi < \pi$, they parametrize the sphere S^3 in $\mathbf{C}^2$ with the radius $r^{1/2}$. The quantities $x_i = \chi^+ \sigma_i \chi$, $i = 1, 2, 3$, depend on r, θ and φ only, and for fixed r they parametrize the sphere S^2 in $\mathbf{R}^3$ with the radius r. We shall denote $x_0 = \chi^+\chi = r$.

In the $SU(2)$-invariant formalism the fields are functions of the form

$$f = f(\chi, \chi^*) = \sum a_{n_1 n_2 m_1 m_2} \chi_1^{*n_1} \chi_2^{*n_2} \chi_1^{m_1} \chi_2^{m_2} , \tag{2.6}$$

where $a_{n_1 n_2 m_1 m_2} \in \mathbf{C}$. To the linear space $\mathcal{H}_k$, $k \in \mathbf{Z}$, belong functions of the form (6), with $k = n_1 + n_2 - m_1 - m_2$ fixed. In particular, $\Phi_1(x), \Phi_2(x)$ in (3) belong to $\mathcal{H}_0$.

By $\mathcal{S}_k$, $k \in \mathbf{Z}$, we denote the linear space of spinor fields on S^2 with the topological charge k:

$$\Psi = \begin{pmatrix} \Psi_1 \\ \Psi_2 \end{pmatrix} = \begin{pmatrix} \chi_1 \\ \chi_2 \end{pmatrix} f + \begin{pmatrix} \chi_2^* \\ -\chi_1^* \end{pmatrix} g , \tag{2.7}$$

where $\Psi_\alpha(x) \in \mathcal{H}_k$, $\alpha = 1, 2$, or equivalently, $f \in \mathcal{H}_{k+1}$ and $g \in \mathcal{H}_{k-1}$. The latter form is a chiral decomposition, as we shall see below.

The multiplications in $\mathcal{S}_k$ by the matrix functions

$$\Delta = y_i \sigma_i , \ \Delta^* = y_i^* \sigma_i , \tag{2.8}$$

where $y_i = -\frac{1}{2}\chi^T \varepsilon \sigma_i \chi \in \mathcal{H}_{-2}$, $y_i^* = -\frac{1}{2}\chi^+ \varepsilon \sigma_i \chi^* \in \mathcal{H}_2$, ($T$ means transposition, and $\varepsilon = (\varepsilon_{\alpha\beta})$), map $\mathcal{S}_k$ to $\mathcal{S}_{k-2}$ and $\mathcal{S}_{k+2}$, respectively.

It can be shown that $P_+ = x_0^{-2}\Delta\Delta^*$ and $P_- = x_0^{-2}\Delta^*\Delta$ are projectors onto the subpaces of $\mathcal{S}_k$ with a given chirality ± 1. Namely, the operator $\Gamma = P_+ - P_-$ is a grading in $\mathcal{S}_k$ satisfying

$$\Gamma\Psi = \begin{pmatrix} \chi_1 \\ \chi_2 \end{pmatrix} f - \begin{pmatrix} \chi_2^* \\ -\chi_1^* \end{pmatrix} g \,. \qquad (2.9)$$

The Dirac operator in the $SU(2)$-invariant formalism maps $\mathcal{S}_k$ into $\mathcal{S}_k$ and acts in the chiral basis as follows:

$$D_0\Psi = \begin{pmatrix} \chi_1 \\ \chi_2 \end{pmatrix} (J^*g) + \begin{pmatrix} \chi_2^* \\ -\chi_1^* \end{pmatrix} (Jf) \,, \qquad (2.10)$$

where $J^*g = \varepsilon_{\alpha\beta}\chi_\alpha^*(\partial_{\chi_\beta} g)$ and $Jf = \varepsilon_{\alpha\beta}(\partial_{\chi_\alpha^*} f)\chi_\beta$. Moreover, D_0 anticommutes with the grading: $D_0\Gamma = \Gamma D_0$.

The quantities $\tilde{\Psi}(x)$, $\tilde{D}_0$ and $\Psi(x)$, D_0 are related by a singular gauge transformation defined separately on the lower and upper hemispheres (see [7]). $\Psi(x)$ and D_0 are globally defined on S^2, which is not true for $\tilde{\Psi}(x)$ and $\tilde{D}_0$.

The gauge field A we take in the form

$$\begin{aligned} A &= (J^*\Phi^*)\,\Delta + (J\Phi)\,\Delta^* \\ &= \sigma_i\,(J_i\ \mathrm{Re}\Phi + \varepsilon_{ijk}x_jJ_k\ \mathrm{Im}\Phi) \,. \end{aligned} \qquad (2.11)$$

Here $\Phi \in \mathcal{H}_0$. Identifying $\Phi_1 = \mathrm{Re}\Phi$, $\Phi_2 = \mathrm{Im}\Phi$, one can show the consistency with (3). We see that the gauge degrees of freedom are described by one complex scalar field Φ.

The action (1) of the model in question can be rewritten as

$$S = \frac{1}{4\pi r^2}\int_{S^2} dx[\Psi^+ i(D_0 + A)\Psi + \frac{1}{4}\mathrm{tr}F^2] \,, \qquad (2.12)$$

where F is defined similarly as in (4). This action:

(i) has the proper space symmetry, i.e. is invariant with respect to the rotations of the sphere generated by

$$J_i = \frac{1}{2}(\chi_\beta^*\sigma_{\alpha\beta}^{*i}\partial_{\chi_\alpha^*} + \chi_\beta\sigma_{\alpha\beta}^i\partial_{\chi_\beta}) \,,\ i = 1,2,3 \,, \qquad (2.13)$$

(ii) is invariant under local gauge transformations:

$$\begin{gathered} f \to e^{i\Lambda}f\,, g \to e^{i\Lambda}g\,, f^* \to e^{-i\Lambda}f^*\,, g^* \to e^{-i\Lambda}g^*\,, \\ \Phi \to \Phi + \Lambda\,,\ \Lambda = \Lambda^* \in \mathcal{H}_0\,, \end{gathered} \qquad (2.14)$$

(iii) is invariant under global chiral transformations:

$$f \to e^{i\alpha}f\,, f^* \to e^{i\alpha}f^*\,, g \to e^{-i\alpha}g\,, g^* \to e^{-i\alpha}g^*\,,\ \alpha \in \mathbf{R}\,. \qquad (2.15)$$

3. The Non-Commutative Case

In the non-commutative case we replace the spinor variables $\chi_\alpha, \chi_\alpha^*, \alpha = 1, 2$, by two pairs of annihilation and creation operators satisfying

$$[\chi_\alpha, \chi_\beta] = [\chi_\alpha^*, \chi_\beta^*] = 0 \ , \ [\chi_\alpha, \chi_\beta^*] = \frac{\lambda}{2}\delta_{\alpha\beta} \ \lambda > 0 \ , \tag{3.1}$$

and acting in the Fock space $\mathcal{F}$ spanned by the vectors

$$|n_1, n_2\rangle = \sqrt{\frac{(2/\lambda)^{n_1+m_2}}{n_1!n_2!}}(A_1^*)^{n_1}(A_2^*)^{n_2}|0\rangle \ , \tag{3.2}$$

where $|0\rangle$ is the vacuum defined by $\chi_1|0\rangle = \chi_2|0\rangle = 0$. By $\mathcal{F}_M$ we denote the subspace

$$\mathcal{F}_M = \{|m_1, m_2\rangle \ \in \mathcal{F} \ , m_1 + m_2 = M\} \ . \tag{3.3}$$

Furthermore, let $\mathcal{H}_k^M$ denote the linear space of the operators of the form (6), with $m_1 + m_2 \leq N$ and $k = n_1 + n_2 - m_1 - m_2$ fixed. Since $f : \mathcal{F}_N \to \mathcal{F}_M \ , N = M + k$, the operator f is represented by an $(M+1) \times (N+1)$-matrix. Similarly, the operators $J_k \ , \ k = 1, 2, 3$, map $\mathrm{Hom}(\mathcal{F}^M, \mathcal{F}^N)$ into itself, and their $su(2)$ action is derived from the $su(2)$ action on $\mathcal{F}^M$ and $\mathcal{F}^N$.

The action we now define as

$$S \ = \ \frac{1}{\lambda NM} \ \mathrm{Tr}_M[\Psi^+ i(D_0 + A)\Psi \ + \ \frac{1}{4}\mathrm{tr}F^2] \ . \tag{3.4}$$

It is useful to fix λ by the condition $\lambda^{-1} = \sqrt{\frac{M}{2}(\frac{M}{2}+1)}$ (this choice fixes the radius of the sphere to 1).

The formulas given above for the standard (commutative) model in the chiral basis can be used directly in the non-commutative case too. Obviously, 1) the model has only finite number of modes, in fact it is a matrix model, and 2) the model has all fundamental symmetries of the commuting model: (i) $SU(2)$-space symmetry, (ii) local gauge invariance, and (iii) global chiral invariance.

Models with spinors have in general a very natural supersymmetric setting, as it was shown (for $k = 0$) in [8]. As the number of degrees of freedom is finite, this leads to a non-perturbative UV-regularization . The details for the purely scalar theories were given in [9].

References

1. A. Connes, Publ. IHES **62** (1986) 257.
2. A. Connes, *Geometrie Noncommutative* (Inter Editions, Paris 1990).

3. M. Dubois-Violette, C. R. Acad. Sci. Paris **307**, Ser. I (1988) 403.
4. M. Dubois-Violette, R. Kerner and J. Madore, Journ. Math. Phys. **31** (1990) 316.
5. C. Jayewardena, Helv. Phys. Acta **61** (1988) 638.
6. H. Grosse and J. Madore, Phys. Lett. **B283** (1992) 218.
7. H. Grosse and P. Prešnajder, Lett. Math. Phys. **33** (1995) 171.
8. H. Grosse, C. Klimčík and P. Prešnajder, *Field theories on the truncated superspace* - to be published.
9. H. Grosse, C. Klimčík and P. Prešnajder, *Towards finite quantum field theory in non-commutative geometry*, CERN-TH/95-138, UWThPh-1995-19, hep-th/9505175.

Local Operators in Massive Quantum Field Theories

A. Koubek

Department of Applied Mathematics and Theoretical Physics, CB3 9EW Cambridge, UK

1. Introduction

A fundamental problem in quantum theory is to establish a connection between its local description (quantum field theory) and measurable quantities (particle masses, scattering amplitudes). In elementary particle physics one relies mostly on approximation techniques due to the non-integrable structure of the interactions. In order to gain a deeper understanding of quantum theory in general, these issues are examined for integrable systems, where one hopes to gain exact relations between the two descriptions. Though many integrable theories in four dimensions are known (for a review see e.g. [1]), much more knowledge has been obtained so far for two dimensional theories, and it is thus this class of models where my investigations focus.

In the particle picture the interaction is encoded into the scattering matrix. The asymptotic states are described by a linear superposition of free one-particle states $|Z_\epsilon(\beta)\rangle$, which are characterised by the particle species ϵ and their momentum, parametrised as $p^{(0)} = m\cosh\beta$, $p^{(1)} = m\sinh\beta$ (m denotes the mass and β the rapidity). They are related through the S-matrix by

$$\begin{aligned} &|Z_{\epsilon_1}(\beta_1)\dots Z_{\epsilon_n}(\beta_n)\rangle_{in} \\ &= S_{\epsilon_1\dots\epsilon_n}^{\epsilon'_1\dots\epsilon'_m}(\beta_1,\dots,\beta_n|\beta'_1\dots\beta'_m)\,|Z_{\epsilon'_1}(\beta'_1)\dots Z_{\epsilon'_m}(\beta'_m)\rangle_{out} \quad . \end{aligned} \tag{1.1}$$

On the other hand, the local description of a theory consists of the space of local operators $\mathcal{A} = \{\mathcal{O}_i\}$ and the set of multi-point correlation functions of them,

$$\langle 0|\mathcal{O}_1(x_1)\dots\mathcal{O}_n(x_n)|0\rangle \quad .$$

These two descriptions are linked through the Lehmann-Symanzik-Zimmermann reduction, since the particles can be obtained as asymptotic limits of the local fields. Another connection is given through the form factors. Consider an arbitrary two-point correlation function

$$G_{ij}(x) \; = < \mathcal{O}_i(x)\, \mathcal{O}_j(y) >$$

of Hermitian operators. Inserting the Identity between the two operators and expanding it into the basis of asymptotic states, it can be expressed as an infinite series over multi-particle intermediate states:

$$\langle \mathcal{O}_i(x)\, \mathcal{O}_j(y) \rangle \tag{1.2}$$

$$= \sum_{n=0}^{\infty} \int \frac{d\beta_1 \dots d\beta_n}{(2\pi)^n} < 0|\mathcal{O}_i(x)|Z_{\epsilon_1}(\beta_1), \dots, Z_{\epsilon_n}(\beta_n) >_{\text{in}}$$
$$_{\text{in}} < Z_{\epsilon_1}(\beta_1), \dots, Z_{\epsilon_n}(\beta_n)|\mathcal{O}_j(y)|0 > \quad . \tag{1.3}$$

The matrix elements

$$< 0|\mathcal{O}_i(0)|Z_{\epsilon_1}(\beta_1), \dots, Z_{\epsilon_n}(\beta_n) >= F_{\epsilon_1 \dots \epsilon_n}(\beta_1, \dots, \beta_n)$$

are called form factors, and in the following I will try to explain, how they can be used in order to establish a link between the local description and the particle picture of a theory.

2. Form Factors and the Space of Local Operators

If one considers two dimensional integrable theories, many simplifying properties occur, which allow to calculate many dynamical quantities exactly. The most remarkable fact is the factorisation of the S-matrix, which determines a general scattering process as a product of two-particle scattering amplitudes. Furthermore these two-particle S-matrices are pure phase-shifts, that is, the incoming and outgoing momenta are the same. This simplification allows to calculate the S-matrix exactly (see *e.g.* P. Kulish's lectures in these proceedings).

Also the form-factors can be determined exactly for integrable two dimensional systems. They obey a set of constraint equations, originating from fundamental principles of quantum theory, such as unitarity, analyticity, relativistic covariance, and locality [3, 4]. The important fact is that the S-matrix is the only dynamical information needed. In the following I will discuss just two examples of form-factor equations, in order to determine their overall structure, and also to explain the solution techniques.

Since the theories here considered are defined in only one space dimension, a scattering process can be viewed as to interchange two particles on the real line:

$$Z_{\epsilon_1}(\beta_1)Z_{\epsilon_2}(\beta_2) = S_{\epsilon_1\epsilon_2}(\beta_1-\beta_2)Z_{\epsilon_2}(\beta_2)Z_{\epsilon_1}(\beta_1) \quad . \tag{2.1}$$

This exchange property will lead to a constraint equation for the form-factors:

$$\begin{aligned} &F^{\mathcal{O}}_{\epsilon_1\ldots\epsilon_i\epsilon_{i+1}\ldots\epsilon_n}(\beta_1,\ldots,\beta_i,\beta_{i+1},\ldots\beta_n) \\ &= S_{\epsilon_i\epsilon_{i+1}}(\beta_i-\beta_{i+1})\,F^{\mathcal{O}}_{\epsilon_1\ldots\epsilon_{i+1}\epsilon_i\ldots\epsilon_n}(\beta_1,\ldots,\beta_{i+1},\beta_i,\ldots\beta_n) \quad . \end{aligned} \tag{2.2}$$

Another constraint equation is derived from the bound state structure of the theory under consideration. If particles Z_i, Z_j form a bound state Z_k, the corresponding two-particle scattering amplitude exhibits a pole at $\beta = iu^k_{ij}$ with the residue

$$-i\lim_{\beta\to iu^k_{ij}}(\beta - iu^k_{ij})S_{ij}(\beta) = (\Gamma^k_{ij})^2 \quad ; \tag{2.3}$$

Γ^k_{ij} is the three–particle on–shell vertex. Corresponding to this bound state, the form–factor exhibits a pole with the residue

$$\begin{aligned} -i\lim_{\beta'\to\beta}(\beta'-\beta)&F^{\mathcal{O}}_{\epsilon_1\ldots ij\ldots\epsilon_n}(\beta_1,\ldots,\beta'+i(\pi-u^j_{ik}),\beta-i(\pi-u^i_{jk}),\ldots,\beta_n) \\ &= \Gamma^k_{ij}F^{\mathcal{O}}_{\epsilon_1\ldots k\ldots\epsilon_n}(\beta_1,\ldots,\beta,\ldots,\beta_n) \quad . \end{aligned} \tag{2.4}$$

As mentioned before, (2.2) and (2.4) are only two examples of form-factor equations. Nevertheless they are two exponents of the two categories of the constraint equations:

1. Equations with fixed n (e.g. (2.2)): they involve form factors with the same number of particles on both sides of the equation.
2. Recursive equations (e.g.(2.4)): They link form factors with different particle numbers with each other - in the example above $n+1$ particle form factors to n particle form factors.

For theories with scalar particles there exists a well established solution method [3]. It consists of the ansatz

$$F_{\epsilon_1\ldots\epsilon_n}(\beta_1,\ldots,\beta_n) = Q_{\epsilon_1\ldots\epsilon_n}(e^{\beta_1},\ldots,e^{\beta_n}) \prod_{i<j}^{n} F_{\epsilon_i\epsilon_j}(\beta_i - \beta_j) \quad . \qquad (2.5)$$

The two-particle form factors $F_{\epsilon_i\epsilon_j}$ can be calculated easily from the form factor equations. The product term satisfies all equations of the first type (with fixed particle number), and also is designed in order to have the correct pole structure of an n-particle form factor.

Through this parametrisation the form-factor equations are reduced to recursive relations for the functions $Q_{\epsilon_1\ldots\epsilon_n}(e^{\beta_1},\ldots,e^{\beta_n})$. Further properties of these functions can be extracted from the form factor equations: they are homogeneous polynomials, symmetric in repeated indices, with a total degree in its arguments fixed by relativistic covariance, and the partial degree determined from the recursion relations.

This information is sufficient for simple models to obtain explicit expressions for the form factors. In all cases though it is possible to determine the space of local operators by just considering these general properties of the functions Q [5]. Each linearly independent solution of the form factor equations corresponds to an independent local operator. Therefore the space of local operators can be determined by counting the number of independent solutions of the form factor equations. This can be done due to the property of the recursion relations, that the dimension of the solution space at level n is the sum of the dimension of the solution space at level $n-1$, and of the dimension of the kernel of the recursion relation at level n. Symbolically this can be written as

$$dim(Q_n) = dim(Q_{n-1}) + dim(\mathcal{K}_n) \quad .$$

An interesting application of this counting method occurs in perturbed conformal field theories. As an example consider the Ising model. Its point of second order phase transition is described by a conformal field theory (the minimal model $\mathcal{M}_3$), and therefore the space of local operators at the critical point is determined by Virasoro irreducible representations. The critical point admits two relevant perturbations which are integrable. The perturbation with the conformal operator with conformal weight $h = \frac{1}{2}$ drives the

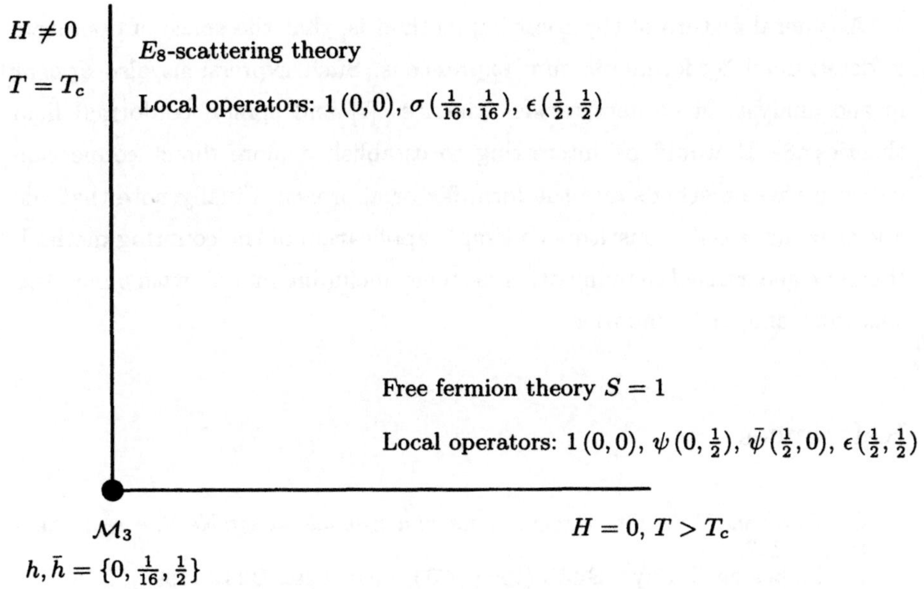

Fig. 2.1. The critical Ising model and its integrable perturbations

model into the regime $T > T_c$ and $H = 0$. It is described by a free fermion theory. The other perturbation with the operator $h = \frac{1}{16}$ corresponds to the Ising model with $H \neq 0$ and $T = T_c$, and is described by a scattering theory containing 8 scalar massive particles [6]. Virasoro symmetry is obviously broken by both perturbations, and it is therefore an interesting problem to determine the space of operators for these theories.

The situation is summarised in figure 2.1. For both perturbations it is possible to determine the space of local operators. It is given in terms of characters of the minimal model M_3. This is a quite remarkable fact, since conformal symmetry is explicitly broken by the perturbation. Further, since the thermal perturbation is described by a free fermion theory, the local operators are just the fermions $(\psi, \bar{\psi})$, the identity operator (1), and the energy density ϵ. Also the spin operator (σ) and the disorder field (μ) can be analysed by the counting method, but they are semi-local operators with respect to the asymptotic states. The magnetic perturbation breaks the Z_2 symmetry of the model, and only scalar operators appear in this perturbation, namely the identity (1), the energy density (ϵ), and the spin operator (σ).

A general feature of the counting method is, that the space of operators is determined by fermionic sum expressions. Such expressions also appear in the analysis of corner transfer matrices [7] and spinon conformal field theories [8]. It would be interesting to establish a more direct connection between these methods and the form factor approach. Finally note that the above example only constitutes a simple application of the counting method. It can be generalised to many other systems, including models with a massless spectrum and/or boundaries.

References

1. C. Devchand, V. Ogievetskii, *Four-dimensional integrable theories*, hep-th/9410147.
2. A. Koubek Nucl. Phys. **B435** (1995), 703; Phys. Lett. **B346** (1995), 275.
3. M. Karowski, P. Weisz, Nucl. Phys. **B139** (1978), 445.
4. F.A. Smirnov, *Form–Factors in Completely Integrable Models of Quantum Field Theory*, (World Scientific 1992); F. A. Smirnov, in *Introduction to Quantum Group and Integrable Massive Models of Quantum Field Theory*, Nankai Lectures on Mathematical Physics, World Scientific 1990.
5. A. Koubek Nucl. Phys. **B435** (1995), 703; A. Koubek Phys. Lett. **B346** (1995), 275.
6. A.B. Zamolodchikov, in Advanced Studies in Pure Mathematics **19** (1989), 641; Int. J. Mod. Phys. **A3** (1988), 743.
7. R. Kedem, T.R. Klassen, B.M. McCoy and E. Melzer, Phys. Lett. **B304** (1993), 263; Phys. Lett. **B307** (1993), 68.
8. P. Bouwknegt, A.W.W. Ludwig and K. Schoutens, Phys.Lett. **B338** (1994), 448; hepth/9504074.

On Anomalies and Noncommutative Geometry

E. Langmann

Theoretical Physics, Royal Institute of Technology, S-10044 Sweden

1. Introduction

In the following I discuss examples, where basic structures from noncommutative geometry (NCG) [1] naturally arise in quantum field theory. The discussion is based on work with the ultimate aim to get better mathematical understanding of quantum gauge theory models like QCD(3+1). (There is also a close connection with the representation theory of infinite dimensional Lie groups, which I shall not discuss.) The examples are restricted to external field problems, i.e. fermions coupled to non–dynamical Yang–Mills fields. This simplification makes possible a complete mathematical analysis. Though rather drastic, it already allows to study in detail several non–trivial aspects of QFT, like the structure of UV divergences in the fermion sector, and how they lead to anomalies. Moreover, it motivates the development of new efficient calculation tools, which, as I believe, should also be useful for analyzing the fully quantized theories.

The philosophy of NCG — to generalize the differential geometric machinery to situations without underlying manifold but rather algebras of Hilbert space operator — seems to be the natural way to understand the relation between the rich differential geometric structure of anomalies (anomalies as de Rham forms, characteristic classes, descent equations relating anomalies in different dimensions etc.) and their explicit QFT derivation ('dirty' calculations using Feynman diagrams, perturbation theory etc.). A general idea here is to interpret Feynman diagrams as regularized traces of certain operators on some Hilbert space, and to try to identify NCG structures based on the algebra of these operators. The regularized traces are of operators which are not trace class. In the examples discussed anomalies can be identified as regularized traces of commutators $[a,b]=ab-ba$ of certain operators a and b (I believe this is true in general). Even though such an expression is always zero, if e.g. a is trace class and b bounded, it can still be defined in more general cases and be non–zero then. Such regularized traces $\mathrm{Tr}_{reg}([a,b])$ are also closely related to the Wodzicki residue and the Dixmier trace, playing a fundamental role in NCG [1].

2. NCG and Schwinger Terms

Graded Differential Algebra (GDA). A basic object in NCG is a GDA, generalizing the notion of de Rham forms. To motivate this notion we recall the following purely algebraic characterization of de Rham forms on $\mathbb{R}^d$ (for simplicity we restrict ourselves to manifolds $\mathbb{R}^d$, and all our mappings are C_0^∞, i.e. smooth and compactly supported). One starts with the algebra $\mathcal{C}_d^{(0)} \equiv C_0^\infty(\mathbb{R}^d, \mathrm{gl}_N)$ of $N \times N$–matrix valued functions on $\mathbb{R}^d$. With d the usual exterior differentiation, one defines $\mathcal{C}_d^{(n)}$ as the space of all n–forms which are linear combinations of $\omega_n = X_0 \mathrm{d}X_1 \cdots \mathrm{d}X_n$ with $X_i \in \mathcal{C}_d^{(0)}$, and $\mathcal{C}_d = \bigoplus_{n=0}^\infty \mathcal{C}_d^{(n)}$ ($\mathcal{C}_d^{(n)} = \emptyset$ for $n > d$ here). Then d defines a mapping $\mathcal{C}_d^{(n)} \to \mathcal{C}_d^{(n+1)}$ with $\mathrm{d}^2 = 0$, and

$$\omega_n \omega_m \in \mathcal{C}_d^{(n+m)}, \quad \mathrm{d}(\omega_n \omega_m) = \mathrm{d}(\omega_n)\omega_m + (-)^n \omega_n \mathrm{d}\omega_m \tag{2.1}$$

$\forall \omega_n \in \mathcal{C}_d^{(n)}$, $\omega_m \in \mathcal{C}_d^{(m)}$, $n, m \in \mathbb{N}_0$. Moreover, there exists a linear map $\int$ — integration of de Rham forms —

$$\int \omega_n = \begin{cases} \int_{\mathbb{R}^d} \mathrm{tr}_N(\omega_n) & \text{for } n = d \\ 0 & \text{otherwise} \end{cases}, \tag{2.2}$$

(tr_N is the usual trace of $N \times N$–matrices), and Stokes' theorem holds, $\int \mathrm{d}\omega = 0$ for all $\omega \in \mathcal{C}_d$. Such a triple $(\mathcal{C}_d, \mathrm{d}, \int)$ is called a GDA.

An important example for a GDA based on algebras of Hilbert space operators is as follows. Consider a separable Hilbert space $\mathcal{H}$, which is decomposed in two orthogonal subspaces, $\mathcal{H} = \mathcal{H}_+ \oplus \mathcal{H}_-$. The operator ε, which is ± 1 on $\mathcal{H}_\pm$, is a grading operator, $\varepsilon^* = \varepsilon^{-1} = \varepsilon$, and $\mathcal{H}_\pm = \frac{1}{2}(1 \pm \varepsilon)\mathcal{H}$ ($*$ is the Hilbert space adjoint). Denoting B and B_1 the bounded and trace class operators on $\mathcal{H}$, respectively, and $B_{2p} = \{a \in B | (a^*a)^p \in B_1\}$ (these are the so–called Schatten classes), one defines the algebras

$$g_p \equiv \{u \in B | [\varepsilon, u] \in B_{2p}\} \tag{2.3}$$

for $2p$ a positive integer. Then

$$\hat{\omega}_n = (i)^n u_0 [\varepsilon, u_1] \cdots [\varepsilon, u_n] \quad \equiv u_0 \hat{\mathrm{d}} u_1 \cdots \hat{\mathrm{d}} u_n \quad \forall u_i \in g_p, n = 0, 1, \ldots, \tag{2.4}$$

can be regarded as generalized differential forms. Indeed, denoting by $\hat{\mathcal{C}}_p^{(n)}$ the space of all linear combinations of n–forms (2.4) ($\hat{\mathcal{C}}_p^{(0)} = g_p$),

$$\hat{\mathrm{d}}\hat{\omega}_n = \mathrm{i}(\varepsilon \hat{\omega}_n - (-)^n \hat{\omega}_n \varepsilon) \tag{2.5}$$

defines a mapping $\hat{\mathcal{C}}_p^{(n)} \to \hat{\mathcal{C}}_p^{(n+1)}$ such that $\hat{\mathrm{d}}^2 = 0$. $\hat{\mathrm{d}}$ can therefore can be regarded as exterior differentiation. One can easily check that the relations (2.1) hold. An integration $\hat{\int}$ can be defined as

$$\hat{\int} \hat{\omega}_n = \begin{cases} \mathrm{Tr}_C \left(\Gamma^{2p} \hat{\omega}_n \right) & \text{for } n = 2p-1 \\ 0 & \text{otherwise} \end{cases} \tag{2.6}$$

($\Gamma^{2p} = \Gamma / 1$ for $2p$ odd / even), where Γ is a grading operator on $\mathcal{H}$ such that $\varepsilon\Gamma = -\Gamma\varepsilon$, and

$$\mathrm{Tr}_C (a) \equiv \frac{1}{2} \mathrm{Tr} (a + \varepsilon a \varepsilon) \tag{2.7}$$

is a conditional Hilbert space trace. Stokes' theorem holds here due to cyclicity of trace.

Then $(\hat{\mathcal{C}}_p, \hat{\mathrm{d}}, \hat{\int})$ with $\hat{\mathcal{C}}_p = \bigoplus_{n=0}^{\infty} \hat{\mathcal{C}}_p^{(n)}$ is a GDA. As discussed below, it actually is a natural generalization of the de Rham complex $(\mathcal{C}_d, \mathrm{d}, \int)$, if $2p = d + 1$.

Quasi–free Second Quantization (QFSQ). The abstract mathematical framework, which has been found useful for studying external field problems of fermions, is very much in the spirit of NCG, even though historically it has been developed independently. (For a more detailed discussion see e.g. [2]; this approach mainly uses methods from functional analysis [3]. For an alternative approach based on differential geometric methods see [4]. The generalization of the latter from g_1 to $g_{p\geq 1}$ was first given in [5]. My discussion of this is based on [6].)

In an external field problem, the starting point is a 1-particle description of the fermions, and the aim is to 'second quantize', i.e. to find the corresponding QFT description. One has a Hilbert space $\mathcal{H}$ describing the possible 1-particle states, a Hamiltonian H, and other observables which are given by self–adjoint operators on $\mathcal{H}$. Note in the following that, even though in most applications $\mathcal{H}$ is a L^2–space over some space manifold, no reference to this manifold or the explicit form of the observables is made, and this makes this framework very general and flexible.

The Hamiltonian H naturally defines a splitting of $\mathcal{H}$ in the subspaces of positive and negative energy states, $\mathcal{H} = \mathcal{H}_+ \oplus \mathcal{H}_-$. The corresponding grading operator ε with $\varepsilon \mathcal{H}_\pm = \pm \mathcal{H}_\pm$ can be written as sign(H) (using the spectral theorem of self–adjoint operators, where sign$(x) = 1$ (-1) for $x \geq 0$ $(x < 0)$).

Given these data one can construct the corresponding QFT model. The fermions field algebra is defined as C^*–algebra generated by the field operators $\psi^*(f)$ linear in $f \in \mathcal{H}$, and $\psi(f) = \psi^*(f)^*$, obeying the CAR $\left(\psi^*(f) + \psi(g)\right)^2 = (f, g)$ (inner product in $\mathcal{H}$). The physical representation of this algebra is then on the fermion Fock space $\mathcal{F}$ over $\mathcal{H}$ and is uniquely determined by the 'vacuum' Ω, such that $\psi(f_+)\Omega = \psi^*(f_-)\Omega = 0$ for all $f_\pm \in \mathcal{H}_\pm$, which corresponds to the Dirac sea.

The aim then is to 'second quantize' observables and construct for operators u on $\mathcal{H}$ the corresponding multiparticle observables, i.e. operators $\mathrm{d}\Gamma(u)$ on $\mathcal{F}$ such that $[\mathrm{d}\Gamma(u), \psi^*(f)] = \psi^*(uf)$. One then finds, that it is not always possible to construct such an *operator* $\mathrm{d}\Gamma(u)$, but only if u is in g_1 introduced above. Thus the Schatten ideal conditions of NCG naturally appear here. The $\mathrm{d}\Gamma(u)$ for $u \in g_1$ form an algebra of operators on $\mathcal{F}$, and one has relations

$$[\mathrm{d}\Gamma(u), \mathrm{d}\Gamma(v)] = \mathrm{d}\Gamma([u,v]) + \hat{c}_1(u,v), \tag{2.8}$$

where

$$\hat{c}_1(u,v) = \frac{1}{2}\mathrm{Tr}_C\left(u[\varepsilon, v]\right), \tag{2.9}$$

is a term arising from the regularization (normal ordering) required in this construction. It is called Schwinger term in the physics literature and is a non–trivial 2–cocycle of the Lie algebra g_1.

As discussed below, this general framework is sufficient only for QFT in 1+1 dimensions, since there the observables of interest actually are in g_1. In higher dimensions the interesting observables u are only in $g_{p\geq 2}$. Then $\mathrm{d}\Gamma(u)$ cannot be defined as operator. It is, however, still possible to define it as *sesquilinear form.* (Recall that for a s.l.f. A, only transition amplitudes (f, Ag) are defined for f, g in some dense set of the Hilbert space.) To obtain a suitable generalization of (2.8), one has to consider also splittings of $\mathcal{H}$ given by other grading operators F, namely those for which $F-\varepsilon$ is in g_p. We denote this set as Gr_p. This is because the unitary operators $U = \exp(\mathrm{i}u) \in g_p$, generated by self–adjoint operators $u \in g_p$, act as transformations changing ε to $F = U^{-1}\varepsilon U$, which are in Gr_p, $F-\varepsilon = U^{-1}[\varepsilon, U]$. Infinitesimally this action is described by the Lie derivative $\hat{\mathcal{L}}_u f(F) \equiv -idf(\exp(-itu)F\exp(itu))/\,dt|_{t=0}$ on quantities f depending on F. The construction of $\mathrm{d}\Gamma(u;F)$ can then be done by an additional 'wave function renormalization'. One obtains an algebra of operators $G(u,F) = \hat{\mathcal{L}}_u + \mathrm{d}\Gamma(u;F)$ similar to (2.8), with a Schwinger term $\hat{c}_{2p-1}(u,v;F)$ depending on $F \in Gr_p$, e.g. for $p=2$ [5],

$$\hat{c}_3(u,v;F) = -\frac{1}{8}\mathrm{Tr}_C\left((F-\varepsilon)[[\varepsilon,u],[\varepsilon,v]]\right). \tag{2.10}$$

The case $p=1$ is special, since one can choose $\mathrm{d}\Gamma(u;F)$ and $\hat{c}_1$ independent of F and therefore can forget about Gr_1.

Gauss law anomalies. We now describe, how the abstract framework of quasi–free second quantization above is used to derive the anomalous commutators of Gauss law generators for chiral QCD, i.e. chiral fermions coupled to a Yang–Mills field. Our setting is YM theory on $\mathbb{R}^d$ with structure group $\mathrm{SU}(N)$ represented by $N \times N$–matrices (for simplicity we do not distinguish $\mathrm{SU}(N)$ from its representation, i.e. we assume $\mathrm{SU}(N) \subset \mathrm{gl}_N$). The space dimension is $d = 1, 3, 5\ldots$. We denote by $\mathcal{A}_d$ the set of all YM field configurations, i.e. 1-forms $A = \sum_{i=1}^d A_i \mathrm{d}x^i$ with $A_i \in C_0^\infty(R^d, \mathrm{su}(N))$ ($\mathrm{su}(N) \subset \mathrm{gl}_N$ the Lie algebra of $\mathrm{SU}(N)$). The gauge group is $C_0^\infty(\mathbb{R}^d, \mathrm{SU}(N))$ and acts on $\mathcal{A}$ as $A \mapsto U \circ A \equiv U^{-1}AU - \mathrm{i}U^{-1}\mathrm{d}U$. Its Lie algebra is $C_0^\infty(\mathbb{R}^d, \mathrm{su}(N))$, acting on functionals f of A by the Lie derivative, $\mathcal{L}_X f(A) \equiv -idf(\,\mathrm{e}^{\mathrm{i}tX} \circ A)/dt|_{t=0}$.

We now consider chiral fermions coupled to external Yang–Mills fields. Cohomological arguments show that for odd dimensions d there are 2–cocycles $c_d(X,Y;A)$, e.g.

$$c_1(X,Y) = \frac{1}{2\pi}\int_{\mathbb{R}^1} \mathrm{tr}_N\,(X\mathrm{d}Y) \tag{2.11}$$

for $d = 1$ [4], and

$$c_3(X,Y;A) = \frac{\mathrm{i}}{24\pi^2}\int_{\mathbb{R}^3} \mathrm{tr}_N\,(A[\mathrm{d}X,\mathrm{d}Y]) \tag{2.12}$$

for $d = 3$ [7]. It has been suggested on cohomological grounds, that these 2–cocycles should arise as Schwinger terms commutators of the Gauss' law generators of chiral QCD [7]. This can be shown using the general formalism of QFSQ described above. (For a different solution to this problem for $d = 3$ see [8].)

The starting point is the 1-particle description of chiral fermions. The states at fixed time are described by the Hilbert space $h = L^2(\mathbb{R}^d) \otimes \mathbb{C}^\nu_{spin} \otimes \mathbb{C}^N_{color}$, where $\nu = 2^{(d-1)/2}$ is the number of spin indices. For a given YM configuration A, the 1-particle Hamiltonian is $\mathrm{d}_A = \sum_{i=1}^d \gamma^i\,(-\mathrm{i}\partial_i + A_i)$, with γ^i the usual γ–matrices acting on $\mathbb{C}^\nu_{spin}$ and obeying $\gamma^i\gamma^j + \gamma^j\gamma^i = 2\delta^{ij}$. This naturally defines a self–adjoint operator on h for all $A \in \mathcal{A}_d$, and so do all $X \in C_0^\infty(\mathbb{R}^d;\mathrm{gl}_N)$ (containing $C_0^\infty(\mathbb{R}^d,\mathrm{SU}(N))$ and $C_0^\infty(\mathbb{R}^d,\mathrm{su}(N))$; we recall that every $X \in C_0^\infty(\mathbb{R}^d,\mathrm{gl}_N)$ defines a bounded operator on h, $(Xf)(x) = X(x)f(x)$ for all $f \in h$, which we denote by the same symbol).

The essential property now is that for h and $\varepsilon = \mathrm{sign}(\mathrm{d}_0)$ above, there are natural embeddings of $C_0^\infty(\mathbb{R}^d,\mathrm{gl}_N)$ in g_p, and of $\mathcal{A}_d$ in Gr_p:

$$X \in C_0^\infty(\mathbb{R}^d,\mathrm{gl}_N),\ A \in \mathcal{A}_d \Rightarrow X \in g_p,\ F_A \equiv \mathrm{sign}(\mathrm{d}_A) \in Gr_p \text{ for } 2p = d+1. \tag{2.13}$$

Thus QFSQ gives by restriction the algebra of Gauss law generators of chiral QCD with Schwinger terms $\hat{c}_d(X,Y;F_A)$, and the question of whether the Schwinger terms (2.11) and (2.12) arise in chiral QCD reduces to the question, whether the 2–cocycles (2.11) and (2.9) (for $d = 1$) and (2.12) and (2.10) (for $d = 3$) are cohomologous. This is very nontrivial, since the abstract Schwinger terms $\hat{c}_d$ are given by highly non–local expressions, whereas the Schwinger terms c_d are local integrals of de Rham forms. For $d = 1$ it has been known to be true since quite some time [2], and for $d = 3$ it was shown by explicit calculation recently [9]. However, given our discussion of graded differential complexes above, this result becomes very plausible, since $\hat{c}_d$ is (up to a constant) just the noncommutative generalization of c_d. For $d = 1$ this is obvious, and for $d = 3$ it follows, if we regard $F - \varepsilon$ as the noncommutative generalization of A. The latter, however, is very natural: If the YM field is a pure gauge, $A = -\mathrm{i}U^{-1}\mathrm{d}U$, we have $F_A - \varepsilon = U^{-1}\varepsilon U - \varepsilon = -\mathrm{i}U^{-1}\hat{\mathrm{d}}U$. Equivalence of Schwinger terms thus is just a special case of a general embedding theorem of the de Rham complex $(\hat{\mathcal{C}}_p, \hat{\mathrm{d}}, \hat{\int})$ in the complex $(\mathcal{C}_d, \mathrm{d}, \int)$, especially that the noncommutative integral $\hat{\int}$ generalizes integration of de Rham forms $\int$,

$$(\mathrm{i})^d \mathrm{Tr}_C\left(\Gamma X_0[\varepsilon, X_1]\cdots[\varepsilon, X_d]\right) = c_d \int_{\mathbb{R}} \mathrm{tr}\left(X_0 \mathrm{d}X_1 \cdots \mathrm{d}X_d\right) \qquad (2.14)$$

$\forall X_i \in C_0^\infty(\mathbb{R}^d; \mathrm{gl}_N)$, with some constants c_d, and $\Gamma = 1$ for d odd, and $\Gamma = \gamma_{d+1}$ for d even. A simple proof of this (motivated by the calculation in [9]) was recently given in [10] (one gets $c_d = (2\mathrm{i})^{[d/2]} 2\pi^{d/2}/d(2\pi)^d \Gamma(d/2)$). For $d = 1$ this proves that the Schwinger terms are in fact identical (as $c_1 = 1/\pi$), for $d = 3$ it proves identity for pure YM field configurations which are pure gauges (as $c_3 = \mathrm{i}/3\pi^2$).

3. Efficient Anomaly Calculations

In [9] we found that the calculus of pseudodifferential operators is an extremely powerful calculational tool in anomaly calculations: it gives a simple way of calculating regularized traces Tr_{reg} of commutators of operators (these methods were used earlier in a similar context in [8]). The same tool was essential in [10]. Recently we used this very tool for a very short QFT derivation of the axial anomalies for all even dimensions [11]. The strategy was as usual, to write the effective action for fermions in an external YM field as Tr_{reg} of a Dirac operator d_A, and to calculate its variation under axial gauge transformations. Using only elementary rules for manipulating Hilbert space operators, we could write the latter as a sum of terms $\mathrm{Tr}_{reg}([a, b])$. To my opinion, this very short derivation makes mathematically precise traditional perturbative calculations of the anomaly and is very much in the spirit of NCG.

4. Final Remarks

In my discussion in **2.** I tried to convince the reader, that the NCG viewpoint (i.e. generalizing from the differential complex $(\mathcal{C}_d, \mathrm{d}, \int)$ to $(\hat{\mathcal{C}}_p, \hat{\mathrm{d}}, \hat{\int})$) is very useful for general QFT calculations concentrating on the essential QFT aspects, namely the nature of the UV divergences characteristic for a specific dimension. It naturally leads to generalizations of Schwinger terms to the noncommutative setting. This and the anomaly calculation discussed in **3.** suggest, that all YM fermion anomalies should have noncommutative generalizations. I recently found that this is indeed the case. In fact, there is a noncommutative generalization of the whole tower of descent equations [12]. It involves a generalization of the noncommutative integration formula (2.14) to $(d-n)$–dimensional submanifolds of $\mathbb{R}^d$, such that Stokes' theorem holds (details will appear elsewhere). To my opinion, this noncommutative version of the descent equations provides a nice explanation of how the rich geometric structure of anomalies arises from QFT: it is present already on the level of Hilbert space operators entering the Feynman diagrams.

Acknowledgment
I would like to thank J. Mickelsson for a pleasant collaboration, and H. Grosse and S. Rajeev for helpful discussions.

References

1. A. Connes, *Noncommutative Geometry*, Academic Press (1994)
2. L.-E. Lundberg, Commun. Math. Phys. 50, 103 (1976); S. N. M. Ruijsenaars, J. Math. Phys. 18, 517 (1977); A. L. Carey and S. N. M Ruijsenaars., Acta Appl. Mat. **10**, 1 (1987); H. Grosse and E. Langmann, Int. Jour. of Mod. Phys. **21**, 5045 (1992); E. Langmann, Jour. Math. Phys. **35**, 96 (1994)
3. R. Reed and B. Simon, *Methods of Modern Mathematical Physics I. Functional Analysis*, Academic Press, New York (1968)
4. A. Pressley and G. Segal, *Loop Groups*, Oxford Math. Monographs, Oxford (1986); J. Mickelsson, *Current Algebras and Groups*, Plenum Monographs in Nonlinear Physics, Plenum Press (1989)
5. J. Mickelsson and S. G. Rajeev, Commun. Math. Phys. 116, 400 (1988); J. Mickelsson, Commun. Math. Phys. 117, 261 (1988)
6. E. Langmann, *On Schwinger terms in* $(3+1)$*-dimensions*, Proc. of the "XXVII Karpacs Winter School of Theoretical Physics" (Karpacz 1991), World Scientific, Singapore (1991); E. Langmann, Commun. Math. Phys. 162, 1 (1994)
7. L. D. Faddeev, Phys. Lett. 145B, 81 (1984); L. D. Faddeev and S. L. Shatiashvili, Theor. Math. Phys. **60** (1984) 206; J. Mickelsson, Commun. Math. Phys. 97, 361 (1985)
8. J. Mickelsson, Lett. Math. Phys. 28, 97 (1993); J. Mickelsson, *Wodzicki residue and anomalies of current algebras*, Proc. of "Integrable Models and Strings" (Helsinki 1993), Lecture Notes in Physics 436, Springer (1994)
9. E. Langmann and J. Mickelsson, Phys. Lett. B **338**, 241 (1994)
10. E. Langmann, Jour. Math. Phys. (in press)
11. E. Langmann and J. Mickelsson, Lett. Math. Phys. (in press)
12. B. Zumino, Nucl. Phys. **253**, 477 (1985)

Quantum Deformations of Space-Time Symmetries with Fundamental Mass Parameter

J. Lukierski

Institute for Theoretical Physics, University of Wrocław, pl. Maksa Borna 9, 50-204 Wrocław

1. Introduction

The quantum deformations of space-time symmetries lead to the replacing of classical Poincaré algebra as well as classical Poincaré group by their quantum conterparts described by noncommutative and noncocommutative Hopf algebras [1–3]. We recall that the fourmomenta P_μ as well as the translation parameters x_μ are endowed with nontrivial scale transformations:

$$P_\mu \longrightarrow \lambda^{-1} P_\mu \,, \qquad x_\mu \longrightarrow \lambda x_\mu \,. \tag{1}$$

One can distinguish the following two different types of quantum deformations:

1. The q-deformations with scale-invariant deformation parameter q ($q = 1$ corresponds to the classical symmetry). In such a case the nonlinearity appearing in quantum-deformed Poincaré algebra is given by the function of dimensionless Lorentz generators.
2. The κ-deformations, with the dimensionfull deformation parameter, usually described by a masslike parameter κ ($\kappa = \infty$ corresponds to the classical symmetry). In this case the generic nonlinearity of quantum-deformed Poincaré algebra is provided by functions of the fourmomentum generators.

In this talk we shall describe the κ-deformations.

2. κ-Deformed $D = 4$ Poincaré Algebra

The κ-deformation of $D = 4$ Poincaré algebra was introduced firstly in [4, 5] (see also [6–8]) and is one of the weakest possible deformations of the classical Poincaré algebra. In particular recently there was obtained for $D = 4$ κ-Poincaré algebra the so-called bicrossproduct basis with the following properties [8]:

1. The only deformed classical κ-Poincaré algebra relations are the commutators between boosts N_i and the three-momenta P_i ($i = 1, 2, 3$):

$$[N_i, P_j] = \delta_{ij} \left\{ \frac{\kappa}{2} \left(1 - e^{-\frac{2P_0}{\kappa}}\right) + \frac{1}{2\kappa} \mathbf{P}^2 \right\}. \tag{1}$$

Remaining relations of classical Poincaré algebra are not modified under κ-deformation.

2. The coproducts $\Delta(\,\cdot\,)$, defining the realizations of κ-deformed Poincaré algebra on the tensor product of two representation spaces, have the following form[1]:

$$\begin{aligned} \Delta(N_i) &= N_i \otimes 1 + e^{-\frac{P_0}{\kappa}} \otimes N_i + \frac{\epsilon_{ijk}}{\kappa} P_j \otimes M_k \,, \\ \Delta(P_i) &= P_i \otimes 1 + e^{-\frac{P_0}{\kappa}} \otimes P_i \,. \end{aligned} \tag{2}$$

Remaining coproducts, for the generators M_i (spatial rotations) and P_0 (energy generator), are classical, i.e.

$$\Delta(P_0) = P_0 \otimes 1 + 1 \otimes P_0 \,, \qquad \Delta(M_i) = M_i \otimes 1 + 1 \otimes M_i \,. \tag{3}$$

The relations (3) tell us that the energy and nonrelativistic angular momenta of two systems described by the tensor product representation of κ-Poincaré algebra are additive as in the classical theory.

3. The Hopf algebra is the bialgebra with antipode. Antipode generalizes the notion of the inverse element for classical groups and classical Lie algebras. We obtain that

$$\begin{aligned} S(N_i) &= -e^{\frac{P_0}{\kappa}} N_i + \tfrac{1}{\kappa} \epsilon_{ijk} e^{\frac{P_0}{\kappa}} P_j M_k \,, \\ S(P_i) &= -e^{\frac{P_0}{\kappa}} P_i \,. \end{aligned} \tag{4}$$

The explicite form (1)-(2) of the deformed relations permits to describe the κ-deformed Poincaré algebra $\mathcal{U}_\kappa(\mathcal{P}_4)$ as the bicrossproduct Hopf algebra, which can be denoted as follows [8]:

$$\mathcal{U}_\kappa(\mathcal{P}_4) = \mathcal{U}(SO(1;3)) \mathcal{U}_\kappa(T_4) \,. \tag{5}$$

The general theory of bicrossproduct Hopf algebras was described by Majid [9] and provides the general scheme for the quantum deformations of semidirect products of Lie algebras. The κ-Poincaré algebra became one of two nontrivial thoroughly studied explicite examples ilustrating the general theory of bicrossproduct Hopf algebras[2].

---

[1] We recall here that the coproduct $\Delta : A \longrightarrow A \otimes A$ is a homomorphism of the algebra A.

[2] The second example is given in [10] and can be described as the κ-deformation of $D = 3$ Euclidean group of motion.

3. κ-Deformed Poincaré Group

In classical Lie algebra theory the corresponding classical Lie group can be obtained by the exponential map. This map does not have a direct counterpart for quantum Lie algebras, but is replaced by the dualization procedure. We recall that two Hopf algebras H, $\widetilde{H}$ are dual to each other, if (se e.g. [11])

1. there exists a nondegenerate bilinear form $\langle \widetilde{a}, a \rangle$ $(a \in H, \widetilde{a} \in \widetilde{H})$, describing the mapping $\widetilde{H} \otimes H \longrightarrow C$;
2. the form $\langle \cdot \, , \, \cdot \rangle$ has the following properties:

$$\langle \widetilde{a}, bc \rangle = \left\langle \widetilde{\Delta}(\widetilde{a}), b \otimes c \right\rangle , \qquad \left\langle \widetilde{a} \cdot \widetilde{b}, c \right\rangle = \left\langle \widetilde{a} \otimes \widetilde{b}, \Delta(c) \right\rangle , \tag{1a}$$

$$\left\langle \widetilde{S}(\widetilde{a}), b \right\rangle = \langle \widetilde{a}, S(b) \rangle \, . \tag{1b}$$

The formulae (1a) express the property that the multiplication and comultiplication are related by the duality operation, i.e. from the algebra structure of H follows the coalgebra structure of $\widetilde{H}$, and analogously the coalgebra relations in H determine the multiplication sector of the Hopf algebra $\widetilde{H}$. It appears that the bicrossproduct form (5) of κ-deformed Poincaré algebra is very useful in deriving the Hopf algebra structure of the dual κ-deformed Poincaré group (see [13]). Denoting the generators of κ-Poincaré group by $(\Lambda^{\mu}{}_{\nu}, v^{\rho})$, one obtains the following relations for κ-Poincaré group [12, 8]:

$$[\Lambda^{\mu}{}_{\nu}, \Lambda^{\rho}{}_{\tau}] = 0 \, , \tag{2a}$$

$$[\Lambda^{\mu}{}_{\nu}, v^{\rho}] = -\frac{i}{\kappa}\{(\Lambda^{\mu}{}_{0} - \delta^{\mu}{}_{0})\Lambda^{\rho}{}_{\mu} + (\Lambda^{0}{}_{\nu} - \delta^{\nu}{}_{0})\eta^{\nu\rho}\} \, , \tag{2b}$$

$$[v^{\mu}, v^{\nu}] = \frac{i}{\kappa}(\delta^{\mu}{}_{0}v^{\nu} - \delta^{\nu}{}_{0}v^{\mu}) \, , \tag{2c}$$

with the additional constraints $\Lambda^{\mu}{}_{\nu} \cdot \Lambda^{\nu}{}_{\rho} = \delta^{\mu}{}_{\rho}$. The coproducts take the form

$$\Delta(\Lambda^{\mu}{}_{\nu}) = \Lambda^{\mu}{}_{\rho} \otimes \Lambda^{\rho}{}_{\nu} \, , \tag{4a}$$

$$\Delta(v^{\mu}) = \Lambda^{\mu}{}_{\rho} \otimes v^{\rho} + v^{\mu} \otimes 1 \, . \tag{4b}$$

Denoting $\Lambda^{\mu}{}_{\rho} \otimes 1 \equiv \Lambda^{\mu}{}_{\rho}$, $1 \otimes v^{\mu} \equiv x_{\mu}$, and $v^{\mu} \otimes 1 \equiv a^{\mu}$, one obtains from (4b) the classical Poincaré transformations of the noncommutative coordinates $x^{\mu} = (x^0, x^i)$, satisfying due to (2c) the commutation relations

$$[x^i, x^j] = 0 \, , \qquad [x^0, x^i] = \frac{1}{\kappa}x^i \, . \tag{4c}$$

The relations (4c) describe the algebra of κ-Minkowski coordinates. Because the κ-Poincaré algebra and κ-Poincaré group are dual to each other, there is a natural action of the κ-Poincaré algebra on x_{μ} given by the formula

$$a \triangleright x^{\mu} = \sum \left\langle a_{(1)}, x^{\mu} \right\rangle a_{(2)} , \tag{4d}$$

where $a \in \mathcal{U}_{\kappa}(\mathcal{P}_4)$ and $\Delta(a) = \sum a_{(1)} \otimes a_{(2)}$. The formula (4d) for all ten generators of $a \in \mathcal{U}_{\kappa}(\mathcal{P}_4)$ was given explicitly firstly in [8].

4. κ-Deformed $D = 4$ Supersymmetry

During last two years also the κ-deformed $D = 4$ Poincaré superalgebra [14] as well as the κ-deformed Poincaré supergroup [15] have been obtained. The basic supersymmetry relations in so-called standard basis take the form [14] $(A, B = 1, \ldots, 4;\ \overline{Q} = Q^T \gamma_0)$:

$$\{\overline{Q}_A, Q_B\} = (\gamma_0)_{AB} \cdot 2\kappa \sinh \frac{P_0}{2\kappa} - (\gamma_i)_{AB} P_i \,. \tag{4a}$$

The κ-deformed coproduct for the supercharges looks as follows:

$$\Delta(Q_A) = Q_A \otimes e^{\frac{P_0}{4\kappa}} + e^{-\frac{P_0}{4\kappa}} \otimes e^{-\frac{P_0}{4\kappa}} \,. \tag{4b}$$

There were obtained several interesting results in the framework of κ-deformed $D = 4$ supersymmetries, i.e.

1. the supersymmetric extension of the κ-deformed Minkowski space (4c) [15],
2. the bicrossproduct form of κ-deformed $D = 4$ Poincaré superalgebra [16] as well as of the $D = 4$ Poincaré supergroup [17].

It appears that the κ-deformed Poincaré algebra can be obtained from the bosonic sector of κ-deformed Poincaré superalgebra only after putting $Q = Q_{\dot{\alpha}} = 0$.

5. Final Remarks

The application of the above presented formalism to physical models can follow the following two lines:

1. One can consider the κ-deformation only as the regularization of "physical" standard relativistic theories. In standard regularization procedure the relativistic symmetries remain unmodified. The quantum group regularization introduces the modification of theory accompanied by simultaneous modification of the symmetries, in such a way that the theory remains symmetry-invariant. Unfortunately such a regularization procedure requires the knowledge of κ-Poincaré invariant interaction vertices. At present only the κ deformation of free fields (e.g. Klein-Gordon field [5], Dirac field [7, 18]) has been succesfully studied.
2. One can consider the deformation parameter κ as introducing geometrically the third fundamental constant in physics (besides light velocity c and the Planck constant $\hbar$) equal to the Planck mass ($\kappa \sim 10^{13}$ GeV). In such a case the present relativistic physics (classical as well as quantum) is an approximation at the distances greater than the Planck length ($\sim 10^{-33}$ cm) of more exact, κ-relativistic theories. Some preliminary calculations of effects related with finite value of κ were even tested experimentally (see e.g. [19], where the lower bound for κ ($\kappa > 10^{12}$ GeV) has been estimated).

References

1. W.G. Drinfeld, in Proc. Intern. Congress of Mathematics (Berkeley, USA, 1986), Vol. 1 (Academic Press, NY, 1987), p. 798.
2. L. Woronowicz, *Comm. Math. Phys.* **111**, 613 (1987).
3. L. Faddeev, N. Reshetikhin, and L. Takhtajan, *Algebra i Analiz* **1**, 178 (1989) (in Russian; English translation in *Leningrad Math. J.* **1**, 193 (1990)).
4. J. Lukierski, A. Nowicki, H. Ruegg, and V.N. Tolstoy, *Phys. Lett.* **B264** 331 (1991).
5. J. Lukierski, A. Nowicki, H. Ruegg, *Phys. Lett.* **B293** 344 (1992).
6. S. Giller, J. Kunz, P. Kosiński, M. Majewski, and P. Maślanka, *Phys. Lett.* **286**, 57 (1992).
7. J. Lukierski, H. Ruegg, and W. Rühl, *Phys. Lett.* **313**, 357 (1993).
8. S. Majid and H. Ruegg, *Phys. Lett.* **334**, 348 (1994).
9. S. Majid, *Journ. Algebra* **130**, 17 (1990).
10. S. Majid, *J. Funct. Anal.* **95**, 291 (1991).
11. S. Majid, *Int. Journ. Phys.* **A5**, 1 (1990).
12. S. Zakrzewski, *Journ. Phys.* **A27**, 2075 (1994).
13. J. Kosiński and P. Maślanka, *The duality between the κ-Poincaré algebra and κ-Poincaré group*, hep-th-9411033.
14. J. Lukierski, A. Nowicki, amd J. Sobczyk, *Journ. Phys.* **A26**, L1099 (1993).
15. P. Kosiński, J. Lukierski, P. Maślanka, and J. Sobczyk, *Journ. Phys.* **A27**, 6827 (1994).
16. P. Kosiński, J. Lukierski, P. Maślanka, and J. Sobczyk, *Journ. Phys.* **A**, in press.
17. P. Kosiński, J. Lukierski, P. Maślanka, and J. Sobczyk, *Bicrossproduct Hopf Superalgebras and $D = 4$ κ-deformed Poincaré Supergroup*, Wrocław Univ. preprint IFT UWr 887/95.
18. A. Nowicki, E. Sorace, and M. Tarlini, *Phys. Lett.* **B302**, 419 (1993).
19. G. Domokos and S. Kovesi-Domokos, *J. Phys.* **G20**, 1989 (1994).

Magnetic Field-Induced 2D Wigner Solid versus Incompressible Quantum Liquid

G. Meissner

Theoretische Physik, Universität des Saarlandes, D-66041 Saarbrücken, Germany

1. Introduction

The main purpose of this contribution is a discussion of some quantitative predictions of a unified many-body approach for Coulomb-interacting two-dimensional (2D) electrons (charge: $Q = -e$), competing between two condensed phases, a liquid and a solid one. Strong correlations of the considered many-body system are of significance in the limit of high magnetic fields applied perpendicular to the 2D plane. Due to the well-known massive degeneracy associated with 2D free-electron motion of cyclotron frequency ω_c, the electron-electron interaction within the lowest Landau level plays an important role. In actual systems disorder potentials, as e.g. at hetero-junction interfaces, may also be essential [1]. The highly correlated motion of these electrons at sufficiently small random disorder, however, should favor the formation of the two novel phases of condensed matter at temperatures (T) low enough compared to the cyclotron energy, i.e., $k_B T \ll \hbar\omega_c$, and at magnetic fields (B) sufficiently high, such that the mean number $\nu^{-1} = B/(\phi^0 n_e)$ of magnetic flux quanta $\phi^0 = ch/e$ per electron of areal density $n_e = N_e/A$ exceeds one.

The incompressible quantum liquid (IQL)-phase of Bose condensed charge-vortex composites $e_p\phi_q^0$ for certain rational filling factors $\nu = p/q < 1$ (with p and q denoting mutual primes and q usually being odd) exhibits the fractional quantum Hall effect (FQHE), i.e., a quantized Hall conductance $\sigma_{12} = \nu e^2/h$ is accompanied by minima in the longitudinal conductance σ_{11} , surviving moderate disorder [2].

The Wigner solid (WS)-phase with a lattice-periodic structure of the guiding centers of the electrons, minimizing their Coulomb repulsion $e^2/\epsilon r_L$ repeatedly at non-rational filling factors below a certain critical value ν_c, exhibits a non-vanishing shear modulus $\mu > 0$ ($r_L = (c\hbar/eB)^{1/2} = \hbar/(m\omega_c)^{1/2}$: Larmor radius; m: effective elctron mass; ϵ: background dielectric constant). In the dilute electron ($\nu \rightarrow 0$) or high-magnetic field ($r_L \rightarrow 0$) limit, the ground-state of that sort of a reentrant system may finally be identified with the triangular electron lattice formed by a classical 2D Wigner crystal of lattice constant $d = (\sqrt{3}n_e/2)^{-1/2}$ [3]. A random potential will induce a finite correlation length ξ in the Wigner solid, whose shear stiffness μ remains meaningfull, however, as long as $\xi/d \gg 1$.

2. Outline of the Many-Body Approach

The fundamental quantity for investigating, e.g., ground-state energies and collective excitations in our many-body approach [4], is the wave-vector and frequency dependent response function

$$\chi_{\Delta\Delta}(\boldsymbol{k}, \Omega = \omega + i\eta) = \frac{i}{\hbar}\int_0^\infty dt\, e^{i\Omega t} \left\langle [\Delta(\boldsymbol{k}, t), \Delta(-\boldsymbol{k}, 0)]\right\rangle \tag{2.1}$$

for non-commuting density fluctuations $\Delta(\boldsymbol{k}) = \sum_l exp(-i\boldsymbol{k}\cdot \boldsymbol{X}(l))$ of the guiding centers $\boldsymbol{X}(l)$ with non-commuting Cartesian components, being reflected in the commutation relations

$$\left[\Delta(\boldsymbol{k}), \Delta(\boldsymbol{k}')\right] = -2i\sin((\boldsymbol{k}\times \boldsymbol{k}')r_L^2/2)\Delta(\boldsymbol{k}+\boldsymbol{k}').$$

These density fluctuations $\Delta(\boldsymbol{k})$, hence, obey a W_∞ algebra [5]. Closely related are the spectral function $\chi''_{\Delta\Delta}(\boldsymbol{k}, \omega)$, being given by the discontinuity of the complex response function $\chi_{\Delta\Delta}(\boldsymbol{k}, \Omega)$ at the real frequency axis, and the static structure factor $S_{\Delta\Delta}(\boldsymbol{k}) = \langle \Delta(\boldsymbol{k})\Delta(-\boldsymbol{k})\rangle / N_e$ of N_e electrons. Sum-rule techniques are employed, in order to put the inverse of the response function $\chi^{-1}_{\Delta\Delta}(\boldsymbol{k}, \Omega)$ in the form of a rigorous spectral representation, which can be used for actually calculating ground-state energies $E(\nu)$ and collective exitations $\omega(\boldsymbol{k})$ on the basis of these quantities.

Using the concept of broken magnetic translation invariance, an implicit equation for the melting temperature $T_M(\nu)$ as a function of the filling factor $\nu = 2\pi r_L^2 n_e$ can finally be obtained quite generally. This equation contains the polarization operator and a generalized particle-hole interaction of the lowest Landau level (LLL). It can be derived from a homogeneous integral equation, holding for $\boldsymbol{D}\langle\overline{\rho}(\boldsymbol{r})\rangle$, i.e., the covariant derivative $\boldsymbol{D} = (\partial/\partial\boldsymbol{r} + iQ\boldsymbol{A}/\hbar c)$, of the expection value $\langle\overline{\rho}(\boldsymbol{r})\rangle$ of that part of the electron density operator

$$\overline{\rho}(\boldsymbol{r}) = \int d^2k\, exp(i\boldsymbol{k}\cdot\boldsymbol{r} - \boldsymbol{k}^2 r_L^2/4)\Delta(\boldsymbol{k})/(2\pi)^2, \tag{2.2}$$

which no longer contains the fast cyclotron motion. The quantity $\boldsymbol{A}$ in $\boldsymbol{D}$ denotes the vector potential of the magnetic field $\boldsymbol{B} = \boldsymbol{\nabla}\times\boldsymbol{A}$.

The competition between Coulomb repulsion and exchange attraction apparently is a sufficient condition for this system to become unstable against broken magnetic translation invariance at non-rational filling factors $\nu < \nu_c$, where $\boldsymbol{D}\langle\overline{\rho}(\boldsymbol{r})\rangle \neq 0$. Due to a specific kind of particle-hole symmetry in the lowest Landau level, the Fourier transform of the effective interaction between two guiding centers,

$$v_{eff}(\boldsymbol{k}) = \frac{1}{2}\left[v(\boldsymbol{k}) - \frac{r_L^2}{2\pi}\int d^2k'\, e^{i\boldsymbol{k}\cdot\boldsymbol{k}' r_L^2} v(\boldsymbol{k}')\right], \tag{2.3}$$

with the repulsive part $v(\boldsymbol{k}) = (2\pi e^2/\epsilon k)\, exp(-\boldsymbol{k}^2 r_L^2/2)$ in the expression of the ground-state energy per electron

$$E(\nu)/N_e = \frac{1}{2}\int \frac{d^2k}{(2\pi)^2} v(\boldsymbol{k})\,[S_{\Delta\Delta}(\boldsymbol{k}) - 1]\,, \tag{2.4}$$

indeed reveals this sort of competition, i.e. $v_{eff}(\boldsymbol{k}) < 0$, if $kr_L > 1$.

3. Summary of Results

Based on the outlined many-body approach, quantitative predictions concerning the ground-state energies $E(\nu)$, the collective excitations $\omega(\boldsymbol{k})$, and the phase separation curve $T_M(\nu)$ of the two competing phases are conceivable.

The critical filling factor ν_c can be estimated from comparing ground-state energies calculated both for the solid and the liquid phase, i.e. $E_S(\nu_c) = E_L(\nu_c)$. Our estimate obtained for $\nu_c \simeq 0.202 \geq 1/5$ together with reentrant aspects around $\nu = 1/5$ seems to be in reasonable agreement with experimental findings [1] and theoretical expectations [6]. For the WS-phase the one-loop approximation of high magnetic field expansion,

$$E_S(\nu)/N_e = \{-0.78213\nu^{1/2}+0.24101\nu^{3/2}+0.02992\nu^{5/2}+\ldots\}(e^2/\epsilon r_L), \tag{3.1}$$

has been used, containing the static lattice energy, the harmonic zero-point energy, and anharmonic corrections [7]. A fit of Monte-Carlo simulations with Laughlin's wave function was used [8] for

$$E_L(\nu)/N_e = -0.78231\nu^{1/2}\{1 - 0.221\nu^{0.74} + 0.012\nu^{1.20}\}(e^2/\epsilon r_L) \tag{3.2}$$

in the IQL-phase at filling factors $\nu = 1/q$ with odd $q = 3, 5, \ldots$.

Concerning the phase separation curve $T_M(\nu)$, we would just like to mention, that the familiar mean-field result [9], $k_B T_M/(e^2/\epsilon r_L) = 0.557\nu(1-\nu)$, is easily recovered in our approach as a leading approximation, replacing the generalized particle-hole interaction of the LLL by v_{eff} of Eq.(3), and using the Hartree-Fock approximation (HFA) for the polarization operator. Strong correlations could be included by evaluating the polarization operator in single-mode approximation (SMA).

Results for collective excitations $\omega(\boldsymbol{k})$ can be obtained from $\mathrm{Re}\,\chi^{-1}_{\Delta\Delta}(\boldsymbol{k}, \Omega = \omega(\boldsymbol{k})) \equiv 0$, employing sum-rule techniques [10]. The finite gap $\omega_L(\boldsymbol{k} = 0) \neq 0$ of the magneto-rotons, versus gapless behavior $\omega_S(\boldsymbol{k} \to 0) \sim \boldsymbol{k}^{3/2}$ of the magneto-phonons (Goldstone mode) in the long wavelength limit, has been shown to hold rigorously. Thus, one finds in the WS-phase

$$\omega_S^2(\boldsymbol{k}) = \det|M_{\alpha\beta}(\boldsymbol{k}, 0)|\omega_c^{-2} = \frac{2\pi e^2}{m^2\omega_c^2\epsilon}\mu \boldsymbol{k}^3 + O(\boldsymbol{k}^4), \tag{3.3}$$

where $M_{\alpha\beta}(\boldsymbol{k},0) = 2\pi e^2 n_e k_\alpha k_\beta/(k\epsilon m) + Z_{\alpha\beta\gamma\delta}k_\gamma k_\delta + 0_{\alpha\beta}(\boldsymbol{k}^4)$ denotes the static self-energy, and $\mu = mn_e Z_{1122}$ is the isothermal shear modulus. In the dilute electron ($\nu \to 0$) limit, e.g., $\mu = 0.24506\, e^2 n_e^{3/2}\epsilon^{-1}$ [3]. The gap at $\boldsymbol{k} = 0$ in the dispersion relation of the magneto-rotons of the IQL-phase,

$$\omega_L^2(\boldsymbol{k}=0) = \lim_{\boldsymbol{k}\to 0} \overline{\omega}(\boldsymbol{k})\, \chi_{\Delta\Delta}^{-1}(\boldsymbol{k},0) \neq 0, \tag{3.4}$$

follows rigorously from the long wavelength $\boldsymbol{k}^4$-behavior, holding for the first frequency moment $\overline{\omega}(\boldsymbol{k}) \sim \boldsymbol{k}^4$ of the spectral function $\chi''_{\Delta\Delta}(\boldsymbol{k},\omega)$, and for the static susceptibility $\chi_{\Delta\Delta}(\boldsymbol{k},0) \sim \boldsymbol{k}^4$, reflecting the strongly suppressed density fluctuations due to the incompressibility of this sort of a quantum liquid. Quantitatively, one obtains, e.g., at $\nu = 1/3$ in SMA an estimate for $\hbar\omega_L(\boldsymbol{k}=0) \simeq 0.15\,(e^2/\epsilon r_L)$, which taking into account finite-thickness effects is reduced by about 25% in fair agreement with the value $0.11\,(e^2/\epsilon r_L)$ from inelastic light scattering experiments [11].

The liquid-solid transition thus is finally accompanied by a softening of the magneto-roton minimum of the IQL-phase (at a certain finite wave-vector in $\omega_L(\boldsymbol{k})$ becoming a reciprocal lattice vector in the WS -phase), and the solid-liquid transition by a softening of the shear modulus μ, respectively. Indications of reentrance between the two competing phases are found from a comparison of quantitative results obtained for the ground-state energies and the collective excitations.

References

1. Jiang, H.W., R.L. Willet, H.L. Stoermer, D.C. Tsui, L.N. Pfeiffer, and K.W. West, Phys. Rev. Lett. **65**, 633 (1990).
2. Tsui, D.C., H.L. Stoermer, and A.C. Gossard, Phys, Rev. Lett. **48**, 1599 (1982); Laughlin, R.B., Phys. Rev. Lett. **50**, 1395 (1983).
3. Meissner, G., H. Namaizawa, and M. Voss, Phys. Rev. B **13**, 1370 (1976); Bonsall, L. and A.A. Maradudin, Phys. Rev. B **15**, 1959 (1977).
4. Meissner, G., Physica B **184**, 66 (1993).
5. See, e.g., Capelli A., C.A. Trugenberger, and G.R. Zemba, Phys. Lett. **306** B, 100 (1993).
6. Froehlich, J., Lectures on: *Many-Body Theory and Two-Dimensional Field Theory*.
7. Meissner, G. and U. Brockstieger, in G. Landwehr Ed. (Springer, Berlin, Heidelberg) Series in Solid State Sciences **71**, 85 (1987).
8. Levesque, D., J.J. Weis, and A.H. MacDonald, Phys. Rev. B **30**, 1056 (1984).
9. Fukuyama, H., P.M. Platzman, and P.W. Anderson, Phys. Rev. B **19**, 5211 (1979).
10. Meissner, G., in G. Landwehr Ed. (Springer, Berlin, Heidelberg) Series in Solid State Sciences **101**, 605 (1992).
11. Pinczuk, A., B.S. Dennis, L.N. Pfeiffer, and K. West, Phys. Rev. Lett. **70**, 3983 (1993).

Random Walks over Wilson Loop Space

A. Protogenov

Institute of Applied Physics, Russian Academy of Sciences, 46 Ul'yanov St., 603600 Nizhny Novgorod, Russia

1. Introduction

Strong quantum fluctuations in low dimensional spin systems change the structure of the ground state and the classification of low-lying excitations. Breaking of local magnetic order due to doping and creation of charge degrees of freedom makes the role of many-body correlations more essential. The spin disorder of any number of particles can be expressed in terms of fluxes Φ of a statistical gauge field, "measuring" a degree of the spin noncollinearity along an arbitrary large loop C. In the case of Gaussian distribution of fluxes with the dispersion proportional to the total area $S(C)$, the mean value satisfies the confinement law: $< \exp(i\Phi) >\sim \exp(-\alpha S(C))$. The choice of the gauge and diffeomorphic invariant eigenvalues $\exp(i\Phi)$ of the Wilson operator simplify the structure of the phase space of planar spin systems under consideration. The finite volume of the phase space makes the dimension of the Hilbert space finite [1], due to the periodicity of the canonical conjugated variables $\exp(i\Phi_a)$ ($a = n$ or $a = m$), for example in the simplest case of the phase space being a torus. The wave function $\Psi(\exp(i\Phi))$ will be a k-dimensional vector, because the eigenvalue of the Wilson operator is equal to the k-th root of unity, i. e. $\Phi_n = 2\pi in/k$ and $n = 0,\ldots,k-1$. Here k is related to the coefficient K of the $SU(2)$ Chern-Simons term in a usual long-wave description by $k = K + 2$ [2]. In the general case the quantum number K has a sense of the linking number of the particle world lines. In particular, within the $SU_q(2)$ quantum group theory of fractional statistics, the semion excitations have the value $k = K + 2 = 4$, i. e., the d-wave quantum number K. The quantum number n has a sense of the puncture number inside of the loop C.

The problems of properties of the electronic system in the phase with Gaussian-random or homogeneous distribution of magnetic fluxes are discussed in the recently published papers [3, 4]. On the other hand, the quantum group approach to the description of the diffusion in spin chains has been used in [5]. In this paper we study the q-diffusion dynamics of the states characterized by the random distribution of the fluxes with fixed value of k and an arbitrary value of n. We are interested in the time behavior of transition processes $n \to n \pm 1$.

2. q-Random Walks

Let us consider the transition from the state $\Psi(w_n) \equiv \Psi_n$ to the neighboring states $\Psi(w_{n\pm1})$. We use the parametrization of the wave function argument by the eigenvalue $w_n = \exp(\Phi_n)$ of the Wilson operator

$$\hat{W} = Tr_{\mathcal{R}} \mathcal{P} \exp(\oint_C \mathbf{a}\, \mathbf{dl}).$$

Here $\mathcal{R}$ denotes an irreducible representation; $\mathbf{a}$ is the non-Abelian gauge potential. Using these variables the processes of the creation $\Psi(w_n) \to \Psi(w_n q)$ and the annihilation $\Psi(w_n) \to \Psi(w_n/q)$ of the fluxes can be considered as hops on the lattice constructed by the geometrical progression rules. The denominator $q = \exp(2\pi i/k)$ of this geometrical progression is the k-th root of unity. The wave function is now determined on a discretized circle with the points locating in the positions of the k-th roots of unity: $\{1, q, \ldots, q^{k-1}\}$. In this case the procedure of the comparison of function values in neighboring points is represented by the difference operator D_q^+:

$$D_q^+ \Psi(w_n) = \frac{\Psi(q w_n) - \Psi(w_n)}{q w_n - w_n}\,, \tag{1}$$

and by Hermite-conjugated ones D_q^-:

$$D_q^- \Psi(w_n) = \frac{\Psi(w_n/q) - \Psi(w_n)}{(w_n^{-1}/q - w_n^{-1})}\,. \tag{2}$$

The action of the Laplace operator Δ_q is given by [6]

$$\Delta_q \Psi_n \equiv D_q^- D_q^+ \Psi_n = \frac{\Psi_{n+1} + \Psi_{n-1} - 2\Psi_n}{q + q^{-1} - 2}\,. \tag{3}$$

The equation of motion in the Euclidean time,

$$\frac{\partial \Psi}{\partial t} = \Delta_q \Psi, \tag{4}$$

has the form of the q-diffusion equation [7]. It is easily seen that the operator (3) coincides with the Hamiltonian for a rotator in the limit $k \to \infty$ ($q \to 1$), and Eq. (4) is a usual diffusion equation in this limit.

The eigenfunctions and the eigenvalues of Δ_q are given by

$$\Psi_{n,m} = \frac{1}{\sqrt{k}} q^{mn}\,, \qquad n, m = 0, 1, \ldots, k-1, \tag{5}$$

$$\epsilon_m = \frac{\sin^2(\pi m/k)}{\sin^2(\pi/k)} \,. \tag{6}$$

By using Eqs. (5), (6), the general solution of Eq. (4) may be written as

$$\Psi_n(t) = \sum_{m=0}^{k-1} a_m \Psi_{n,\,m} e^{-\epsilon_m t} = \sum_{m=0}^{k-1} g_m G(n-m,\,t)\,, \tag{7}$$

where $g_n = \Psi_n(0)$ is an initial distribution. The Green function

$$G(n,\,t) = \frac{1}{k} \sum_{s=0}^{k-1} \exp\{-\epsilon_s t + 2\pi i s n/k\} \tag{8}$$

generalizes the form of the usual Gaussian distribution function.

3. Discussion

First of all, we would like to draw attention to the fact, that the spectrum (6) has the sense of the squared q-dimension $[n]_q^2$ of an irreducible representation of the algebra $U_q(sl_n)$. Here

$$[n]_q \equiv \left(q^{n/2} - q^{-n/2}\right) / \left(q^{1/2} - q^{-1/2}\right).$$

From this viewpoint the time evolution looks like *quantum group random walks in the space of different hooks of the Young tableaux* connected with representations of the q-deformed algebra. This motion can also be considered [8] as random walks on a Dynkin diagram, and t as a discrete time parameter. The statistical models related with random walks on Dynkin diagrams have been studied in Ref. [9]. The investigation of discrete time evolution requires separate detailed consideration. We plan to return to this problem in the next paper. The symmetries of Eq. (4) with q-deformation of the l.h.s. and some solutions for the real value of the deformation parameter were studied in Ref. [10].

It is easily shown, using Gaussian-like distribution (8), that in the initial state the mean value $\langle \epsilon_n \rangle$, i.e., the variance

$$\langle [n]_q^2 \rangle = \frac{1}{k} \sum_{m=0}^{k-1} \epsilon_m = \frac{1}{2\sin^2(\pi/k)}\,, \tag{1}$$

is proportional to the q-analog of the Casimir operator for odd representations [11]. At $t \to 0$ the variance agrees with the diffusion law $\langle [n]_q^2 \rangle = \mathcal{D}(k)\,t$, with the diffusion coefficient $\mathcal{D}(k)$, which equals $1/2$ in the fermion case $k = 2$. For semions $\mathcal{D}(4) = 5$.

The q-deformed equation of the random walk motion was first obtained in Ref. [7]. We show here that Madjid's assumption [7], that there is not reason to think that the solutions of q-diffusion equation "should exist along familiar Gaussian lines", is valid in the following case.

The determination of the operator (3) is not unique. Its determination [7] as D_q^2, where

$$D_q\Psi(w) = \{\Psi(qw) - \Psi(w/q)\}/\{w(q-q^{-1})\},$$

leads to the existence of *only zeroth modes* in the diffusion equation. In that case Eq. (4) has the following form:

$$\begin{aligned}\frac{\partial\Psi}{\partial t} = \; & w^{-2n}[q^{-1}\Psi(w_{n+2}) + q\Psi(w_{n-2}) \\ & -(q+q^{-1})\Psi(w_n)] \equiv \hat{A}\Psi.\end{aligned} \tag{2}$$

The multiplier $(q-q^{-1})^{-2}$ in the right-hand side of this equation is included in the definition of t. The formal solution of Eq. (10),

$$\Psi_n(t) = e^{\hat{A}t}\Psi_n(0), \tag{3}$$

is a polynomial in t and q,

$$\Psi_n(t) = \sum_{s=0}^{r-1} \frac{t^s}{s!}\hat{A}^s\Psi_n(0), \tag{4}$$

because the operator $\hat{A}$ is the nilpotent operator of degree r: $\hat{A}^r = 0$. The dependence of $r(k)$ is shown in the following table.

k	odd	even, $4 \le k \le 10$	12	14, 16
r	$(k+1)/2$	2	3	4

In the steady-state Eq. (10) is satisfied by solutions (5), which are equal to the functions $\Psi_n(0)$ in Eq. (12). To make this fact more clear, let us rewrite Eq. (10) in the equivalent form

$$\frac{\partial\Psi}{\partial t} = a^2\Psi. \tag{5}$$

The annihilation and creation operators a, a^+ in the finite-dimensional q-holomorphic representation are defined by [12]

$$a = g^{-1}\frac{h-h^{-1}}{q-q^{-1}}, \qquad a^+ = g, \tag{6}$$

where $h = \exp[(2\pi i/k)P]$, and $g = \exp(iQ)$. The action of the last operators is given by

$$\begin{aligned}(h\Psi)(w_n) &= \Psi(w_{n+1}), \\ (g\Psi)(w_n) &= w_n\Psi(w_n).\end{aligned} \tag{7}$$

They satisfy the following commutator relations:

$$h^k = g^k = I, \quad hh^+ = gg^+ = I, \quad hg = qgh. \tag{8}$$

Using the latter property it is possible to find [12] the commutation relation [13, 14]

$$aa^+ - qa^+a = q^{-P}. \tag{9}$$

The relation (17) shows that the angular momentum operator P and the particle number operator coincide. The eigenstates $\Psi_m(w_n)$ of the operator P ($P\Psi_m = m\Psi_m$, $m = 0, 1, \ldots, k-1$) coincide with the functions Ψ_n in Eq. (5).

All eigenvalues ε being equal to zero means, that Eq. (10) describes "a motion" in the degeneracy space of the ground state. This allows us to formulate the following hypothesis associated with the dependence of the entropy on the braid linking index k. The operator of the time evolution of states in the frames of the quantum-group approach can be generalized [15] with the conservation of energy and nonconservation of entropy, i.e. with the possibility of transition from pure states to mixed ones, as this takes place in open systems. The presence or absence of the transitions can depend on the invariance of the considered mixed state with respect to the action of the q-generalized evolution operator.

In conclusion, we have performed a study of the q-Brownian motion in the flux space. For the deformation parameter being a root of unity, it is found out that the general solution of the q-diffusion equation is the q-power polynomial, and the distribution of the flux numbers is characterized by the variance, that is proportional in the initial state to the q-analog of the Casimir operator.

Acknowledgements

I would like to thank M. A. Antonets, and I. A. Shereshevsky for many helpful and stimulating discussions. This work was supported in part by the Russian Fundamental Research Foundation, Grant No. 95-02-05620, by the High School Committee, Grant No. 94-31.4-20, and by the International Science Foundation, Grant No. R8K300.

References

1. E. Witten, Comm. Math. Phys. **121**, 351 (1989).
2. L. Alvarez-Gaume, J. M. F. Labastida, and A. V. Ramallo, Nucl. Phys. **B334**, 103 (1990).
3. C. Pryor and A. Zee, Phys. Rev. B **46**, 3116 (1992).
4. T. Sugiyama and N. Nagaosa, Phys. Rev. Lett. **70**, 1980 (1993).
5. F. C. Alcaraz and V. Rittenberg, Phys. Lett. B **314** 377 (1993).
6. T. Kobayashi and T. Suzuki, J. Phys. A **26**, 6055 (1993).
7. S. Majid, Int. J. Mod. Phys. A **8** 4521 (1993).
8. R. Coquereaux, preprint CPT-94/P.3020.

9. V. Pasquier, Nucl. Phys. **B285** 162 (1987); J. Phys. A **20**, 5707 (1987).
10. R. Floreanini and L. Vinet, Lett. Math. Phys. **32**, 37 (1994).
11. P. B. Wiegmann and A. V. Zabrodin, Phys. Rev. Lett. **72**, 1890 (1994).
12. E. G. Floratos, J. Phys. A **24**, 4739 (1991).
13. L. Biedenharn, J. Phys. A **22**, L873 (1989).
14. A. MacFarlane, J. Phys. A **22**, 4581 (1989).
15. P. Schupp, preprint LBL-34942, UCB-PTH-93/35, hep-th/9312075.

Polynomial Identities on Reflection Equation Algebra

P.N. Pyatov[1] and P.A. Saponov[2]

[1] Bogoliubov Laboratory of Theoretical Physics, JINR, 141980 Dubna, Moscow Region, Russia
[2] Theoretical Department, IHEP, 142284 Protvino, Moscow Region, Russia

1. Introduction

In this note we will deal with an algebra called the Reflection Equation Algebra (REA), which is closely connected with an R-matrix formulation of the quantized universal enveloping algebras (see [1]). The reader can find a compact and comprehensive introduction to the subject in Ref. [2] and also in lectures of P.P.Kulish in the present volume. Here we only mention that the REA appears in integrable models with nontrivial boundary conditions and in non-commutative geometry.

More exactly, we will consider REAs corresponding to linear quantum groups only. These algebras are generated by a set of N^2 operators L^i_j $(i, j = 1, \ldots N)$. Being combined into a matrix L (with operator entries), they satisfy the following commutation relations:

$$R_{12}\, L_1\, R_{12}\, L_1 = L_1\, R_{12}\, L_1\, R_{12}. \tag{1.1}$$

Here R is an $N^2 \times N^2$ R-matrix of $gl(N)$-type, obeying the Yang-Baxter equation and the Hecke condition:

$$R_{12}\, R_{23}\, R_{12} = R_{23}\, R_{12}\, R_{23}, \tag{1.2}$$

$$R^2 = 1 + \lambda R, \qquad \lambda = q - \frac{1}{q}\,. \tag{1.3}$$

The explicit expression for this R-matrix can be found in Ref.[1].

The subscripts in compact matrix notations used in above equations mark the number of corresponding vector space, and the matrix product in repeated vector spaces is assumed. For example $R_{12}L_1 \to \sum_{a_1} R^{i_1 i_2}_{a_1 j_2} L^{a_1}_{j_1}$, and so on.

When the deformation parameter q tends to zero, the R-matrix turns into the permutation matrix of the two vector spaces:

$$R_{12} \to P_{12}: \quad P_{12}\, v \otimes u = u \otimes v,$$

and the L-operator matrix turns into the one with commuting elements (classical matrix). But the connection of REA and classical matrix calculus turns out to be much deeper and allows one to obtain some useful results. Namely, it is possible to define quantum analogues of trace and determinant for the L-matrix, which are central elements of the REA (see e.g. [3, 4]):

$$Tr_q\, L = Tr\, \mathcal{D}\, L, \qquad \mathcal{D} = \mathrm{diag}(q^{1-N}, q^{3-N}, \dots, q^{N-1}),$$
$$\epsilon_q^{1\dots N} \mathrm{Det}_q L = (L_1 R_{12} R_{23} \dots R_{N-1N})^N \, \epsilon_q^{1\dots N} .$$

Here we have used the notation $\epsilon_q^{1\dots N} \equiv \epsilon^{i_1 \dots i_N}$ for the q-deformed Levi-Civita tensor, which is defined up to a normalizing factor by its q-antisymmetry property:

$$R_{ii+1}\epsilon_q^{1\dots N} = -\frac{1}{q}\epsilon_q^{1\dots N}, \quad 1 \le i \le N-1. \tag{1.4}$$

The normalization condition is usually taken to be $\epsilon_q^{i_1 \dots i_N}|_{i_1=1,\dots,i_N=N} = 1$.

In the present note we establish further connections of the REA and the classical matrix algebra. We found a q-generalization of the Cayley-Hamilton theorem, and q-analogues of the so-called Newton relations connecting the two basis sets of generators for the center of REA. One of these sets is formed by "trace-like" elements and contains the quantum traces of powers of the matrix L: $Tr_q L^k$. Another one is formed by the "determinant-like" elements and contains, in particular, the quantum determinant $\mathrm{Det}_q L$. It should be mentioned, that the characteristic identity (generalized Cayley-Hamilton theorem) was obtained in a different approach in Ref.[5] (see Remark 4.8 of that paper), when studying the algebraic structure of quantum Yangians $Y_q(gl(N))$. We reproduce this result in the REA approach.

It will be convenient to recall here some facts from the theory of matrices (see e.g. [6]), which will then be generalized to the quantum case. For an $N \times N$ matrix A with complex entries one can define its characteristic polynomial

$$\triangle\,(x) \equiv \det ||\, x\,\mathbf{1} - A\,|| \equiv x^N + \sum_{k=1}^{N} (-1)^k \sigma(k)\, x^{N-k} . \tag{1.5}$$

The coefficients $\sigma(k)$ of the characteristic polynomial expressed in terms of eigenvalues $\{\lambda_i\}$ of the matrix A form the set of basic symmetric polynomials in N variables:

$$\sigma(1) \equiv \sum_{i=1}^{N} \lambda_i = Tr\, A\,, \quad \sigma(2) \equiv \sum_{i<j} \lambda_i \lambda_j \;, \dots, \quad \sigma(N) \equiv \prod_{i=1}^{N} \lambda_i = \det A\,. \tag{1.6}$$

Being expressed directly in terms of the matrix elements of A, each $\sigma(k)$ is given by the sum of all the principal minors of the k-th order,

$$\sigma(k) = \frac{1}{k!\,(N-k)!}\, \epsilon^{1\dots N} A_1 \dots A_k\, \epsilon^{1\dots N} . \tag{1.7}$$

Here $\epsilon^{1\dots N}$ is the antisymmetric Levi-Civita N-tensor. Another standard set of symmetric polynomials is given by the traces of powers of the matrix A,

$$s(k) \equiv \sum_{i=1}^{N} (\lambda_i)^k = Tr\, A^k\,, \quad 1 \le k \le N\,. \tag{1.8}$$

The two basic sets $\{\sigma(k)\}$ and $\{s(k)\}$ are connected by the so-called Newton relations

$$k\,\sigma(k) - s(1)\,\sigma(k-1) + \ldots + (-1)^{k-1} s(k-1)\,\sigma(1) + (-1)^k s(k) = 0. \tag{1.9}$$

The Cayley-Hamilton theorem asserts, that the matrix value of the characteristic polynomial $\Delta(A)$ vanishes identically.

Now, let us proceed to establishe analogues of these relations for the quantum matrices L.

2. Quantum Newton Relations and Characteristic Polynomial

In the sequel we will need the notation for the q-numbers p_q and the square of the q-deformed Levi-Chivita tensor,

$$p_q \equiv \frac{(q^p - q^{-p})}{\lambda} \qquad |\epsilon_q|^2 \equiv \epsilon_q^{1\ldots N} \epsilon_q^{1\ldots N} = q^{N(N-1)/2} N_q!, \quad N_q! \equiv 1_q 2_q \ldots N_q.$$

Now we turn to derive the q-generalization of the Newton relations. To this end we consider the set of central "trace-like" elements,

$$s_q(k) = q^{1-N} Tr_q L^k, \qquad 1 \le k \le N\,, \tag{2.1}$$

where the normalizing factor is chosen for convenience. Another generating set consists of the "determinant-like" elements

$$\sigma_q(k) = \alpha_k \epsilon_q^{1\ldots N} (L_1 R_1 \ldots R_{k-1})^k \epsilon_q^{1\ldots N}\,, \tag{2.2}$$

where we have introduced the notation $R_i \equiv R_{ii+1}$. The factors α_k are normalizing constants. The first of them is fixed to be

$$\alpha_1 = q^{1-N} N_q / |\epsilon_q|^2$$

by the natural condition $\sigma_q(1) = s_q(1)$. The other ones will be specified below.

The derivation of the Newton relations (1.9) at the classical level is based on the spectral representations (1.6), (1.8) for $\sigma(k)$ and $s(k)$. But in the quantum case the spectrum of the matrix L can be constructed only, if the center of REA is algebraically closed, which, in turn, can be clarified only in a particular representation of REA. To overcome this difficulty we shall develop a little more technique.

Define an operator S_N, which symmetrizes any $N \times N$ matrix X in N matrix spaces:

$$S_N(X) = X_1 + R_1X_1R_1 + \ldots + R_{N-1}\ldots R_1X_1R_1\ldots R_{N-1}. \tag{2.3}$$

The characteristic properties of this symmetrizer,

$$[S_N(X)\,,\,R_i] = 0\,,\quad 1 \le i \le N-1\,, \tag{2.4}$$

are fulfilled due to the relations (1.2), (1.3). Furthermore, the following useful formula

$$\epsilon_q^{1\ldots N} S_N(X) = s_X \epsilon_q^{1\ldots N} \tag{2.5}$$

is a direct consequence of (2.4) and (1.4). Here the scalar factor s_X reads

$$s_X = \frac{1}{|\epsilon_q|^2}\epsilon_q^{1\ldots N} S_N(X)\epsilon_q^{1\ldots N} = q^{1-N} Tr_q X.$$

For $X = L^k$ this factor coincides with $s_q(k)$ (2.1). Now we are able to prove the

Proposition. For $1 \le k \le N$, the generators $\sigma_q(k)$ and $s_q(k)$ are connected by the relations

$$\frac{k_q}{q^{k-1}}\sigma_q(k) - s_q(1)\,\sigma_q(k-1) + \ldots + (-1)^{k-1}s_q(k-1)\,\sigma_q(1) + (-1)^k s_q(k) = 0\,, \tag{2.6}$$

provided that the numerical factors α_p are fixed as follows:

$$\alpha_p = \frac{N_q!}{(N-p)_q!\,p_q!}\,\frac{q^{-p(N-p)}}{|\epsilon_q|^2}\,. \tag{2.7}$$

Proof. Consider the quantities $s_q(k-p)\sigma_q(p)$ for $1 \le p \le k-1$. With the help of (2.5) and the definitions of $s_q(k)$ and $\sigma_q(k)$, one can perform the following transformation:

$$\begin{aligned}
&s_q(k-p)\sigma_q(p)\\
&= \alpha_p \frac{p_q}{q^{p-1}}\epsilon_q^{1\ldots N}(L^{k-p+1}R_1\ldots R_{p-1})(L\,R_1\ldots R_{p-1})^{p-1}\epsilon_q^{1\ldots N}\\
&+\alpha_p \frac{(N-p)_q}{q^{N-p-1}}\epsilon_q^{1\ldots N}(L^{k-p}R_1\ldots R_p)(L\,R_1\ldots R_p)^p\epsilon_q^{1\ldots N}\,.
\end{aligned}$$

Now arbitrary coefficients α_p should be fixed in such a way, that the last term in $s_q(k-p+1)\,\sigma_q(p-1)$ and the first one in $s_q(k-p)\,\sigma_q(p)$ become equal. This is the case, if α_p satisfy the relations

$$\alpha_p = q^{2p-1-N}\frac{(N-p+1)_q}{p_q}\alpha_{p-1}\,. \tag{2.8}$$

Then, taking the alternating sum $\sum_p(-1)^{p-1}s_q(k-p)\sigma_q(p)$, we find that the only terms which survive are the first one in $s_q(k-1)\sigma_q(1)$ and the last one in $s_q(1)\sigma_q(k-1)$, and thus we obtain relations (2.6). Finally, given the value of α_1, one easily shows that the recursion (2.8) is solved by (2.7).

It is worth mentioning that the singular points, where the connection between $\{s_q(k)\}$ and $\{\sigma_q(k)\}$ breaks down, are those of the roots of unity: $k_q = 0$ for $1 \le k \le N$. This is apparently related to the fact, that the isomorphism of the Hecke algebra of the A_{N-1} type and the group algebra of symmetric group $\mathbf{C}S_N$ is also destroyed at these points (see e.g. [7]).

Now let us turn to derive the quantum characteristic identity for the matrix L.

Theorem. The characteristic polynomial of the matrix L looks like

$$\triangle(x) = \sum_{k=0}^{N} (-x)^k \sigma_q(N-k)\,, \tag{2.9}$$

where x is a complex number, and for the L-matrix the characteristic identity

$$\triangle(L) = \sum_{k=0}^{N} (-L)^k \sigma_q(N-k) \equiv 0 \tag{2.10}$$

is satisfied.

Proof. With the use of (1.4), (1.1) it is straightforward to check, that the expression

$$\prod_{k=0}^{N-1} \left[(L - q^{2k}x\mathbf{1})R_1 \ldots R_{N-1}\right] \epsilon_q^{1\ldots N}$$

satisfies the q-antisymmetry property (1.4), and, hence, it is proportional to the q-deformed Levi-Civita tensor

$$\prod_{k=0}^{N-1} \left[(L - q^{2k}x\mathbf{1})R_1 \ldots R_{N-1}\right] \epsilon_q^{1\ldots N} = \epsilon_q^{1\ldots N} \triangle(x)\,, \tag{2.11}$$

where the scalar factor is calculated as

$$\Delta(x) = \frac{1}{|\epsilon|^2} \epsilon_q^{1\ldots N} \prod_{k=0}^{N-1} \left[(L - q^{2k}x\mathbf{1})R_1 \ldots R_{N-1}\right] \epsilon_q^{1\ldots N}\,.$$

This expression can be further simplified with the use of (1.4), (1.1), (1.2). The calculations are straightforward but rather lengthy, and we omit them here presenting only the result in (2.9).

To prove the characteristic identity we shall contract relation (2.11) with $\epsilon_q^{2\ldots N+1}$:

$$(L - x\mathbf{1}) \left[\epsilon_q^{2\ldots N+1} \prod_{k=1}^{N-1} \left((L - q^{2k}x\mathbf{1})R_1 \ldots R_{N-1}\right) \epsilon_q^{1\ldots N}\right]$$
$$= (\epsilon_q^{2\ldots N+1} \epsilon_q^{1\ldots N}) \triangle(x)\,.$$

The r.h.s. of this relation is proportional to the unit matrix and, hence, the bracketed expression in the l.h.s. is proportional to $(L - x\mathbf{1})^{-1}$. The classical limit of this relation is the standard base for proving the Cayley-Hamilton theorem [6]. In the quantum case all the considerations are completely the same, and the resulting statement is that the matrix polynomial $\triangle(L)$ identically vanishes.

As an example of application of the Theorem above we present the compact expression for the inverse matrix of L:

$$L^{-1} = \frac{1}{\sigma_q(N)} \sum_{k=0}^{N-1} (-L)^k \sigma_q(N-k-1) ,$$

which is obtained by the formal multiplying of (2.10) by $[\sigma_q(N)L]^{-1}$.

On passing to concrete REA representations, the order of the characteristic identity may decrease, because the basic symmetric polynomials $\sigma_q(k)$ become dependent. This is illustrated in the recent paper [8], where the L-matrices were realized as pseudo-differential operators acting on the quantum plane, and they were found to possess the characteristic identity of the quadratic order.

References

1. Faddeev L.D., Reshetikhin N.Yu., Takhtadjan L.A.: Algebra i Analiz **1** (1989) 178. English translation in: Leningrad Math. J. **1** (1990) 193.
2. Kulish P.P., Sasaki R.: Prog. Theor. Phys. **89** (1993) 741.
3. Shupp P., Watts P., Zumino B.: Lett. Math. Phys. **25** (1992) 139; Comm. Math. Phys. **157** (1993) 305.
4. Faddeev L.D., Pyatov P.N.: *The differential calculus on quantum linear groups.* hep-th/9402070, to appear in F.A.Berezin memorial volume.
5. Nazarov M. and Tarasov V.: Publications RIMS **30** (1994) 459.
6. Lankaster P.: *Theory of Matrices.* Academic Press, New York – London, 1969.
7. Wenzl H.: Invent. Math. **92** Fasc.2. (1988) 349.
8. Chu C.-S., Zumino B.: *Realization of vector fields for quantum groups as pseudodifferential operators on quantum spaces.* Preprint UCB-PTH-95/04 and q-alg/9502005.

Introduction to Poisson σ-Models

P. Schaller[1] and Th. Strobl[2]

[1] Institut für Theoretische Physik, TU Wien, Wiedner Hauptstr. 8-10, A-1040 Vienna, Austria

[2] Institut für Theoretische Physik, RWTH Aachen, Sommerfeldstr. 26-28, D-52056 Aachen, Germany

The close interplay between geometry, topology, and algebra turned out to be a most crucial point in the analysis of low dimensional field theories. This is particularly true for the large class of topological and almost topological two-dimensional field theories, which can be treated comprehensively in the framework of Poisson-σ models. Examples are provided by pure Yang-Mills and gravity theories and, to some extent, by the G/G gauged WZW-model.

It turned out [1, 2, 3] that the common mathematical structure behind these theories is the one of Poisson manifolds. In the present contribution we review how this structure leads to the formulation of the Poisson-σ models. We will demonstrate that the general features of Poisson manifolds provide powerful tools for the analysis of these models, and we will apply the results to the examples mentioned above.

For pedagogical reasons we will, in the presentation of the examples, restrict ourselves to $SU(2)$ as gauge groups of YM and G/G and, in the case of gravity theories, to $\mathcal{R}^2$-gravity ($\mathcal{R}$ being the Ricci scalar). This has the advantage, that the considered target spaces will turn out to be three-dimensional then, opening possiblities for a simple visualization and explicitness of formulas. The method is, however, by no means restricted to these cases; for YM and G/G, e.g., even arbitrary non-compact gauge groups are comprised by the treatment.

Let us start by reviewing some features of Poisson manifolds [4], which finally will become the target space of our 2D field theories. (To avoid any confusion, let us note right away that the notion of a Poisson manifold is somewhat more general than the one of a symplectic manifold.)

Denote by N a manifold equipped with a Poisson bracket relation $\{.,.\}$ between functions on N. By definition the Poisson bracket is bilinear and antisymmetric. It is, moreover, subject to the Leibnitz rule $\{f,gh\} = \{f,g\}h + g\{f,h\}$, and the Jacoby identity $\{f,\{g,h\}\} + cycl. = 0$.

In local coordinates X^i on N a Poisson bracket may be expressed in terms of a two-tensor $\mathcal{P}^{ij}(X) := \{X^i, X^j\}$. With the Leibnitz rule this relation determines the Poisson structure uniquely:

$$\{f,g\} = \mathcal{P}^{ij} f_{,i} g_{,j} \,, \tag{1}$$

where $f_{,i} \equiv \partial f/\partial X^i$, and summation convention is understood. From (1) it is obvious that $\mathcal{P}^{ij}$ transforms covariantly under coordinate transformations. In terms of $\mathcal{P}^{ij}$ the Jacoby identity reads

$$\mathcal{P}^{ij}{}_{,k}\mathcal{P}^{kl} + cycl.(i,j,l) = 0\,. \tag{2}$$

There is another consequence of the Leibnitz rule: The Poisson bracket of the constant function with anything else vanishes. In the case of a general Poisson manifold there may be also other functions which share this property. They are called Casimir functions.

An example for a $2n$-dimensional Poisson manifold is provided by the phase space of an n-particle system: Denote by Q^α the positions of the particle, and by P_α the momenta. The Poisson structure is then given by $\{Q^\alpha, P_\beta\} = \delta^\alpha_\beta$, and $\{Q^\alpha, Q^\beta\} = 0 = \{P_\alpha, P_\beta\}$.

With (1) it is easy to see that the constant function is the only Casimir function in this example, which obviously holds whenever $\mathcal{P}^{ij}$ is nondegenerate. Actually Poisson manifolds with nondegenerate $\mathcal{P}$ are symplectic manifolds, the symplectic two-form being equal to the inverse of the Poisson tensor $\mathcal{P}$. For them the above example is in some sense generic: Locally *any* symplectic manifold allows so-called Darboux coordinates such that the Poisson tensor takes the form of the $2n \times 2n$ matrix:

$$\begin{pmatrix} 0 & \boldsymbol{1} \\ -\boldsymbol{1} & 0 \end{pmatrix}\,, \tag{3}$$

where $\boldsymbol{1}$ denotes the $n \times n$ unit matrix.

Another example for a Poisson structure is given by the Poisson bracket

$$\{X^i, X^j\} = \sum_{k=1}^{3} \varepsilon^{ijk} X^k \tag{4}$$

on $I\!R^3$, where ε^{ijk} is the completely antisymmetric three-tensor. Obviously the Poisson structure (4) is degenerate: Darboux coordinates cannot exist on a manifold of odd dimension. Indeed $R^2 := \sum_{i=1}^3 X^i X^i$ is a Casimir function.

In the example (4) the Poisson structure can be restricted consistently to any two-sphere given by the choice of a constant value for R^2. The restricted Poisson structure is nondegenerate. It is easy to convince oneself that Darboux coordinates on a two-sphere of radius R_0 are provided by one of the coordinates X^i and the azimutal angle around the respective axis; e.g. with $Z := X^3$ and $\Phi := \arctan(X^2/X^1)$ it follows from (4) that $\{\Phi, Z\} = 1$. The physical phase spaces corresponding to these two-spheres are the classical analogues of spin systems. (We will come back to this point below.)

The picture outlined here generalizes in a straightforward way: A Poisson manifold with degenerate Poisson structure foliates into a family of lower dimensional manifolds (symplectic leaves), each of which is characterized by assigning constant values to the Casimir functions and is equipped with a nondegenerate Poisson structure. To be more precise: Some of the symplectic leaves may have lower dimension than the generic ones, such as the origin $X^i = 0$ in the example (4), and in some of these cases (not so for (4)) level surfaces of Casimir functions correspond to several symplectic leaves.

Moreover, a symplectic leaf need not be a (regular) submanifold of N, i.e. it might wind around densely in some higher dimensional subregion of N (as a higher-dimensional generalization of what one knows from paths on a torus in classical mechanics, cf., e.g., [5]). In that case constant values of Casimir functions characterize parts of such leaves only. So what we are dealing with is a stratifacation of N rather than a foliation. Nevertheless we will stick to the more common nomenclature introduced before.

As an example for a nonlinear Poisson structure let us take the quadratic modification

$$\{X,Y\} = -Z^2 + \frac{1}{4}, \quad \{Y,Z\} = X, \quad \{Z,X\} = Y \tag{5}$$

of (4) on $\mathbb{R}^3$ with coordinates (X,Y,Z), which will become relevant in the analysis of (Euclidean) $\mathcal{R}^2$-gravity (with cosmological constant). It is straightforward to check, that

$$C_{\mathcal{R}^2} := X^2 + Y^2 - \tfrac{2}{3}Z^3 + \tfrac{1}{2}Z \tag{6}$$

is a Casimir function of the bracket (5). $\Phi := \arctan(Y/X)$ and Z, on the other hand, are still canonical conjugates (i.e. $\{\Phi, Z\} = 1$). They may be used as local Darboux coordinates on any of the symplectic leaves, which (generically) coincide with the level surfaces of (6).

It is a nice excercise to visualize the 'foliation' of $N = \mathbb{R}^3$ into symplectic leaves. This is done most easily by rotating the square root of the positive parts of

$$h(Z) := \tfrac{2}{3}Z^3 - \tfrac{1}{2}Z + C_{\mathcal{R}^2} \tag{7}$$

around the Z-axis (since $h(Z) = X^2 + Y^2$). The resulting picture is the following: For values of $C_{\mathcal{R}^2}$ smaller than $-1/6$ the symplectic leaves are diffeomorphic to the (X,Y)-plane. At $C_{\mathcal{R}^2} = -1/6$ there is an additional pointlike leaf at $(X,Y,Z) = (0,0,-\frac{1}{2})$. Indeed, this is one of the two points where the right-hand side of (5) (and thus also the Poisson tensor $\mathcal{P}$) vanishes. Within the range $(-1/6, 1/6)$ of the Casimir, the pointlike leaf turns into an ellipsoid, again accompanied by a 'plane' (situated at larger values of Z). For $C_{\mathcal{R}^2} = 1/6$ the ellipsoid and the plane touch at $(X,Y,Z) = (0,0,\frac{1}{2})$. This value of $C_{\mathcal{R}^2}$ corresponds to three symplectic leaves: An ellipsoid with one pole taken out ($\sim$ plane), this pole $(X,Y,Z) = (0,0,\frac{1}{2})$, and a 'plane without origin'. For larger values of the Casimir, finally, one has only planar leaves again.

Let us come back to the Poisson structure (4) and see its role in the two-dimensional Yang Mills theory. The field content of the latter is given by an $su(2)$ algebra-valued one-form $A = A_i T^i$ on a two-dimensional world sheet M, where the generators T^i of $su(2)$ may be represented by imaginary multiples of the Pauli matrices. Then the action reads:

$$L_{SU(2)} = \frac{1}{2\alpha} \int_M tr\, F \wedge *F, \quad F = \left(dA_i + \tfrac{1}{2}\varepsilon^{jk}{}_i A_j \wedge A_k\right) T^i . \tag{8}$$

Here α is a coupling constant and the Hodge dual $*$ is to be taken with respect to some volume-form μ on M.

To analyze the model one may find it useful to rewrite the action in a first order formalism. Denoting the momenta conjugate to the connection A (or, more precisely, to A_{1i}) by X^i, one obtains

$$L'_{SU(2)} = \int_M X^i(dA_i + \frac{1}{2}\varepsilon^{jk}{}_i A_j \wedge A_k) - \frac{\alpha}{2}\mu X_i X^i . \tag{9}$$

The equivalence to (8) is seen most easily by inserting the field equation $X_i = \frac{1}{\alpha} * F_i$ into (9). Equivalently one might perform the Gaussian integral over X^i in (9). Now with (4) and a partial integration the $SU(2)$-YM-action (9) may be seen to be of the form

$$L = L_{top} + L_{\mathcal{C}},$$

$$L_{top} = \int_M A_i \wedge dX^i + \tfrac{1}{2}\mathcal{P}^{ij}(X) A_i \wedge A_j, \tag{10}$$

$$L_{\mathcal{C}} = \int \mu\, \mathcal{C}(X),$$

where $\mathcal{C}(X)$ denotes the Casimir function $-\frac{\alpha}{2}R^2$ of the Poisson bracket (4).

This form provides a natural generalization of the Yang Mills theory: Let M be the two-dimensional world sheet of a field theory. As dynamical fields we first take coordinates X^i on some arbitrary Poisson manifold N. Additionally we take a field A which is one-form valued on M as well as on N, i.e., with coordinates x^μ on M; A can be written as $A = A_i \wedge dX^i = A_{\mu i} dx^\mu \wedge dX^i$. Then L_{top} provides (the topological part of) the Poisson σ-model action. With the additional input of a volume-form μ on M, and the choice of a Casimir function $\mathcal{C}$, it may be extended further by the addition of $L_{\mathcal{C}}$.

L_{top} is defined without any reference to a background metric on M. This establishes that the model defined by the action L_{top} is a topological field theory of the Schwarz type [6]. In the SU(2)-case this corresponds to the topological Yang-Mills theory with coupling constant $\alpha \to 0$ (BF-theory). The addition of a nontrivial $L_{\mathcal{C}}$ to the action, of course, spoils the topological nature of the model.

The symmetries of the Poisson-σ-model (10) are a straightforward generalization of the SU(2) gauge symmetry, where the structure constants are replaced by derivatives of the Poisson tensor:

$$\delta_\epsilon X^i = \epsilon_i(x)\mathcal{P}^{ij}, \quad \delta_\epsilon A_i = d\epsilon_i + \mathcal{P}^{lm}{}_{,i} A_l \epsilon_m . \tag{11}$$

Proving the invariance of the action under these transformations, the Jacoby identity (2) is most important. So here it becomes clear that this identity plays a crucial role in the theory of Poisson-σ-models.

From the action $L = L_{top} + L_{\mathcal{C}}$ the equations of motion follow immediately:

$$dX^i + \mathcal{P}^{ij}A_j = 0,$$

$$dA_i + \frac{1}{2}\mathcal{P}^{lm}{}_{,i} A_l \wedge A_m + \mathcal{C}_{,i} = 0\,. \tag{12}$$

In the beginning of this talk we asserted that the special features of Poisson structures greatly facilitate the analysis of Poisson-σ-models. This is best illustrated by the classification of the local solutions of the field equations:

Remember that any Poisson manifold foliates into a family of symplectic leaves. The latter may be parametrized by a complete set of independent Casimir functions X^I. On each of the symplectic leaves we may find Darboux coordinates X^α. Together Casimir functions and Darboux coordinates give rise to a coordinate system (X^I, X^α) on N, with $\mathcal{P}^{Ii} = 0 = \mathcal{P}^{ij}{}_{,k}$ and $\mathcal{P}^{\alpha\beta}$ having the standard form (3).

In these coordinates any Casimir function $\mathcal{C}$ on N depends on the X^I only. The equations of motion (12) simplify to

$$dX^I = 0\,, \quad dA_I = -\mathcal{C}_{,I}\,, \quad A_\alpha = (\mathcal{P}^{-1})_{\beta\alpha} dX^\beta\,, \tag{13}$$

where, as a matrix, $(\mathcal{P}^{-1})_{\beta\alpha} = \mathcal{P}^{\alpha\beta}$, according to (3). So the $X^I(x)$ have to be constant on M, but otherwise arbitrary, whereas the $X^\alpha(x)$ remain completely undetermined by the field equations. Any choice of the latter determines A_α uniquely through the last equation in (13). Each of the A_I is determined up to an exact one-form only.

Still one has not made use of the gauge freedom. As is obvious from (11), any choice of the X^α is gauge equivalent, and also $A_I \sim A_I + dh_I$, where the h_I are arbitrary functions. Thus locally any solution to the field equations is uniquely determined by the constant values of the X^I.

Let us illustrate the usefulness of such considerations at the example of 2D $\mathcal{R}^2$-gravity with Lagrangian

$$L_{\mathcal{R}^2} = \tfrac{1}{4} \int_M d^2x \sqrt{\det |g|}\, \left(\tfrac{1}{4}\mathcal{R}^2 + 1\right)\,. \tag{14}$$

Here g is the metric on M, which, for simplicity, we take to be of Euclidean signature, and $\mathcal{R}$ is the Ricci scalar of the corresponding Levi-Civita connection. First we have to clarify how the action (14) fits into our framework. The decisive observation is, that

$$L'_{\mathcal{R}^2} = \int X\,(de^1 - \omega \wedge e^2) + Y\,(de^2 + \omega \wedge e^1) + Z d\omega + (\tfrac{1}{4} - Z^2)\, e^1 \wedge e^2 \tag{15}$$

constitutes an equivalent formulation of (14) in an Einstein-Cartan formalism with zweibein e^1, e^2 (so $g = e^1e^1 + e^2e^2$) and connection one-form ω. Indeed X and Y are seen to be Lagrange multipliers for torsion zero, and the elimination of Z, analogous to the one of the X^i in the transition from (9) to (8), reproduces the $\mathcal{R}^2$-term due to the relation $\mathcal{R} = 2 * d\omega$. Now it is straightforward to verify that, upon the identification $X^i = (X, Y, Z)$ and $A_i = (e^1, e^2, \omega)$, (15) takes the form (10) with $\mathcal{C} = 0$ and the Poisson bracket (5).

According to our study of that bracket, Casimir-Darboux coordinates are provided here by $(C_{\mathcal{R}^2}, \Phi, Z) =: \widetilde{X}^i \equiv (X^I, X^\alpha)$ (cf. Eq. (6)). Then (13) yields $A_{\widetilde{1}} = df$, $A_{\widetilde{2}} = dZ$, and $A_{\widetilde{3}} = -d\Phi$, where f, Z, and Φ are some arbitrary functions on the worldsheet M. Let us choose f and Z to coincide with coordinates $x^0/2$ and x^1 on M, respectively. It then is a simple calculation to verify that, with $A_i = \frac{\partial \widetilde{X}^j}{\partial X^i} A_{\widetilde{j}}$, one obtaines for the metric $g = A_1 A_1 + A_2 A_2$:

$$g = h(x^1)(dx^0)^2 + \frac{(dx^1)^2}{h(x^1)}, \tag{16}$$

where the function h has been defined in (7). A procedure completely analogous to the one above reveals, that (16) holds also for Minkowskian signature, if one merely changes the sign in front of the first term. So in the Euclidean and in the Minkowskian case of (14) *locally* g is of a generalized Schwarzschild form and is parametrized by one quantity $C_{\mathcal{R}^2}$, which plays the role of the Schwarzschild mass.

Additional structures evolve, if global aspects are taken into account. In [7], e.g., a classification of all global solutions to (14) (and many other gravity models) is provided for Minkowskian signature. This classification contains continuous and discrete parameters in addition to $C_{\mathcal{R}^2}$. But already the global classification of Euclidean solutions of (14) is still an open problem to the best of our knowledge. It is hoped that the Poisson σ-formulation is helpful also in this respect.[1]

The features of Poisson manifolds are not only helpful in the classical theory, but also in the quantum theory of Poisson-σ-models. To illustrate this, let us analyze the quantum states in a Hamiltonian formulation.

To this end let us assume M to be of the form $S^1 \times \mathbb{R}$, parameterized by a 2π-periodic coordinate x^1 and the evolution parameter x^0. As the action (10) is already in first order form, the derivation of the corresponding Hamiltonian system is simple. The zero components A_{0i} of the A_i play the role of Lagrange multipliers, giving rise to the system of first class constraints ($\partial = \partial/\partial x^1$)

$$G^i \equiv \partial X^i + \mathcal{P}^{ij} A_{1j} \approx 0\,. \tag{17}$$

The fundamental Poisson brackets (not to be confused with the Poisson brackets on the target space N) are given by

$$\{X^i(x^1), A_{1j}(y^1)\} = -\delta^i_j\, \delta(x^1 - y^1), \tag{18}$$

[1] Actually, it turns out that a classification of all solutions to the Poisson σ-model with Poisson structure (5) and trivial fibration (cf. below) on arbitrary Riemann surfaces $M = \Sigma_\gamma$ is not too difficult. However, since Σ_γ with genus $\gamma \neq 1$ has a nontrivial tangent bundle, X^1 and X^2 are no more mere functions on $M = \Sigma_\gamma$, but they are sections of a nontrivial vector-bundle. So, in order to describe Euclidean $\mathcal{R}^2$-gravity correctly by its powerful reformulation as a Poisson σ-model, a nontrivial fibration of the target space N over the base-manifold M will have to be taken into account.

and the Hamiltonian reads

$$H = \int dx^1 (\mathcal{C} - A_{0i} G^i) . \tag{19}$$

To quantize the system in an X-representation, we consider quantum wave functions as complex valued functionals on the space Γ_N of parameterized smooth loops in N:

$$\Gamma_N = \{\mathcal{X} : S^1 \to N, x \to X^i(x)\} . \tag{20}$$

Following the Dirac procedure, only such quantum states are admissible, which satisfy the quantum constraints ($\hbar := 1$)

$$\hat{G}^i(x)\Psi[\mathcal{X}] = \left(\partial X^i(x) + i\mathcal{P}^{ij}(X)\frac{\delta}{\delta X^j(x)}\right)\Psi[\mathcal{X}] = 0 . \tag{21}$$

To get some insight into the problem of solving (21), let us specify the constraints for the SU(2) case, using the Casimir-Darboux coordinates (R, Φ, Z). We have

$$(\partial R)\Psi = 0 , \qquad (\partial\Phi + i\frac{\delta}{\delta Z(x)})\Psi = 0 , \qquad (\partial Z - i\frac{\delta}{\delta\Phi(x)})\Psi = 0 . \tag{22}$$

The first equation (22) is seen to restrict the support of Ψ to loops, which are contained entirely in some of the two-spheres with constant R. It is easy to see, furthermore, that the remaining equations (22) are solved by

$$\Psi = \exp\left(i \oint_{\mathcal{X}} \partial\Phi(x) Z(x) dx\right) \quad \text{(locally on } \Gamma_S) , \tag{23}$$

where Γ_S is the space of loops on a chosen two-sphere S. As the differential equations on Γ_S are of first order, this local solution is unique up to a multiplicative (generally R-dependent) constant. However, as it stands, (23) is not well-defined for loops intersecting one of the poles $X^i = (0, 0, \pm R)$, and also, generically, it is discontinuous there. (Consider, e.g., an infinitesimal loop winding around the north-pole $X^i = (0, 0, R)$, and another one close to that north-pole having trivial winding number. On the first loop (23) yields (approximately) $\Psi = \exp(i2\pi R)$, on the second one $\Psi = 1$.) Multiplying (23) by $\exp(-iR\oint \partial\Phi dx)$, which, on the patch of validity of (23) in Γ_S, is an R-dependent constant, the discontinuity of (23) at the north-pole may be removed. However, continuity at the south-pole yields a restriction to loops on spheres S with integer or half-integer radius R.

The situation is overviewed best [8], if one reformulates (23) with Stokes' theorem as

$$\Psi[\mathcal{X}] \propto \exp\left(i\int_D dZ \wedge d\Phi\right) , \quad \partial D = \mathcal{X} ; \tag{24}$$

i.e. D is a two-surface, the boundary of which is the argument loop of the wave functional. Now, the two-form $dZ \wedge d\Phi =: \Omega$ is well-defined all over a two-sphere $\mathcal{S}$ of fixed radius R; obviously Ω is nothing but the symplectic two-form on $\mathcal{S}$. According to (24) the phase factor of Ψ is the area enclosed by $\mathcal{X}$ (measured by means of Ω). All the ambiguity is now separated into the ambiguity of choosing D on a two-sphere $\mathcal{S}$ such that its boundary coincides with a given loop $\mathcal{X}$. Obviously there are two inequivalent choices for D. Ψ should be unaffected by this choice. This is the case, iff the total area of the sphere is an integer multiple of 2π, which again leads to $2R \in Z$.

In this formulation the generalization to arbitrary Poisson σ-models is straightforward: For any integral symplecitc leaf there exists one quantum state of the Poisson σ-model. By definition a symplectic leaf $\mathcal{S}$ is called integral, iff the integral of the symplectic form Ω over any closed surface $\Sigma \subset \mathcal{S}$ is an integer multiple of 2π:

$$\oint_{\Sigma} \Omega = 2\pi n \,, \quad n \in Z \,. \tag{25}$$

To be more precise, this result only holds, if the first homotopy of the symplectic leaves $\mathcal{S}$ is trivial. Otherwise it might happen, that a loop $\mathcal{X}$ serving as argument in the wave function does not enclose an area. In this case we have to take the area between $\mathcal{X}$ and a reference loop in the same homotopy class as the phase factor in the wave function. (The construction given above for trivial homotopy corresponds to choosing a constant (pointlike) loop as reference loop.) Effectively this yields that the wave function depends on an additional quantum number, parametrizing the first homotopy group of the symplectic leaves.

To get further insight into the integrality condition (25), observe that the integrality of the symplectic form on a classical phase space is nothing but the Bohr-Sommerfeld quantization condition. So the integrality condition (25) means, that the classical phase space associated with a symplectic leaf is quantizable [9]. We already mentioned, that the symplectic leaves in the SU(2)-example may be seen as the classical analogues of spin systems. In that case the integrality condition expresses the fact, that spin is always integer or half integer. According to the analysis above, to any *quantum* spin system there exists *one* state of the $SU(2)$ YM-theory.

Next let us specify the results for the example of Euclidean $\mathcal{R}^2$-gravity characterized by the Poisson structure (5). We found that for any value of the Casimir $C_{\mathcal{R}^2}$ there exists one symplectic leaf, which is diffeomorphic to $\mathbb{R}^2 \sim T^*\mathbb{R}$ (besides possibly additional ones that will be dealt with below). These leaves are certainly always quantizable phase spaces, (25) is satisfied trivially. So the proportionality constant in (24) becomes an arbitrary function of $C_{\mathcal{R}^2}$ here. This leads to physical wave functions on a line effectively. However, this is not all of the story. We know that for $C_{\mathcal{R}^2} \in [-1/6, 1/6]$ there are further symplectic leaves, and we have to find out which of them are quantizable.

For the pointlike leaf at $C_{\mathcal{R}^2} = -1/6$, the integrality condition (25) is satisfied trivially. So, we obtain one additional quantum state of $\mathcal{R}^2$-gravity for this value of the Casimir. (It is arguable, however, whether this state contributes to the physical Hilbert space: Possibly it should not be taken into account, because the corresponding gauge orbit is of lower dimension than the generic ones.) Next we have to clarify, which of the elliptic leaves are quantizable ($C_{\mathcal{R}^2} \in (-1/6, 1/6)$). This is not too difficult: Integrating the symplectic form $\Omega = dZ \wedge d\Phi$ over the respective ellipsoid, obviously one obtains an integer multiple of 2π, iff the 'height' $Z_{max} - Z_{min}$ of the ellipsoid is an integer. An explicit calculation shows that this is the case only for the ellipsoid at $C_{\mathcal{R}^2} = 1/9\sqrt{6}$ ($\Rightarrow Z_{max} = 1/2 - 1/\sqrt{6}$, $Z_{min} = -1/2 - 1/\sqrt{6}$). This definitely provides an additional quantum state of $\mathcal{R}^2$-gravity, and the spectrum of the Casimir becomes twofold degenerate at that specific value. At $C_{\mathcal{R}^2} = 1/6$ the situation is not that clear: There are three leaves, one of which is not simply connected. The latter fact would lead to an infinite number of states at that value of $C_{\mathcal{R}^2}$. However, the space of gauge orbits is not Hausdorff precisely there. Some continuity requirement thus might lead effectively to an identification of all these additional states. Evidently, at this point there is some ambiguity in how to define a quantum theory.

Studying (14) for Minkowskian metrics g, one again obtains a continuous spectrum for $C_{\mathcal{R}^2}$. Now, however, this spectrum is infinitely degenerate within all of the range $(-1/6, 1/6)$. The reason is that those leaves, that have been elliptic in the Euclidean formulation of the theory, turn into (two-dimensional) non-compact ($\Rightarrow$ condition (25) statisfied always) and non-simply connected ones (cf. [7] for more details).

As a last application of the theory of Poisson σ-models, we consider the G/G WZW model. Its action is

$$L_{GWZW}(g, a_+, a_-) = \frac{k}{8\pi} \int_M tr\, (\dot{g}g^{-1}\dot{g}g^{-1} - \partial g g^{-1} \partial g g^{-1})\, d^2x$$

$$4 + \frac{k}{12\pi} \int_{B \,\text{with}\, \partial B = M} tr\, (dgg^{-1} \wedge dgg^{-1} \wedge dgg^{-1})$$

$$+ \frac{k}{4\pi} \int_M tr\, [a_+(\dot{g} - \partial g)g^{-1} - a_- g^{-1}(\dot{g} + \partial g) - a_+ g a_- g^{-1} + a_+ a_-]\, d^2x\,, \quad (26)$$

where g takes values in some Lie-group G, and $a_\pm$ in the corresponding Lie-algebra. The second line, the Wess-Zumino part of the action, requires a restriction of k to integers; under this condition it becomes single-valued modulo 2π. The first two lines in (26) are a conformal generalization of the action for a free scalar field: Indeed for $g \in U(1) \Leftrightarrow g = \exp(i\phi)$, this action becomes proportional to $\int[\dot{\phi}^2 - (\partial\phi)^2]d^2x$. The third and last line in (26) ensures, that the global symmetry under adjoint transformations, $g \to h^{-1}gh$, of the previous lines becomes a local gauge symmetry of the action (upon $a_\pm \to h^{-1}a_\pm h + h^{-1}\dot{h} \pm h^{-1}\partial h$).

For semisimple G, (26) may be brought into Poisson σ-form. This shall be displayed for $G = SU(2)$. In that case

$$g = \begin{pmatrix} X^1 + iX^2 & X^3 + iX^4 \\ -X^3 + iX^4 & X^1 - iX^2 \end{pmatrix}, \quad (27)$$

where the $X^i \in \mathbb{R}$ are subject to the condition

$$\sum_{i=1}^{4} X^i X^i = 1 \,. \tag{28}$$

This displays best that topologically $SU(2)$ is a three-sphere S^3. Also it is easy to determine the symmetry orbits of (26), i.e. the conjugacy classes $g \sim h^{-1}gh$: They are described by constant values of $X^1 = tr\, g/2$, and according to (26) they are two-spheres for $X^1 \in (-1,1)$, and points for the poles of the S^3, $X^1 = \pm 1$.

Rewriting (26) in first order form and integrating out a_+, which then enters quadratically, one finds [2]

$$L'_{GWZW} = L_{top}(g, A) + L_\delta(g) \,. \tag{29}$$

Here L_{top} is the action (10) with Poisson bracket

$$\{X^i, X^j\} = \frac{2\pi}{k}\,(X^1 + iX^2) \sum_{l=1}^{4} \varepsilon^{1ijl} X^l \,, \tag{30}$$

where ε^{ijlm} is the completely antisymmetric four-tensor. L_δ, on the other hand, is

$$L_\delta = k \int_B \delta(X^1)\delta(X^2)\, dX^1 \wedge dX^2 \wedge d\phi \,, \tag{31}$$

where we have introduced an angular variable ϕ via

$$\exp(i\phi) := \frac{X^3 + iX^4}{\sqrt{(X^3)^2 + (X^4)^2}} \,. \tag{32}$$

L_δ is a relict of the Wess-Zumino term in (26). The corresponding integrands

$$\delta(X^1)\delta(X^2)\, dX^1 \wedge dX^2 \wedge d\phi \tag{33}$$

and $dgg^{-1} \wedge dgg^{-1} \wedge dgg^{-1}/12\pi$, respectively, are from the same (non-trivial) cohomology class on $SU(2)$, and (33) is a representative of this class with minimal support. The support of (33) is merely an S^1 ($X^1 = X^2 = 0 \Rightarrow (X^3)^2 + (X^4)^2 = 1$), which obviously is contained entirely within the conjugacy class $X^1 = 0$. As a consequence we will be allowed to disregard L_δ in the analysis of (29) for all symmetry orbits $X^1 \neq 0$.

The Poisson tensor (30) is not real, and consequently also the A_i are complex-valued now. Still the equations of motion resulting from the action (29) as well as the Hamiltonian structure are completely equivalent to the original formulation (26), at least if appropriate conditions for A at $(X^1, X^2) \to (0,0)$ are taken into account, cf. [2]. As the complex conjugate fields $\bar{A}_i$ do not enter (29), the $\bar{A}_i$ may be eliminated in a Hamiltonian formulation right at the beginning, and, for $X^1 \neq 0$, one is left again with (17) and (18).

Let us remark here also that, due to (28), the dX^i are not linearly independent. Consequently only three of the A_i are independent. Alternatively we could have introduced true local coordinates X^i on $SU(2)$, $i = 1, 2, 3$. However, the embedding of $SU(2)$ into $\mathbb{R}^4$ has advantages with respect to a global description. Note that the restriction of the Poisson tensor (30) on $\mathbb{R}^4$ to $SU(2)$ is consistent, because the left-hand side of (28) is a Casimir function.

As expected X^1 is a Casimir function of (30). However, the direct identification of conjugacy classes of $SU(2)$ with symplectic leaves fails for the two-sphere $X^1 = 0$, as obviously the Poisson tensor $\mathcal{P}^{ij}$ vainishes at its equator $(X^1, X^2) = (0, 0)$. This artifact turns out to be cured, however, when L_δ and the boundary conditions for the A-fields are taken into account [2], so that also $X^1 = 0$ corresponds to one symmetry orbit of (29) only.

Let us determine, which of the symplectic leaves are integral now. Certainly the pointlike ones at $X^i = (\pm 1, 0, 0, 0)$ are. As the angular variable ϕ and $(ik/2\pi)\ln(X^1 + iX^2)$ are canonically conjugates with respect to (30), on the remaining two-spheres $X^1 = const \neq 0$ the symplectic two-form may be written as

$$\Omega_{GWZW} = \frac{ik}{2\pi} d\ln(X^1 + iX^2) \wedge d\phi \,. \tag{34}$$

Introducing another spherical coordinate θ on S^3 by $X^1 := \cos\theta$, $\theta \in [0, \pi]$, the respective S^2 is parametrized by $\phi \in [0, 2\pi]$ and $X^2 \in [-\sin\theta, \sin\theta]$, with θ fixed, and the integral over the symplectic form becomes

$$\oint_{S^2} \Omega_{GWZW} = \begin{cases} 2k\theta & , \quad \theta \in [0, \pi/2) \Leftrightarrow X^1 \in (0, 1] \\ 2k(\theta - \pi) & , \quad \theta \in (\pi/2, \pi] \Leftrightarrow X^1 \in [-1, 0) \end{cases} \,. \tag{35}$$

Here we made use of the fact, that for $X^2 = \pm\sin\theta$ one has $\ln(X^1 + iX^2) = \pm i\theta + n2\pi i$, where the integer n is determined most easily by the requirement that $\int \Omega_{GWZW} = 0$ for the pointlike leaves at $X^1 \equiv \cos\theta = \pm 1$. So, those leaves characterized by a value of $X^1 = \cos\theta$ yielding an integer multiple of 2π for the righthand side of (35) carry a quantum state of the $SU(2)$-GWZW-model.

At the equator of the two-sphere $X^1 = 0 \Leftrightarrow \theta = \pi/2$, clearly the symplectic form (34) becomes ill-defined, and so becomes its integral (35). Here one has to take into account the changes induced by L_δ, and the above-mentioned conditions on the A-fields. However, the correct result may be guessed already by a simple limiting procedure: From (35) we obtain

$$\lim_{X^1 \to \pm 0} \int_{S^2} \Omega_{GWZW} = \pm k\pi, \tag{36}$$

which, as k is integer valued, is unique modulo to 2π. So we conclude that the adjoint orbit $X^1 = tr\, g/2 = 0$ carries a quantum state, (36) is an integer multiple of 2π, i.e. iff k is even. This result may be verified also by a more careful analysis [2].

Summing up, we conclude that the integral orbits (i.e. the orbits or conjugacy classes allowing for nontrivial quantum states of the $SU(2)$-GWZW model) are given by $\theta = n\pi/k \Leftrightarrow X^1 = \cos(n\pi/k)$, $n = 0, 1, ..., k$. Here we counted also the two states localized at the pointlike orbits at $g = \pm \mathbf{1}$. If one disregards these states, one should renormalize k to $k+2$ simultanously, so as to reproduce standard path integral results.

It is instructive to compare the final picture with that of the YM theory. There we had $N = su(2) = \mathbb{R}^3$, now we have $N = SU(2) = S^3$. Both target spaces N are foliated into two-spheres, one respectively two of which degenerate to a point. In the YM-case an infinite number of these two-spheres is integral, in the G/G case it is a finite number only. We have seen, furthermore, that any of the YM-quantum states corresponds to a quantum spin system and thus to a unitary representation of the $su(2)$-algebra. One can show that similarly any of the G/G-quantum states is associated to a representation of the quantum group $su_q(2)$ (with $q = \exp[i2\pi/(k+2)]$). These results generalize to arbitrary compact gauge groups.

A final remark: The YM-theory taught us a lecture of what kind of terms may be added to L_{top} without spoiling its symmetries (11). So does the G/G-model: Given a two-form or a closed three-form on N, which is invariant under all symmetries generated by vector-fields of the form $\mathcal{P}^{ij} d/dX^j$ (as is the case for (33)), one may extend the Lagrangian by adding the pull-back of this form to M. As this does not involve any metric or volume-form on M, the resulting action is still topological.

Already by means of these simple examples the richness of the Poisson σ-model on the quantum level becomes obvious. Still with respect to the calculation of the partition function for the general model, only partial results have been achieved yet [3]. It will be interesting to investigate further, what kind of topological features of the target space of the theory are encoded in the partition function and correlators of the Poisson σ-model. Independently of this, it should have become clear by the exposition above, that Poisson σ-models allow for a comprehensive treatment of many seemingly different 2D field theories.

Acknowledgement

We are grateful to A.Alekseev for collaboration on some of the topics presented here. Moreover we want to thank the organizers of the school for giving us the opportunity for this presentation, T.Asselmeyer for discussions, and H.Pelzer for critically reading the manuscript.

References

1. P.Schaller and T.Strobl, Mod. Phys. Letts. A9 (1994) 3129, and *Quantization of Field Theories Generalizing Gravity-Yang-Mills Systems on the Cylinder*, In: LNP 436, p. 98, *Integrable Models and Strings*, eds. A.Alekseev et al, Springer 1994 or gr-qc/9406027.
2. A.Y.Alekseev, P.Schaller and T.Strobl, *The Topological G/G WZW Model in the Generalized Momentum Representation*, hep-th/9505012.
3. P.Schaller and T.Strobl, *Poisson σ-models: A generalization of Gravity-Yang-Mills Systems in Two Dimensions*, to appear in Proc. of 'Conference on Integrable Systems', Dubna 1994 or hep-th/9411163.
4. For more details confer, e.g., Y. Choquet-Bruhat, C. DeWitt-Morette, *Analysis, Manifolds and Physics, Part II: 92 Applications*, North-Holland Physics 1989.
5. W. Thirring, *Lehrbuch der Mathematischen Physik*, Band 1, Springer Wien New York 1979. R. Abraham and J.E. Marsden, *Foundations of Mechanics*, Benjamin, New York (2nd Edition) 1981.
6. D. Birmingham, M. Blau, M. Rakowski and G. Thompson, Phys. Rep. 209 (1991) 129.
7. T.Klösch and T.Strobl, *Classical and Quantum Gravity in 1+1 Dimensions*, Parts I,II,III, in preparation.
8. D.Amati, S.Elitzur and E.Rabinovici, Nucl. Phys. 418 (1994) 45.
9. Cf. also N.M.J. Woodhouse, *Geometric Quantization*, second Edition 1992, Clarendon Press, Oxford.

On Solution of Self-Duality Equation in Quantum-Group Gauge Theory

B.M. Zupnik

Bogoliubov Laboratory of Theoretical Physics , JINR, Dubna, Moscow Region, 141980, Russia

We discuss a gauge theory for the quantum group $SU_q(2) \times U(1)$ on the quantum Euclidean space. This theory contains three physical gauge fields and one $U(1)$-gauge field with a zero field strength. We construct the quantum- group self-duality equation (QGSDE) in terms of differential forms. A deformed analogue of the BPST-instanton solution is obtained.

An attractive idea of quantum deformations for the gauge theories has been considered in the framework of different approaches [1 - 3]. Formally one can discuss independent deformations of basic spaces, and gauge groups and possible correlations between these deformations. We shall here consider a gauge theory with identical one-parameter deformations of the 4-dimensional Euclidean space and the gauge group $SU(2)$. A consistent formulation of the gauge theory for the semisimple quantum group $SU_q(N)$ is unknown to us, so we shall deal with the quantum group $U_q(2) = SU_q(2) \times U(1)$. It will be shown, that the $U(1)$-gauge field can be treated as a field with a zero field strength.

Consider the standard relations between elements T^i_k of the quantum $U_q(2)$-matrix [4],

$$R\,T\,T' = T\,T'\,R \quad \Leftrightarrow \quad R^{ik}_{lm} T^l_j T^m_n = T^i_l T^k_m R^{lm}_{jn}\ , \tag{1}$$

where I is a unity matrix , R is the constant symmetric matrix with components $R^{ik}_{lm}(q)$ $(i,k,l,m = 1,\ 2)$, and q is a real deformation parameter. We use the R-matrix method in the condensed notations of Ref [3] (see also [4],[7]).

The parameters $q(ik)$ $(q(12) = q, q(21) = q^{-1}, q(11) = q(22) = 1)$ define a q-deformation of the ε-symbol, $\varepsilon_{ik}(q) = \sqrt{q(ik)}\varepsilon_{ik}$, where ε_{ik} is the ordinary antisymmetric symbol.

The R-matrix can be written in terms of projection operators $P^{(\pm)}$: $R = qP^{(+)} - q^{-1}P^{(-)}$, $P^{(+)} + P^{(-)} = I$. The operator $P^{(-)}$ for $U_q(2)$ is proportional to the product of two $\varepsilon(q)$- symbols:

$$[P^{(-)}]^{ik}_{lm} = -\frac{q}{1+q^2}\varepsilon^{ki}(q)\varepsilon_{ml}(q), \qquad \varepsilon^{ki}(q)\varepsilon_{il}(q) = \delta^k_l\,. \tag{2}$$

Here the basic identity for $\varepsilon(q)$-symbols with upper indices is written also.

We shall use the covariant relation for the quantum determinant $D(T)$ of the $U_q(2)$-matrix,

$$\varepsilon_{ml}(q) T^l_i\, T^m_k = \varepsilon_{ki}(q)\, D(T). \tag{3}$$

A covariant expression for the inverse quantum matrix $S(T) = T^{-1}$ can be obtained from this equation.

The $SU_q(2)$-metric $\mathcal{D}(q)$ determines the matrix product of transposed q-matrices[4],

$$T_i^l \mathcal{D}_l^m(q) S(T)_m^k = \mathcal{D}_i^k(q) = -\varepsilon_{mi}(q)\varepsilon^{mk}(q). \tag{4}$$

The unitarity condition for the matrix T can be formulated with the help of involution $\overline{T_k^i} = S_i^k$ [4].

Let us consider the bicovariant differential calculus on the $U_q(2)$ group [5-7]:

$$T\, dT' = R\, dT\, T'\, R, \tag{5}$$
$$D(T)\, dT = q^2 dT\, D(T).$$

Note that the condition $D(T) = 1$ is inconsistent in the framework of this calculus. Consider the relations for the right-invariant differential forms $\omega = dT\, S$ [3]:

$$\omega R \omega + R \omega R \omega R = 0\,, \quad T\, \omega' = R\, \omega\, R\, T. \tag{6}$$

The quantum trace ξ of the form ω plays an important role in this calculus:

$$\xi(T) = \mathcal{D}_i^k(q)\omega_k^i(T) \neq 0, \quad \xi^2 = 0, \quad d\xi = 0, \tag{7}$$

$$dT = \omega T = (q^3 - q)^{-1}[T, \xi], \quad q dD(T) = \xi D(T),$$
$$d\omega = \omega^2 = (q - q^3)^{-1}\{\xi, \omega\}\,. \tag{8}$$

Note that the basic relations of the bicovariant calculus on $GL_q(2)$ and $U_q(2)$ have been analysed in detail [5-7]. We shall denote this calculus as BC-calculus.

The BC-calculus serves as the basis for a consistent formulation of quantum-group gauge theory in the framework of noncommutative algebra of differential complexes [2-3]. Consider formally the quantum group gauge connection $A_b^a(x) = dx^m [A_m(x)]_b^a$ defined on some basic space. Suppose that this differential complex satisfies the following simplest commutation relation:

$$(A\, R\, A + R\, A\, R\, A\, R\,)_{cd}^{ab} = A_e^a R_{gd}^{eb} A_c^g + R_{ef}^{ab} A_g^e R_{hn}^{gf} A_m^h R_{cd}^{mn} = 0. \tag{9}$$

These relations generalize the anticommutativity conditions for components of the classical connection form. Note that the general relation for A contains a nontrivial right-hand side [3].

Coaction of the gauge quantum group $U_q(2)$ has the following standard form:

$$A \to T\, A\, S(T)\, +\, dT\, S(T) = T\, A\, S\, +\, \omega(T), \tag{10}$$
$$\alpha = \mathrm{Tr}_q A \to \alpha + \xi(T).$$

The restriction $\alpha = 0$ is inconsistent with (9), but we can use the gauge-covariant relations $\alpha^2 = 0$ and $\mathrm{Tr}_q A^2 = 0$.

It should be stressed, that we can choose the zero field-strength condition $d\alpha = 0$ for the $U(1)$-gauge field [1]. This constraint is gauge invariant and consistent with (9). The deformed pure gauge field α can be decoupled from the set of physical fields in the limit $q = 1$. We shall consider further the $U_q(2)$-gauge theory with three "physical" gauge fields and one "zero-mode" $U(1)$ field. The curvature 2-form $F = dA - A^2$ is q-traceless for this model.

Quantum deformations of Minkowski and Euclidean 4-dimensional spaces have been studied intensively in many papers. We shall treat the coordinates x^i_α of q-deformed Euclidean space $E_q(4)$ as generators of a noncommutative algebra ($Rx\, x' = x\, x'R$), covariant under the coaction of the quantum group $G_q(4) = SU^L_q(2) \times SU^R_q(2)$.

The q-deformed central Euclidean interval τ can be constructed by analogy with the quantum determinant

$$\tau = x^1_1 x^2_2 - q x^1_2 x^2_1 = -\frac{q}{1+q^2}\varepsilon^{\beta\alpha}(q)\varepsilon_{ki}(q) x^i_\alpha x^k_\beta. \tag{11}$$

We shall use an analogue of the bicovariant $U_q(2)$-calculus for studying differential complexes on $E_q(4)$. The commutation relations between matrices x and dx can be obtained from Eqs(6 - 9) by formal substitution $T \rightarrow x$. One can obtain , for example,

$$x^i_\alpha dx^k_\beta = R^{ik}_{lm} dx^l_\gamma x^m_\rho R^{\gamma\rho}_{\alpha\beta}, \tag{12}$$

$$P^{(+)} dx dx' P^{(+)} = 0 = P^{(-)} dx dx' P^{(-)}.$$

The basic decomposition of 2-forms on $E_q(4)$ has the following form:

$$dx^i_\alpha dx^k_\beta = [P^{(-)} dx dx' + dx dx' P^{(-)}]^{ik}_{\alpha\beta} = \varepsilon^{ki}(q) d^2 x_{\alpha\beta} + \varepsilon_{\beta\alpha}(q) d^2 x^{ik}, \tag{13}$$

where Eq(2) for $P^{(-)}$ is used. By analogy with the classical case we can treat two terms of this decomposition as self-dual and anti-self-dual 2-forms under the action of a duality operator $*$.

Consider the right-invariant 1-forms on $E_q(4)$,

$$\omega^i_k(x) = [dx S(x)]^i_k = dx^i_\alpha S^\alpha_k, \quad dx = \omega x, \tag{14}$$

where $S(x)$ is the inverse matrix for x defined by Eq(13). It is convenient to rewrite the decomposition of 2-forms in terms of the right-invariant self-dual and anti-self-dual forms:

$$dx dx' = \omega x \omega' x' = \omega R \omega R x x', \tag{15}$$

$$P^{(-)} dx\, dx' = q P^{(-)} \omega R \omega x x' P^{(+)} = P^{(-)} \Omega_S P^{(+)} x x', \tag{16}$$

[1] This condition is consistent also for the case of $GL_q(N)$ group.

$$dx\,dx'P^{(-)} = \Omega_A P^{(-)} xx', \tag{17}$$

$$\Omega_S = *\Omega_S = q^4\omega^2 + q\omega\xi, \tag{18}$$

$$\Omega_A = -(*\Omega_A) = q^{-1}\omega\xi - \omega^2. \tag{19}$$

Here the commutation relations of BC-calculus on $E_q(4)$ and properties of the $P^{\pm}$-operators were used. It should be stressed, that the condensed notations simplify significantly these calculations.

Let us introduce the simple ansatz for quantum $U_q(2)$ anti-self-dual gauge fields:

$$A^a_b = dx^i_\alpha A^{\alpha a}_{ib}(x) = \omega^a_b(x) f(\tau), \tag{20}$$
$$A^{\alpha a}_{ib}(x) = \delta^a_i S^\alpha_b(x) f(\tau),$$

where $f(\tau)$ is a function of q-interval (11). Note that this ansatz is a partial case of a more general construction of the differential complex on $GL_q(2)$ [2,3]. Addition of the term $\xi(x)g(\tau)$ results in a relation for the connection A more complicated than (9) .

Consider the q-traceless curvature form for the connection (23), which can be calculated in the framework of the BC-calculus on $E_q(4)$:

$$F = \omega^2 f(\tau)[1 - f(q^2\tau)] + (q^3 - q)^{-1}\omega\xi[f(\tau) - f(q^2\tau)]. \tag{21}$$

An appearance of the finite translation $f(q^2\tau)$ is a general feature of the calculus on the quantum space.

The anti-self-duality equation for our ansatz is equivalent to the nonlinear finite-difference equation:

$$*F = -F \;\Rightarrow\; F \sim \Omega_A f(\tau)[1 - f(q^2\tau)], \tag{22}$$

$$f(\tau) - f(q^2\tau) = (1 - q^2) f(\tau)[1 - f(q^2\tau)], \tag{23}$$

where Ω_A is the anti-self-dual 2-form (19).

This equation has a simple solution analogous to the classical BPST-solution

$$f(\tau) = \frac{\tau}{a + \tau}\,, \tag{24}$$

where a is an arbitrary "constant" that can be treated as a central periodical function: $a(\tau) = a(q^2\tau)$. Note that our solution for the connection A contains the parameter q only through definitions of $\omega(x)$ and τ ; however, the corresponding curvature has more explicit q-dependence (21).

It is easy to obtain the 5-parameter solution via substitution [2] $x^i_\alpha \to \hat{x}^i_\alpha = x^i_\alpha + c^i_\alpha$ in Eqs(23, 24). Stress that our anti-self-dual solution is a function on the braided algebra with the noncommuting generators x, dx, a and c.

$$R\,\hat{x}\,\hat{x}' = \hat{x}\,\hat{x}'\,R, \quad R\,c\,c' = c\,c'\,R, \quad c\,x' = R\,x\,c'\,R, \tag{25}$$

---

[2] The addition of q-matrices was considered early by V.Jain, O.Ogievetsky, and S.Majid.

$$c\,dx' = R\,dx\,c'\,R, \quad [\hat{x}, \tau(\hat{x})] = 0, \tag{26}$$
$$d\hat{x} = dx, \quad dc = 0,$$
$$\tau(\hat{x})dx = q^2 dx \tau(\hat{x}).$$

The harmonic (twistor) interpretation of the quantum-group self-duality equation was considered in Refs[8].

It seems very interesting to study reductions of QGSDE to lower dimensions, and to search for a more general deformation scheme for the self-duality equation.

References

1. I.Ya.Aref'eva, I.V.Volovich, Mod. Phys. Lett.A **6**, 893 (1991)
2. A.P.Isaev, Z.Popowicz , Phys.Lett.B **281**, 271 (1992) ; Phys.Lett.B **307**, 353 (1993)
3. A.P.Isaev, Preprint JINR E2-94-38 , Dubna, 1994
4. N.Yu.Reshetikhin, L.A.Takhtadjan, L.D.Faddeev, Algeb. Anal. **1**, 178 (1989)
5. S.L.Woronowicz, Comm. Math. Phys. **122**, 125 (1989)
6. Yu.I.Manin , Teor.Mat.Fiz. **92**, 425 (1992)
7. P.Schupp, P.Watts, B.Zumino, Comm. Math. Phys.**157**, 305 (1993)
8. B.M.Zupnik, Preprints JINR E2-94-449, E2-94-487 Dubna, 1994; JETPh Lett.61, no.6, 434 (1995)

Springer-Verlag and the Environment

We at Springer-Verlag firmly believe that an international science publisher has a special obligation to the environment, and our corporate policies consistently reflect this conviction.

We also expect our business partners – paper mills, printers, packaging manufacturers, etc. – to commit themselves to using environmentally friendly materials and production processes.

The paper in this book is made from low- or no-chlorine pulp and is acid free, in conformance with international standards for paper permanency.

GPSR Compliance

The European Union's (EU) General Product Safety Regulation (GPSR) is a set of rules that requires consumer products to be safe and our obligations to ensure this.

If you have any concerns about our products, you can contact us on ProductSafety@springernature.com

In case Publisher is established outside the EU, the EU authorized representative is:

Springer Nature Customer Service Center GmbH
Europaplatz 3
69115 Heidelberg, Germany

Batch number: 10620143

Printed by Printforce, the Netherlands